D1297205

The Seven Pillars of Creation

The Seven Pillars
of Creation

*The Bible, Science, and the Ecology
of Wonder*

WILLIAM P. BROWN

OXFORD
UNIVERSITY PRESS

2010

OXFORD
UNIVERSITY PRESS

Oxford University Press, Inc., publishes works that further
Oxford University's objective of excellence
in research, scholarship, and education.

Oxford New York
Auckland   Cape Town   Dar es Salaam   Hong Kong   Karachi
Kuala Lumpur   Madrid   Melbourne   Mexico City   Nairobi
New Delhi   Shanghai   Taipei   Toronto

With offices in
Argentina   Austria   Brazil   Chile   Czech Republic   France   Greece
Guatemala   Hungary   Italy   Japan   Poland   Portugal   Singapore
South Korea   Switzerland   Thailand   Turkey   Ukraine   Vietnam

Copyright © 2010 by Oxford University Press, Inc.

Published by Oxford University Press, Inc.
198 Madison Avenue, New York, NY 10016

www.oup.com

Oxford is a registered trademark of Oxford University Press

All rights reserved. No part of this publication may be reproduced,
stored in a retrieval system, or transmitted, in any form or by any means,
electronic, mechanical, photocopying, recording, or otherwise,
without the prior permission of Oxford University Press.

Library of Congress Cataloging-in-Publication Data
Brown, William P., 1958–
The seven pillars of Creation : the Bible, science,
and the ecology of wonder / William P. Brown.
p. cm.
Includes bibliographical references (p.) and index.
ISBN 978-0-19-973079-7
1. Creation—Biblical teaching.
2. Bible. O.T.—Criticism, interpretation, etc.
3. Bible and science. I. Title.
BS651.B7878 2010
231.7'65—dc22   2009023025

9 8 7 6 5 4 3 2
Printed in the United States of America
on acid-free paper

Contents

Preface

In the course of my research, I had the privilege of consulting with a number of scholars in science and theology. One esteemed theologian (a Gifford lecturer, no less) confided to me that his own efforts at incorporating the findings of science into his theological work continue to be regarded by his peers as something done "on the side." I was shocked. I shudder to think that my own work could be regarded as nothing more than a diversion. And I am not even a theologian! So what compels a biblical exegete to read up on both theology and science, to venture far afield from one's own expertise, not to mention comfort zone? Admittedly, I have often asked myself that question. Am I doing something merely "on the side"? Or worse, am I neglecting my own discipline? Now that this project is completed, at least for this book, all I can say is that the "outside" work I've done has thoroughly enriched and renewed my sense of the discipline to which I feel called. Sometimes it takes a long, winding "detour" to feel more at home in one's own element.

My reasons for writing this book are primarily personal. (The cultural and theological reasons are given in the introductory chapter.) I was raised on science. My father was a farm boy who decided to leave the farm to pursue scientific research. He taught Animal Science at the University of Arizona until he retired and was my early mentor in all things academic. Having focused on science in high school, I devoted my first two years of college to pursuing a degree in engineering. However, through factors entirely unanticipated and much to my parents' dismay, I graduated with a degree in philosophy. Even more surprising, I subsequently found myself studying the Bible in seminary. But up until that time I had built my own reflecting telescope, helped conduct research in an oncology lab, and drafted rudimentary mining shafts. Thus, writing this book was a way of retrieving something of the inquisitive spirit of those formative years when I would scan the clear Arizona night skies for star clusters and nebulae and measure the ominous growth of cancer cells in a Petri dish. I was enchanted with nature, both cosmic and microscopic. I still am.

More recently, as I was toying with the idea of pursuing this project, I learned of a past professor at the institution where I teach whose tenure was cut short. Dr. James Woodrow, uncle of Woodrow Wilson, was appointed the Perkins Professor of Natural Science in Connection with Revelation at Columbia Theological Seminary in 1861. He received his PhD in Chemistry from the University of Heidelberg and taught various scientific disciplines at Oglethorpe University before joining Columbia's faculty. When Darwin's *On the Origin of Species* was published in 1859, Professor Woodrow's chair eventually became a focal point of controversy. In 1883 he was formally asked to "set forth his views upon evolution in order that the church might have the benefit of his opinions."[1] Having received his answer, the church found no benefit. In 1886 Woodrow was dismissed from the seminary, and the chair that was created especially for him was never filled again. (Indeed, it no longer exists.) But not to lament: Woodrow simply went down the street to teach at the University of South Carolina, later becoming its president, and served as Moderator of the Synod of South Carolina. Even a street in Columbia is named after him. Although Woodrow's dismissal was prompted by a number of factors, from personality clashes among the faculty to the seminary's fiscal crisis, it was the perceived threat of Darwinism to Christian faith that carried the day.

One of Woodrow's students asserted, "He was a universal genius, one of the greatest scholars the South ever produced. He taught me more than all the other professors combined, and so grounded me in the truth of the Bible that no power on earth can successfully assail my faith."[2] But another said, "I can never forget that it was the lectures in Dr. Woodrow's classroom that checked me in a wild, downward career to infidelity and atheism and cheerless blank despair."[3]

The lesson? First, you cannot rely on student evaluations alone. Second, the still largely unexplored terrain between biblical faith and modern science remains a minefield, at least on the cultural level. Variations of the Scopes trial continue to be replayed throughout the country, polarizing communities and fostering a conceptual apartheid that keeps faith and science not simply at arm's length but at odds with each other. For me, however, learning of Professor Woodrow's plight provided sufficient impetus to see this project through. Fortunately, the theological climate is changing. My church, like most mainline denominations, is much more open to the findings of science than it was over a century ago. There are now several academic chairs established throughout the country for theologians who engage science. In conversation with some of them, as well as with scientists who do not identify with any faith tradition, I have only begun to discover the treasures buried in the terrain. Someday, I hope, that minefield will become a playfield.

That hope is borne by a new generation. At a recent conference on evolution held at Emory University, whose keynote speakers included the renowned biologist E. O. Wilson and primatologist Frans de Waal, a high school freshman, Caitlin Wade, remarked, "I can pray, read my Bible, and study science. I don't have to choose."[4] I hope for Caitlin's sake and for countless other students that studying the Bible and science *together* is an exercise in wonder, an experience in joy to the World and to the Word made flesh and fresh.

William P. Brown

2009

150th anniversary of the publication

of *On the Origin of Species*

200th anniversary of the birth of Charles Darwin

400th anniversary of Galileo's telescope

500th anniversary of John Calvin

Acknowledgments

This study would not have happened without the generous support of individuals and institutions. First, a research project of this sort required significant funding, which was generously provided by the Henry Luce III Foundation. The Luce fellowship for the 2007/08 academic year enabled me to travel and consult various experts in both science and theology. I also thank Columbia Theological Seminary, particularly its Board, for granting me a sabbatical year. Without full-time leave from teaching and committee responsibilities, this project would never have gotten off the ground. My colleagues, particularly Professors Christine Roy Yoder and Kathleen O'Connor, have had to shoulder extra responsibilities as a result, but they have done so with cheerful grace. I hope to return the favor.

As for the project itself, numerous individuals are to be acknowledged. First, many thanks go to my official consultants on the project.

1. George W. Fisher, Emeritus Professor of Geology, Johns Hopkins University
2. Christopher G. De Pree, Associate Professor of Astronomy and Physics, Director of the Bradley Observatory, Agnes Scott College
3. Nancey C. Murphy, Professor of Christian Philosophy, Fuller Theological Seminary

4. J. Wentzel van Huyssteen, James I. McCord Professor of
Theology and Science, Princeton Theological Seminary

Each showed hospitality, patience, and great insight as I sent them page after page of my ongoing work. Not only did they help bring me up to speed on matters of science and theology; they also shared genuine interest and guidance in the direction of my work. Without them, the project would have floundered. While having extended my horizons to the nth degree, their insights have, I hope, also kept me from making too many egregious mistakes. For those that remain I take full responsibility. I am still learning.

Other readers are to be thanked, namely, those who agreed informally to read a preliminary chapter or two. Professor Edward O. Wilson, Pellegrino Professor of Biology Emeritus of Harvard University, was kind enough to read a draft of my introductory chapter and provide important feedback and encouragement. Other scientists include Professor Charles F. Denny, Distinguished Professor of Biology Emeritus of the University of South Carolina Sumter, and Professor William R. Stoeger, SJ, of the Vatican Observatory and the University of Arizona. Both generously agreed to read select chapters from a neophyte when it comes to science. Closer to home is Dr. Albert Anderson, who earned his PhD in physics and became a successful patent lawyer in Atlanta. The near mural-size poster of Albert Einstein hanging in his office epitomizes the infectious sense of wonder he has shared with me throughout the project. Finally, thanks go to Dr. David R. Vinson, a research physician whose mission is to bring evangelical Christians into constructive conversation with science (http:// science.drvinson.net).

There are many others who have cast an open ear and eye to my work, providing rich insight and moral support. I thank the members of the Earth Covenant Ministry (ECM), a grassroots organization committed to educating local churches about creation care: Alan Jenkins (director), Bobbie Wrenn Banks, Jon Houghton, Sarah Tomaka, Rebecca Watts Hull, and the late Ann Morris, to mention a few, have dedicated themselves to raising the ecological consciousness of the church, congregation by congregation. Support and insight have also come from members of North Decatur Presbyterian Church, including Dr. Cynthia Warner, biochemist at the Center for Disease Control in Atlanta, with whom I had the privilege of teaching a Sunday School series on evolution and global warming. She demonstrates in practice that faith and science can fit hand in glove. So also ProfessorDabney Dixon, a biochemist at Georgia State University and an Episcopalian, who knows well the scientific "data of despair" yet remains a beacon of hope and action. Thanks also go to Dr. Wes McCoy, the high school biology teacher and fearless Presbyterian from

Marietta, Georgia, who worked successfully to remove the infamous stickers ("evolution is a theory, not a fact") that the Cobb County Board of Education placed on all high school biology textbooks.

On the biblical and theological side of this study, I express thanks to John Goldingay, James T. Butler, Mark S. Smith, and Douglas F. Ottati for their helpful feedback on various presentations and versions of my chapters, as well as my Columbia colleagues Stan Saunders and Mark Douglas for their continued encouragement and insights. Throughout the course of my research, they have become cherished conversation partners. Other readers of my work include several pastors who embody the best of what it means to be part of a learned ministry. They include Tom Lindeman and Rick Neale. I also thank Mark Davis, Katy Sachse, and Paul Miller, members of a larger network of "pastor theologians" once supported by the Center of Theological Inquiry in Princeton under the visionary direction of Wallace Alston. My conversations with them and many others in this lively network helped plant the seeds of this transdisciplinary project.

On the publication side, I wish to express my thanks to Cynthia Read of Oxford University Press for her support and encouragement throughout the project, to Joellyn Ausanka for efficiently and cheerfully guiding the project to its publishable conclusion, and to Mary Sutherland for her keen editorial eye. I should also mention the two anonymous reviewers of my early work for Oxford. They know who they are. Their constructive comments helped reshape certain sections of my work. Finally, I thank the late Stephanie Egnotovich of Westminster John Knox Press, who helped me to think clearly about what was truly important about the project as opposed to what was destined for the trash bin. Without her initial guidance, this project would have been significantly longer (and no doubt more tedious) than it is, and for this the reader should give thanks!

Never least and most important is my family. Certain issues addressed in this project have come from questions raised by our ever-inquisitive daughter Ella, who wrote an exploratory paper on religion and science as a high school freshman that elicited wonderful discussions. Throughout my work, the winsome disposition of our younger daughter Hannah saved me from taking my work and myself too seriously. Finally, words cannot express the love and gratitude I have for Gail, my partner in marriage, not only for putting up with my frequent trips but most importantly for convincing me that the project needed to be done. She has given me the gift not only of time but also of inspiration and wisdom. I dedicate the book to her.

Abbreviations

AB   Anchor Bible

ANET   *Ancient Near Eastern Texts Relating to the Old Testament*. Edited by J. B. Prichard. 3rd edition. Princeton, NJ: Princeton University Press, 1954

AOAT   Alter Orient und Altes Testament

ATANT   Abhandlungen zur Theologie des Alten und Neuen Testaments

BA   *Biblical Archaeologist*

BHS   *Biblical Hebraica Stuttgartensia*. Edited by K. Elliger and W. Rudolph. Stuttgart: Deutsche Bibelgesellschaft, 1983.

BM   *Before the Muses*. By Benjamin Foster. 2 vols. 2nd edition. Bethesda, MD: CDL Press, 1996.

BWL   *Babylonian Wisdom Literature*. Edited by W. G. Lambert. Oxford: Clarendon Press, 1960.

BZAW   Beihefte zur Zeitschrift für die alttestamentliche Wissenschaft

CBQ   *Catholic Biblical Quarterly*

CBQMS   Catholic Biblical Quarterly Monograph Series

CS   *The Context of Scripture, Volume One*. Edited by William W. Hallo and K. Lawson Younger Jr. Leiden: Brill 2003.

FAT   Forschungen zum Alten Testament

FOTL   Forms of Old Testament Literature

HALOT	*The Hebrew and Aramaic Lexicon of the Old Testament.* By Ludwig Koehler and Walter Baumgartner et al. 2 vols. Leiden: Brill, 2001.
HBT	*Horizons in Biblical Theology*
HTR	Harvard Theological Studies
IDB	*The Interpreter's Dictionary of the Bible.* Edited by G. A. Buttrick. 4 vols. Nashville, Abingdon, 1962.
IEJ	*Israel Exploration Journal*
J-M	*A Grammar of Biblical Hebrew.* By Paul Joüon. 2 vols. Rome, Editrice Pontificio Istituto Biblico, 1991.
JBL	*Journal of Biblical Literature*
JPC	*Journal of Psychology and Christianity*
JR	*Journal of Religion*
JSOTSup	Journal for the Study of the Old Testament: Supplement Series
KTU	*Die keilalphabetischen Texte aus Ugarit.* Edited by M. Dietrich, O. Loretz, and J. Sanmartín. AOAT 24/1. Neukirchen-Vluyn: Neukirchener Verlag, 1976. 2nd enlarged edition of *KTU: The Cuneiform Alphabetic Texts from Ugarit, Ras Ibn Hani, and Other Places.* Edited by M. Dietrich, O. Loretz, and J. Sanmartín. Münster: Ugarit-Verlag, 1995.
LXX	Septuagint
MT	Masoretic Text
NJPS	*Tanakh: The Holy Scriptures: The New JPS Translation according to the Traditional Hebrew Text*
OBO	Orbis biblicus et orientalis
OTS	Old Testament Studies
PACS	Philo of Alexandria Commentary Series. Edited by Gregory E. Sterling.
PSB	*Princeton Seminary Bulletin*
SAACT	State Archives of Assyria Cuneiform Texts
SBLWAW	Society of Biblical Literature Writings from the Ancient World
ThLZ	*Theologische Literaturzeitung*
VT	*Vetus Testamentum*
VTSup	Vetus Testamentum Supplements
ZAW	*Zeitschrift für die alttestamentliche Wissenschaft*

The Seven Pillars of Creation

I

Introduction

From Wonder to Wisdom

Wisdom has built her house; she has hewn her seven pillars.

—Proverbs 9:1

The larger the island of knowledge, the longer the shoreline of wonder.

—Ralph W. Sockman[1]

The religion that is married to science today will be a widow tomorrow. . . .
But the religion that is divorced from science will leave no offspring
tomorrow.

—Holmes Rolston III[2]

I would not say interpretations, Mr. Darrow, but comments on the lesson.

—William Jennings Bryan[3]

The inspiration behind this study stems from a childhood epiphany
that took place in a science museum in Seattle. Set off in a far corner was a small video screen whose button was waiting to be
pushed by a curious child. I happily obliged. A serene scene un-
folded of a young couple lounging on a picnic blanket in a Chicago
park. But before my eyes could linger, the screen took me out above
the park, the city, Lake Michigan, the continent, and the globe. Vast
stretches of outer space quickly came into view. The solar system
resembled something of an atom. Next came other stars, the swirling Milky Way, and finally empty space dotted with tiny galaxies.
Then in a matter of seconds I found myself once again suspended

over the lounging couple, pausing only briefly before closing in on a patch of skin and proceeding all the way to the cellular, molecular, atomic, and quark-scale levels. I was mesmerized by this dizzying ride through the cosmos and the microcosmos. I saw things I thought were privy only to God.

That presentation, I discovered years later, was *Powers of Ten*, the ingenious creation of Charles and Ray Eames.[4] Both a visual thought experiment and a virtual rollercoaster ride, this now "ancient" video dramatically covered the extremities of scale, from the unimaginably vast (10^{25}) to the inscrutably tiny (10^{-16}), from the cosmic to the subatomic. The sum effect on me was nothing short of transcendent.

Gloriously "weird" is how physicist Brian Greene describes the quantum world.[5] "Too wonderful" is how the biblical sage responds to creation's marvels (Prov 30:18–19). The psalmist trembles before the vastness of the universe (Ps 8:3–4). Biologist Ursula Goodenough celebrates the "sacred depths of nature."[6] What do they all have in common? I wonder. Though separated by over two and a half millennia, the authors of ancient Scripture and numerous scientists of today find themselves caught up in a world of abiding astonishment. Like the ancients, many scientists admit to being struck by an overwhelming sense of wonder—even "sacredness"—about nature and the cosmos.[7] What a far cry from Francis Bacon's objectification of the natural realm as humanity's slave!

The wonder of it all prompts one—anyone—to wonder about it all. Bioanthropologist Melvin Konner regards the capacity to wonder as "the hallmark of our species and the central feature of the human spirit."[8] Although *Homo sapiens* ("wise human") may be too self-congratulatory, there is no doubt that we are *Homo admirans*, the "wondering human." Wonder is what unites the empiricist and the "contemplator," the scientist and the believer.[9] "Everyone is naturally born a scientist," admits astrobiologist Chris Impey.[10] We can no more deny that of our distant ancestors than we can deny that of ourselves. Together, the ancient cosmogonist and the modern cosmologist, the biblical sage and the urbane biologist form a "cohort of wonder."[11]

Lost in Wonder, Losing Wonder

Sadly, the cohort is dissolving. The language of wonder has become riddled with the rhetoric of adversity. In their fight against "soulless science," creationists champion a view of creation so narrow that it is decidedly unbiblical. At the other extreme, certain scientists construe faith in God as the enemy of scientific progress and human well-being.[12] As one might expect, misunderstandings and distortions abound as each side reduces the other to laughable caricatures. Illiteracy, both scientific and biblical, reigns.

Is science really hell-bent on eroding humanity's nobility and eliminating all sense of mystery? Not the science I know. Is faith simply a lazy excuse to wallow in human pretension? Not the faith I know. What if invoking God was a way of acknowledging the remarkable intelligibility of creation? What if science informed and enabled persons of faith to become more trustworthy "stewards of God's mysteries" (1 Cor 4:1–2)? What if faith fostered a "radical openness to the truth, whatever it may turn out to be."[13] The faith I know does not keep believers on a leash, preventing them from extending their knowledge of the world. The science I know is not about eliminating mystery. To the contrary, the experience of mystery "stands at the cradle of true art and true science," as Albert Einstein famously intoned. "Whoever does not know it can no longer wonder, no longer marvel, is as good as dead."[14]

"Mystery," of course, can mean anything from the incomprehensible born of ignorance to the surprising anomaly that invites explanation. For me, mystery inspires awe and inquiry. Examples of mystery are the "unreasonable effectiveness of mathematics,"[15] the remarkable intelligibility of nature, something instead of nothing, the emergence of life, and God's love for the world. Mystery acknowledges that, while we cannot know absolutely everything about, say, a particular ecosystem, there is nothing to stop us from knowing *more* about it, infinitely so. Mystery recognizes the provisional nature of our explanations and the inexhaustibility of our investigations. The world will always be more than what we know. Mystery is being grasped by something larger than ourselves, ever compelling us to stretch, rather than limit, the horizons of our awareness. Under the rubric of wonder, mystery has its place alongside understanding.[16]

To recapture something of the awe that fostered the spirit of inquiry among the ancients and today ignites the "vital spark of wonder that drives the best science,"[17] I want to embark on my own tour of sorts, not so much a roller-coaster ride as a leisurely excursion. I propose a tour of the biblical contours of creation conducted in conversation with science, an expedition that boldly charts the now *un*common ground of wonder. The terrain is rugged. Certain theologians, joined by scientists of deep theological conviction, have labored hard to put religion and science on a positive footing. But their discussions have been largely ignored by the front-line combatants and have so far slipped under the radar screen of the general public.[18] Biblical scholars, moreover, have largely avoided entering the conversation, perhaps because the discussion has remained too technical or would take the biblicist far afield of his or her own areas of expertise. Across the table, the theologian and the scientist would have good reason to question what a Bible scholar, steeped in antiquity, could ever contribute to such forward-looking discussions.

This is unfortunate, for the Bible—particularly its authority and interpretation—has been an intractably persistent issue in both the popular "debate" (read "culture war") and the academic dialogue. From the 1925 Scopes trial in Dayton, Tennessee, to the 2005 court case of Dover, Pennsylvania,[19] the Bible has been either the point of contention or the unnamed elephant. The root of the problem, or better challenge, has always been the Bible and how to read it. Thus, my central aim is to explore how to read what the Bible says about creation. What follows is meant not to win the "war" but, more modestly, to reduce the number of casualties, on both sides.[20] I want to move beyond the "debate" by introducing something of the Bible's own inexhaustible richness, its profound wonder.

I begin by recognizing that there is much more to the biblical account of creation than the seven days of creation. There are at least seven *ways* of creation featured in the Bible, seven separate traditions, each worthy of reflection but each incomplete by itself.[21] They are:

1. Genesis 1:1–2:3
2. Genesis 2:4b–3:24
3. Job 38–41
4. Psalm 104
5. Proverbs 8:22–31
6. Ecclesiastes 1:2–11; 12:1–7
7. Isaiah 40–55 (excerpts)

As no single observer can ever gain complete information about the world on any given level, so no one biblical tradition has the complete and final word on the world. Moreover, each tradition has a depth dimension that is all too easily missed, if not dismissed, in the current discussions. The full scope of human life cannot be reduced to blind chance and selfish genes anymore than the Bible's perspectives on creation can be whittled down to seven days and an apple (which is nowhere mentioned in Genesis). Such reductionism is made to the impoverishment of all participants, Darwinists and creationists, physicists and theologians, Sunday school teachers and exegetes alike.

Word and World Made Flesh

This book invites the non-expert who yearns to know more about engaging biblical faith and science in constructive, as opposed to confrontational, ways. This study also welcomes the scientist who desires to know more about what the ancient Scriptures say about cosmology, nature, and humanity's place. In short, I want to help readers become more literate in Scripture *and* science, as I have become in the course of my research. Specifically, I want to bring together two distinct disciplines, biblical theology and modern science,[22] and explore points of conversation in ways that I hope generate more synergy than sparks. My conviction is that one cannot adequately interpret the Bible today, particularly the creation traditions, without engaging science. Otherwise, the Bible's "strange new world" would become an old irrelevant word.

Central to the Christian faith is a doctrine that resists the temptation to distance the biblical world from the natural world: the incarnation. Barbara Brown Taylor puts it well: "[F]aith in an incarnational God will not allow us to ignore the physical world, nor any of its nuances."[23] Such faith calls us to know and respect the physical, fleshy world, whose "nuances" are its wondrous workings: its delicate balances and indomitable dynamics, its life-sustaining regularities and surprising anomalies, its remarkable intelligibility and bewildering complexity, its order and its chaos. Such is the *World* made flesh, and faith in the *Word* made flesh acknowledges that the very forces that produced me also produced microbes, bees, and manatees. As much as Christians cannot ignore the incarnate God, they cannot dismiss the discoveries of science. Theologically, there is no other option: faith in such a God calls people of faith to understand and respect the natural order, the world that God deemed "extremely good" (Gen 1:31) and saw fit to inhabit. The God in whom "we live and move and have our being" (Acts 17:28) has all to do with the world in which we do indeed live and move and have our being. The world subsists *in* God even as God remains present *in* the world. It is, admittedly, a mystery. But through science we become more literate in the mysteries of creation and, in turn, more trustworthy "stewards" of those mysteries.

Wonder and Wisdom

To be literate is more than being informed. "We are drowning in information, while starving for wisdom," laments biologist Edward O. Wilson.[24] Born from wonder, wisdom may very well be the basis and purpose for gaining "cultural competence" and, thus, bridging faith and science. By no coincidence, wisdom assumes a prominent place in Scripture. According to biblical lore, Wisdom personified claims to have been personally present at the creation of the cosmos (Prov 8:22–31).[25] After bearing witness to God's construction of the universe, she creates something of her own, namely, a house with "seven pillars" (9:1). By any standard building code of Near Eastern antiquity, such a house would have been remarkably spacious. But certain rabbinic interpreters saw something more. They discerned a correspondence between Wisdom's seven pillars and the seven days of creation.[26]

This insight, regardless of its exegetical validity, has inspired the title of this project. Wisdom's "edifice complex" is, I submit, an appropriate framework for studying biblical creation in conversation with science. Biblical wisdom was nurtured by a spirit of inquiry.[27] It acknowledges creation's multifaceted integrity, complexity, and mystery. For our world, threatened as it is with environmental degradation, a wisdom that embraces both science and faith is sorely needed. From climate chaos[28] to species extinctions, the data of despair afflict us. But from such data wisdom seeks understanding, feeds hope, discerns solutions, and inspires action. As biblical Wisdom invites her students to enter her spacious home and partake of her varied fare (Prov 9:2–4), so the reader is invited to enter the Bible's various perspectives on creation, to wander and to wonder, and from wonder to gain wisdom.

Faith Seeking Further Understanding

If theology is, to quote St. Anselm (1033–1109), "faith seeking understanding"[29] and science is a form of understanding seeking further understanding, then theology has nothing to fear and, in fact, much to learn from science. Theology cannot advance the scientific quest for the underlying constituents of matter and the physical nature of causation. Science, in turn, cannot lay claim to know God and God's purposes.[30] Both disciplines represent independent fields of inquiry. But, I ask, does their independence preclude cross-disciplinary conversation? Because both seek truth, because each discipline is driven by an "onto-logical thirst, by the thirst to know reality as it is,"[31] each can learn from the other, especially theology from science. If theology is about relating the world to God but does not take into account the world as known through science, then it fails.[32] And such failure strikes at the very heart of the theological task, for among theology's anathemas is the stigma of irrelevance or "the lack of cultural competence."[33] Science, more so now than ever before, constitutes a critical feature of cultural competence. Indeed, according to one astronomer, science is "the one truly global culture."[34]

For the sake of theological competence, I ask the following question, the one that drives this study: What is it like to read the Bible in one hand and the journal *Science* in the other? The question is not simply a matter of interpretation. It broaches a larger question: What is it like to be both a contemplator and an empiricist? Or to borrow directly from the Bible: What is like to be both a sage and a psalmist, a steward of creation's mysteries and a servant of Christ? In my hermeneutical quest I have found that science holds the promise of deepening the Bible's own perspectives on creation. Astronomy, geology, and biology have put to rest all unbiblical notions that the world is a static given, a ready-made creation dropped from heaven. No, nature has its own story to tell. To talk comprehensively about the story of God's creative and redemptive work is to overturn the woefully narrow view that treats the world as merely the stage for humanity's salvation. The world that God so loved in John 3:16 is nothing less than cosmic. The extent of God's provident love reaches the whole of creation (Rom 8:19–23). If Earth's story is deemed at all important for our time, then it must find a place within or at least alongside God's story for all time, whose very bookends are, in fact, creation and new creation. And at either end of either story—whether the Bible's or Earth's—the scope of life and, thus, God's purposes extend far beyond humanity.

In a nutshell, this study is aimed at engaging science in the theological interpretation of Scripture. It is written for those who desire to know both what the Bible possibly *says* about creation in light of its ancient historical and literary contexts and what the Bible can *mean* within our context as informed by science.[35] However one distinguishes the various layers of meaning with which the Bible engages the reader (and vice versa!), all attempts at interpreting the Bible happen in dialogue,[36] and every dialogue begins by seeking points of contact.

As the physicist searches for the most fundamental constituents of reality and the biologist investigates the "ultimate causes" of phylogenetic change, so the theologian seeks to articulate the ultimate nature of reality as witnessed in Scripture, tradition, and experience. To quote Barth, the Bible reveals a "strange new world."[37] So do astrophysics and quantum mechanics. Ultimately, the separate disciplines of science and theology ought to provide complementary maps of the same strange world we inhabit. Mary Midgley suggests that in this mutual quest for truth "consonances" or correlations between such maps are to be expected.[38] But finding clear and unambiguous connections may be too much to expect. Given the methodological differences between science and theology, such consonances cannot be obvious or clear cut. If they are there, they are at best subtle and partial. Ted Peters recasts such consonances as "shared domains of inquiry."[39] Perhaps most suggestive is James Gustafson's metaphor of "intersections": common points of interest shared by various disciplines in which the "traffic" is able to flow smoothly in both directions, as opposed to being stuck in gridlock.[40] Living in Atlanta, I appreciate this metaphor.

Because the discipline in which I live and move is textually based, I propose the term "parallel" as a way of identifying a point of intersection or shared domain. Used widely (and loosely) by biblical scholars, a "parallel" typically designates some kind of literary association such as a motif, narrative juncture, type-scene, literary pattern, or similar words shared among written sources. But in the case of science and the Bible, any "parallel" would need to be significantly qualified. Because a science report and a confession of faith are two very different genres with two very different goals, any alleged point of contact is at most *virtual*.[41] "Virtual parallels," thus, are not correspondences or "consonances" in any literal sense. They are at most analogous points of contact or imaginative associations, tangents at best. Parallelomania, the tendency to see parallels wherever one looks, has no place in this cross-disciplinary dialogue, for it runs the danger of collapsing the conversation. True parallels are few and far between, particularly across disciplinary boundaries, but because they are similar *and* different, they arrest our attention, prompting us to pause and to ponder, and, most important, to enter into fruitful dialogue.

In light of the ongoing nature of scientific discovery and theological inquiry, virtual parallels can only be proposed tentatively.[42] Nevertheless, to cite Warren Brown, constructive connections between science and faith exhibit a degree of "resonance," at least temporarily so. While they do not strike an identical tone, resonances produce a heuristic harmony that invites further exploration and interpretive improvisation.[43] To shift from the aural to the visual, I call such resonances "virtual" because of the real distinctions that exist between science and theology in their method and manner of discourse. "Virtual" connotes a sense of approximation, as when one claims that something is "virtually" identical to something else. More fundamentally, "virtual" designates the generative work of the imagination, as indicated by the now common phrase "virtual reality." A "virtual parallel" signals a sense of connection formed within the informed imagination of the interpreter. Identifying and exploring "virtual parallels" is a way of imaginatively interrelating science and faith without surrendering the integrity of one to the other. It is, in short, a way of jump-starting the conversation, of revving the engines.

But there is much more to the interpretive venture than simply identifying virtual parallels or harmonious resonances. Though the terrain between science and biblical faith may be filled with intriguing connections awaiting exploration, it is also riddled with disjunctions or collisions, claims made by the biblical text about the world that conflict with the findings of science. But as I hope to show, acknowledging the disjunctions is also indispensable for interpreting the biblical text. By whatever name we give them, *both* the resonances and the dissonances, the connections and collisions, the parallels and disjunctions, lead to opportunities for fresh dialogue and for appropriating the ancient creation traditions.

Scoping the Biblical World

As a biblical theologian, I take as my point of departure the Scriptures, striving mightily to make sense of them within their own contexts, literary and historical. But as a person of faith, I have the additional privilege of making sense of these ancient texts within our contemporary contexts. As noted earlier, much of the present conflict surrounding science and faith has to do with the Bible and how to read it. Thus, a way forward can only begin by addressing what the Bible is, and to do that we must begin with what the Bible was.

Scripture did not appear in the twinkling of an eye or fall from heaven on golden plates. The Hebrew Bible underwent its own convoluted evolution from oral beginnings to edited endings. Its various traditions emerged throughout nearly a millennium of turbulent history and theological toil. Scripture is the product of a community whose identity was shaped by the exigencies of history, on the one hand, and an abiding conviction of divine providence, on the other. Regarding the Hebrew Bible or Old Testament, the community in question was ancient "Israel," whose name, according to the enigmatic story of Genesis 32:22–32, has something to do with "striving with God" (v. 28),[44] bestowed, not coincidentally, upon Jacob, the Bible's most notorious underdog.[45] Consisting of stories and legal codes, poetry and narrative, genealogies and parables, laments and praise, Scripture reflects the sacred, painful struggle of a community in lively dialogue with itself, the larger world, and God. As the psalmist proclaims about creation, "O LORD, how manifold are your works!" (Ps 104:24a), so something comparable can be said of the Bible: O LORD, how manifold are your books!

Canonical Authority

As Delwin Brown observes, "Canon is not simply a collection; it is a force."[46] Like gravity, the Bible draws in many of its readers. For believers, Scripture's attractive pull is the basis of its authority. But what kind of authority does Scripture exercise?[47] From the Latin *auctoritas*, "authority" bears the sense of "origination," to which the word "author" is related. "Authority" is also related to the verb *auctorare*, "to bind." Such richness of meaning, however, is lost in common, contemporary usage. "Authority" frequently functions in legal or academic discourse, particularly when a specific decision or precedent is sought. People, for example, seek an authoritative reason that results in a binding legal decision. Journalists seek an authoritative or credible source for their research and reporting. Certain individuals assume authoritative status because they are considered experts in their fields. Authority, in short, is domain specific.

When "authority" is applied to the Bible, certain questions emerge: To what domain(s) does biblical authority pertain? Does it apply to scientific matters as much as it does to issues of faith and moral conduct? That may be easy to answer, but what about murkier matters such as sexuality, global warming, and stem-cell research, where ethical and theological reflection needs to engage the natural and social sciences? To complicate matters further, whereas recourse to authority frequently involves seeking specific decisions or answers to specific issues, how does one seek such things from biblical narratives and lament psalms? How does one "squeeze" authority out of the love poetry of the Song of Songs? Or is only the legal or instructive material of the Bible to be deemed authoritative?

Perhaps the best place to begin is by acknowledging that Scripture is authoritative first and foremost with respect to its theological subject, God, who lies beyond the purview of scientific and historical inquiry. Nevertheless, because God is the creator of all things, respect for the authority of Scripture requires respect for the authority of science. The Bible fully acknowledges that many aspects about our world can be discerned through empirical observation—a premise of the biblical wisdom tradition, as we shall see.

"Authority" in the biblical sense is clearly different from its usage in contemporary discourse. Solomon's decree to cut the infant in half was not in itself the right *legal* decision (1 Kgs 3:25)—indeed it would have been horrifically wrong had it been *literally* carried out! Rather, the king's ruling was meant to evoke a response that resolved the conflict. The Bible's "authority," thus, points to its generative power to evoke reflection and shape the conduct, indeed the very identity, of the reader and the reading community.[48] Biblical authority is as formative as it is normative, and it is made manifest in Scripture's creative, authorial power for its readers.[49] As the attractive, binding force of gravity has helped to shape the universe in all its complexity, so the authority of Scripture forms the community of faith in all its variety.

The Refracting Lens of Hermeneutics

Interpreting the Bible, even reading it, requires a lens. The theory and practice of interpretation is the stuff of hermeneutics, a term derived from the mythical Hermes, the messenger of the Olympian gods known for his cunning and swiftness. Regarding the Bible, or any ancient text, the provisional work of hermeneutics is neither swift nor cunning. Slow and stumbling is more like it, for the ancient text is never easily or fully accessible to modern understanding. A chasm separates the modern interpreter and the ancient author, filled only partially and sometimes erroneously by the venerable history of interpretation.

We cannot fully grasp what an ancient text said to its intended audience any-more than we can transport ourselves back in time and conduct interviews. The historian, even the biblical historian, cannot raise the dead.

Still, by attempting to peer ever so slightly over the hermeneutical divide, we can catch a glimpse of what the ancient text *could* have meant through care-ful philological study, literary analysis, knowledge of history, and comparative study. Such tools help us to develop a matrix of possible meanings, some more plausible than others. But we can never cross the divide; the full meaning of the ancient text remains ever elusive. When a telescope probes the night sky, it not only brings into view objects from great distances, it also looks back in time, measured in light-years. Given the great distances traversed by light, telescopes are the only time machines we have. The more distant the galaxy, the older and more obscure it is to the observer and the faster it is moving away.[50] Such is the built-in limitation of sight, thanks to the finite speed of light. Similar is the limitation of hermeneutics, thanks to the span of time and our own cultural myopia. Call it the finite speed of life.

There is, moreover, something broader about the hermeneutical enterprise than simply determining what the text could have said in its earliest context(s). Were that all there is to hermeneutics, biblical interpretation would be a strictly historical enterprise, an antiquarian's dream. For persons of faith, the Bible is much more than an ancient artifact; it bears relevance, and in ways not neces-sarily reducible to the original intent of the author(s), however partially that can be retrieved. What the text *means*, in other words, is as critically important as what it once meant or said.[51] Discerning the text's meaning involves interpret-ing the text in the light of one's experience and within one's community. One cannot interpret the biblical text without interpreting oneself within one's con-text (cultural, religious, and personal).

As for the Bible's contemporary meaning and relevance, the telescope anal-ogy shifts its focus: through interpretation Scripture itself becomes a lens to view the world and oneself. The Bible is more than a set of theological proposi-tions or historical facts. It is more like a lens through which one sees the world differently, bringing into focus what would otherwise be overlooked.[52] If we press the analogy further, studying the shape and curvature of the lens corre-sponds to discerning what the Bible might have said in its earliest context(s). To explore the Bible's meaning today is to peer *through* the lens of Scripture to discern the world around. Put more precisely, Scripture is like the refracting lens of a telescope that gathers and directs incoming light, the light of the world, in new ways. On the other end of the telescope is the eyepiece, which provides focus and magnifying power. It is what the reader provides from his or her own context.

Yes, Scripture is a world unto itself. Its various genres and traditions all populate this dense and strangely diverse world we call the Bible. But its faithful readers are called not to live *in* the Bible, as if that were even possible, but to live *through* it into the world in which they move and have their being. Through interpretation, the Bible becomes the part of the interpreter's lens to perceive and thereby engage his or her world. What, then, does it mean to view the world through the lens of Scripture alongside the powerful telescopes and microscopes, spectrometers and seismometers that have advanced our understanding of creation in ways unimaginable to the ancient authors of Scripture? Such is the impetus of this study.

The lens of Scripture, however, is not uniform. Every creation tradition offers its own distinctive lens. Some of the accounts are systematically presented. Others are rougher cut with poetic pathos. Some adopt the jagged genre of narrative; others read almost like a dispassionate treatise. Consonant with their variant forms, these accounts offer divergent views of the world, of creation. The ancient editors of Scripture chose not to homogenize them or to combine them into a single, comprehensive account of how and why the world came to be what it is. There is no GUT in Scripture, no Grand Unified Theory of creation (or, better, no Grand Unified Theology). Instead, there are diverse traditions reflecting differing theological perspectives. Why? One can only wonder. So begins our journey.

Guide for the Journey

As we explore the cosmologies of Scripture, we need a field guide of sorts, a method of inquiry that will provide a way of comparing these traditions and setting them in conversation with science. I propose three interrelated steps:

Step 1. Elucidate the text's perspective on creation within the text's own contexts.

Step 2. Associate the text's perspective on creation with the perspective of science.

Step 3. Appropriate the text in relation to science and science in relation to the text.

Step 1: Elucidate

This first step is observational; it aims at being descriptive. Here, I hope to elucidate each biblical tradition on its own terms as much as possible. This involves examining the text's literary shape and historical context, as well as its

theological orientation, all of which contribute to the text's construal of creation. As we will see, each text in its context yields a distinct "model" of creation.
I use the term "model" because, as Ian Barbour notes, models have been indispensable to *both* science and theology.[53] Or to adopt the language of Genesis:
each tradition casts creation in the "image" of something.

In the biblical context, a model or image reveals something about God, the
One who stands behind creation, and about ancient Israel, the community that
stands behind the text. And as much as a given tradition says something about
God, it also says something about humanity and its place in the world. Step 1,
in short, takes seriously the world *of* the text as it stands in its literary and theological complexity, as well as the world *behind* the text, its historical background,
to help account for the distinctive emphases of each tradition.

Step 2: Associate

This step probes the world *in front of* the text, specifically the reader's world as
informed by science, in relation to the text. It is here that I identify certain associations, what I call "virtual parallels," between a biblical perspective or model
and scientific outlook. Nevertheless, there are also "collisions" to be acknowledged. They, too, have hermeneutical value. The connections are virtual; the
collisions are real. But both are formative in shaping the text's meaning for
today. In certain cases, science underlines, even extends and deepens, the world
of the biblical text. In others, it provides a counterpoint by which the meaning
of a biblical tradition can be understood in new ways. In any case, identifying
both the connections and collisions, the "virtual parallels" and the irreconcilable differences, offers fresh opportunities to mine the ancient traditions for
their wisdom and relevance, hence the final step.

Step 3: Appropriate

This "final" step returns to the biblical text, but with the insights gained on our
"detour" into science. Central is the question: How do we understand the biblical traditions in the light of current scientific understanding and vice versa?
Cast more generally: How do persons of faith living in a scientifically informed
world appropriate the ancient creation traditions of Scripture? Appropriation
involves not just a new way of interpreting the text but a new way of living it.
It recognizes that, as John Goldingay aptly notes, "interpretation is a moral
issue."[54]

In sum, this interpretive exercise is one that moves from the biblical text to
science and back, a hermeneutical roundabout that yields new understandings

of what the biblical text means for today. The process, broadly conceived, is nothing new among biblical interpreters: the interpretation of any text oscillates between the rhetorical world of the ancient text and the contemporary world of the interpreter. This movement is often called the "hermeneutical circle," or better "spiral," whereby the ancient text and the interpreter's context are brought together into dialogue.

I have found, however, a more incisive way of describing the hermeneutical process, and it is borrowed from the field of science: the "feedback loop." A feedback loop moves from one level of description or inquiry to another and returns to the first level with enriched understanding,[55] for example, from the psychological to the neurobiological and back. The result is a greater understanding of psychological phenomena. The various levels of scientific inquiry acknowledge different levels of physical reality, from the subatomic to the chemical and the biological to human cognition and collective, social behavior. Each level of physical reality operates within its own parameters, dynamics, and laws, yet interdependently so as one level builds on or is "nested in" the other. A hermeneutical feedback loop traverses these levels without reducing one to another.

I propose by analogy a hermeneutical feedback loop between biblical faith and scientific understanding whereby the former is enriched by the latter. If biblical creation faith is to be intelligible today, then it requires the feedback of science. This does not mean that science should dictate the direction of biblical interpretation and theological reflection. That would turn the dialogue into a hostile takeover. Rather, the process allows science to nudge the work of biblical theology in directions it has not yet ventured and, in so doing, add another layer to Scripture's interpretive "thickness" (as my colleague Walter Brueggemann would say) or wondrous depth.

Though the steps described above seem straightforward and clearly delineated on paper, they never are in practice. I have found that even as I regard the text as my point of departure, the hermeneutical journey actually begins with me, the interpreter. As I hold the world of the text in focus (Step 1), I also remain in focus. I view the text through *my* lens. To the text I bring my concerns and convictions, my filters and prejudices. I come to the text not as a blank slate but as a reader informed by science and shaped by my culture, which in turn shapes my interpretation, even my translation. The text is not a container of meaning waiting to be unlocked and opened, but an object of focus with which I interact and whose meaning emerges only by my interaction with it.[56] But even as I read the text through my own myopic lens, I hope my eyes at least remain wide open, enough so that I can see things that lie just outside my own small interpretive world. Other eyes will have to judge.

The model of biblical interpretation that I propose does not strive for mere coexistence between biblical faith and science.[57] It does not settle for a "correlationist" or connect-the-dots approach. To strive only for compatibility is to set the bar of lively engagement far too low. As E. O. Wilson rightly observes, alliances among disparate disciplines are forged "through the medium of interpretation."[58] And, I would add, they are cemented by feedback, some of which may be negative (i.e., "collisions") but much of it positive ("virtual parallels"). Boiled down, this study is aimed at forming and sustaining an alliance through the deceptively simple exercise of interpretation. The result is no futile God-of-the-scientific-gaps exercise. To the contrary, this is science of the divine vistas.

Wisdom's Quilt

The late eminent biologist Stephen Jay Gould is well known for categorically dividing science and religion into "non-overlapping magisteria" (aka NOMA). But largely overlooked is his remark on the need to "unite the patches built by our separate magisteria into a beautiful and coherent quilt called wisdom."[59] Taking my cue from Gould, I submit that more important than the separation of magisteria is the larger quilt whose various "patches"—quilters call them "blocks"—*together* make sense of the world around us. Indeed, more than Gould perhaps realized, wisdom can turn the dividing wall between science and faith into a porous membrane through which selected ideas can flow. Call it TOMA or "tangentially overlapping magisteria."

There is, in fact, strong biblical precedence for holding Scripture and science in tangential, cross-disciplinary relation. The ancients, though lacking the tools of scientific analysis, were keen observers of the natural world. Solomon, remembered in biblical tradition as the consummate royal sage, is said not only to have been prodigious in composing songs and proverbs, but also to have demonstrated an intimate familiarity with the world of botany and zoology, prompting international acclaim.

> He would speak of trees, from the cedar in Lebanon to the hyssop
> that grows out of the wall; he would speak of animals, birds, reptiles,
> and fish. People would come from all the nations to hear Solomon's
> wisdom; they came from all the kings of the earth who had heard of
> his wisdom (1 Kgs 4:32–34).[60]

In ancient Mesopotamia, we find strong evidence of scholarly, if not scientific, observations of the natural world.[61] Some modern scholars have argued that ancient lists of phenomena, both natural and divine, were the product of an emerging *Listenwissenschaft* (literally "list study"),[62] a precursor to scientific study. Although this point is debatable,[63] it comes as no surprise that certain biblical creation accounts feature something akin to catalogues of natural phenomena (e.g., Job 38–41; Ps 104), indicating the "scientific," or at least taxonomic, impulse to classify.[64]

In an apocryphal text also attributed to Solomon, the "spirit of wisdom" is said to impart what could be called *scientific* knowledge of creation:

> For it is God who gave me unerring knowledge of what exists, to
> know the structure of the world and the activity of the elements; the
> beginning and end and middle of times, the alternations of the
> solstices and the changes of the seasons, the cycles of the year and
> the constellations of the stars, the natures of animals and the
> tempers of wild animals, the powers of spirits and the thoughts of
> human beings, the varieties of plants and the virtues of roots; I
> learned both what is secret and what is manifest, for wisdom, the
> fashioner of all things, taught me.
>
> —Wisdom of Solomon 7:17–21 (NRSV)

According to this text, God's wisdom imparts knowledge of astronomical patterns and biological forms. Though such knowledge is considered a matter of revelation, it is still knowledge of *nature*, a natural knowledge. Wisdom, thus, blurs the boundary between divine revelation and natural discovery, both of which are part of the same package of knowledge and neither of which settles for ignorance.

The ancients observed the world with what they had—eyes to see and ears to hear, accompanied by an eagerness to understand it all in relation to God. The biblical sages in particular studied the natural world, and their perspectives, one could say, helped set the intellectual stage for later scientific study, from the classical Greeks to the present. To them, the world was orderly yet tinged with mystery. Nature was beneficial but also formidable and perilous; hence, the Israelites, along with all ancient peoples, had to learn how to develop creation's salutary side. For them that included breeding animals, terracing slopes, and conserving water, all which had to be learned by trial and error, by testing.[65] They were agriculturalists, keenly aware of how to maximize the land's long term productivity. The land was their laboratory of survival. Agronomy was the first science, next to medicine.

To study nature, moreover, was an existential matter, not unlike Brian Greene's point that even if greater scientific understanding does not yield greater utility, it does yield "its own empowerment."[66] Such empowerment, however, points not just to the capacity of the human intellect. There was something more for these ancient inquirers. They felt gripped by something beyond themselves. They bore witness to a world bursting with the manifold nature of life; they marveled over the cosmic expanse of the universe; they trembled before the world's terrible beauty. They asked questions:

> When I gaze upon your heavens, the work of your fingers,
>> the moon and the stars that you have established;
> What are human beings that you are mindful of them,
>> mortals that you care for them? (Psalm 8:3–4)

Awestruck, the ancients actively inquired. Skip ahead two and a half millennia, and you find that the question of human identity persists even as our understanding of the cosmos has dramatically changed. All the more so: as our understanding of the universe has increased astronomically, the question posed by the psalmist is intensified exponentially. Astrobiologist Chris Impey puts it well: "The history of astronomy has been a steady march of awe and ignominy," awe over the unimaginable size and age of the universe, ignominy in realizing that we are not the only game in town.[67] And so the psalmist's question is posed with even greater urgency; the answer, even from the biblical perspective, affirms both the awe and the ignominy (see Pss 8:5 and 144:3–4).

To be sure, the *specific* questions posed by the ancients are not the questions normally posed by scientists, a point that cannot be overstated. Nevertheless, they were no less probing, and that too is a point worth noting. Their various models of creation can profitably be regarded as "thought experiments" (à la Einstein), probing forays into the world of *realia* as inquiring and conceptually powerful as any theoretical formulation proposed by physicists today. If string theorists can comfortably work with five interdependent formulas, why can't the Bible have seven models of creation? But more than thought experiments, these traditions were, and continue to be, theological "life experiments" that not only evoke wonder but also cultivate wisdom. The seven pillars of creation are Wisdom's seven pillars.

Speaking of wisdom: the need to engage science and biblical faith has never been more urgent. We desperately need a new way in the world that is both empirically and biblically credible. One testimony to such need is given in the recent popular book by E. O. Wilson, an alleged representative of "soulless scientism." Cast as a series of letters addressed to a Southern Baptist pastor, *The Creation* reaches out to the religious community to recapture the wonder of nature and, thereby, build an alliance to help mitigate the rapid destruction of the earth's biodiversity.[68] I, for one, want to strengthen this emerging alliance. The "cohort of wonder" must form a partnership of stewardship, and none too soon. For creation's sake—for God's sake—we need a new Great Awakening, a Green Awakening.

2

Revolution and Evolution

Ancient Near Eastern Backgrounds to Creation

We are all trying to construct cosmologies of survival.

—Donna Moylan[1]

Before embarking on our tour of biblical cosmologies, one thing more needs to be considered, namely, our specific point of departure. Would that be the biological theory of evolution or the astronomical account of the Big Bang? No. Scientific cosmology and Neo-Darwinian theory, for example, will provide fruitful avenues of dialogue, but not right now, not in the beginning. Why not? For the simple reason that modern science is not where the biblical cosmologists began. Instead of entering the biblical traditions through the "front door," that is, through the threshold of *our* world—the world that we know through modern science and culture—we must find the text's "backdoor" entrance, namely, the text's own context. It is through this back entrance that we can detect traces of the path our biblical authors blazed in their own journey of theological discernment.

The framers of creation in the Bible inherited a treasure trove of venerable traditions from their cultural neighbors. Instead of creating their accounts *ex nihilo*, the composers of Scripture developed their traditions in dialogue with some of the great religious traditions of the surrounding cultures, particularly those originating from Mesopotamia and Egypt, as well as those of their more immediate Canaanite neighbors. This should come as no surprise: geographically ancient Palestine was sandwiched between the superpowers of the ancient Near East, serving as both their land bridge and buffer zone. Before Israel ever entered the historical scene, some of these kingdoms, both large and small, imperial and petty, had developed their own accounts of creation, which the biblical writers were at least nominally aware. Indeed, many of these traditions constituted what could be called the cultural "canon" of Near Eastern antiquity,[2] and it was in relation to them that ancient Israel developed its own.

An excursion into the creation traditions of ancient Israel's neighbors is, therefore, required before embarking on our more ambitious trek through the Hebrew Scriptures. This "backdoor" approach takes seriously the Bible's ancient context. We must, in other words, understand something of the genesis of the Bible's Genesis accounts in order to understand more fully what these accounts say. A sampling will suffice.

War of the Gods: Mesopotamia and Canaan

First stop: Mesopotamia. One of Israel's most formidable oppressors was Babylon, situated beside the river Euphrates. Babylon's kings include the great Hammurabi (1848–1806 BCE) of law-code fame, and of imperial fame there is Nebuchadnezzar II (605–562 BCE), who conquered Judah and deported much of its populace. From Babylon, beginning in the late second millennium under the reign of Nebuchadnezzar I (ca. 1133–1116 BCE), comes an epic narrative of cosmic intrigue that includes a story of creation.[3] Titled *Enūma elish* ("when on high") for its first two words (in Akkadian), this epic was recited publicly, perhaps even dramatically, every year during the New Year's Festival in April as part of a grand ritual designed to maintain hegemony throughout the land, to re-establish world order. The origin of this practice can plausibly be traced to the recovery of the statue of Marduk, Babylon's patron deity, from Elam in the south, an appropriate occasion for Nebuchadnezzar to proclaim Marduk's supremacy over the gods. The epic was likely written in honor of the occasion and served as a foundation story for the exaltation not only of Marduk but of the Babylonian king himself, renewing his mandate to rule the country and, thus, to assert Babylon's dominion over Mesopotamia and beyond. As performed and witnessed annually by a host of governmental officials, the story did much more than entertain. Its purpose was to maintain loyalty throughout the vast extent of the burgeoning empire. *Enūma elish*, in short, presents an eminently imperial cosmogony.[4]

Enūma elish

The account begins not with a flash of blinding light or with the blaring of trumpets, but with the murky emergence of primordial powers, setting the stage for a cosmic *War and Peace*, whose characters seem to proliferate from one episode to the next.

The epic begins not with a cosmogony (i.e., a creation of the cosmos) but with a theogony, an account of the gods' origins.

> When on high neither the heavens had been named,
>> nor the earth below pronounced by name.
> There was only primordial Apsu, their progenitor,
>> and creator Tiamat, who bore them all.
> Because their waters were intermingling,
>> no pasture land was yet formed, no marshes yet found—
> When none of the gods had yet appeared,
>> no names yet received, no destinies decreed,
> the gods were created therein.[5]

The primordial condition depicted in these opening lines is a state of deficiency: no heavens (skies), no earth, no pastures, no marshes, even no gods to speak of. But pervading this temporary state of lack are two primordial powers. They bear names but remain inchoate; they are at this point more domains than discrete personalities: Apsu, the primordial power of fresh water, and Tiamat, the elemental power of sea or salt water. Together, they constitute a primordial soup from which all creation is ultimately derived, gods included. Geographically, Apsu and Tiamat constitute the alluvial plain of southern Mesopotamia, of the fresh waters of the Tigris and the Euphrates flowing into the Persian Gulf. And so it is in the co-mingling that divine life is conceived. Within Apsu and Tiamat the gods are born and the "begats" begin.

The first half of the tale is structured as a series of generations, six total. Two pairs of deities, representing the deposits of silt from the rivers (Lahmu and Lahamu) and the circular horizons of the heavens and the earth (Anshar and Kishar), emerge in successive generations. Next is the sky-god Anu, born to the last theogonic pair, who in turn sires one of the central characters of the plot, the god of wisdom, Ea or Nudimmud, who "was superior to his forefathers." Later in the narrative, Ea fathers Marduk, the hero of the epic. As generation follows generation, divine power develops in ever greater concentration.

The stage is set for conflict of the most elemental, indeed parental, sort. Like exuberant teenagers, the gods of Ea's generation are inclined to party ("dance") too much, stirring up Tiamat's belly and upsetting father Apsu to the point of insomnia: "By day I cannot rest, by night I cannot sleep!" complains the primordial progenitor. Because these most primordial of deities are also the most inert, they favor equilibrium over movement and change. But Apsu's complaint is met with harsh rebuke. Tiamat's maternal instincts prevail; she would rather indulge the godlings. Undeterred, Apsu hatches a scheme for their destruction. But the young gods catch wind of it, and Ea casts a sacred spell, putting Apsu into a deep sleep (which, ironically, fulfills Apsu's wish!). Ea dons Apsu's "mantle of radiance," kills him, and builds his home "on top of Apsu." Giving a cry of triumph, Ea then rests quietly (!) inside his new home, named after the great-great grandfather he murdered.

Life continues apace: Ea and his consort Damkina dwell together "in splendor" and beget a son, Marduk, "the cleverest of the clever, sage of the gods." Anu, his grandfather, rejoices and creates for the child's amusement "four, fearful winds." This has the unfortunate consequence of causing a flood-wave that stirs up Tiamat. The older gods, "unable to rest," plead to Tiamat, appealing to her maternal instincts: "Are you not a mother? You heave restlessly. But what about us, who cannot rest? Don't you love us?" This time her instincts are aroused for destruction. They convince Tiamat to destroy this rowdy generation. She consents, and they whip themselves up into a frenetic rage and convene a council to make preparations for war. For weapons, monstrous dragons are created, bearing "mantles of radiance" as they assume divine status. Tiamat also appoints a military leader, her new "lover" Qingu, to lead this army of monsters into battle. To Qingu is given the "tablet of destinies," granting him unrivaled authority.

Meanwhile back at the Apsu, the all-knowing Ea once again catches wind of the preparations, but this time he is caught off guard. He appeals to Anshar, his grandfather, to do something, but "his roar . . . was quite weak." It is Ea's problem, Anshar bluntly states. Ea is the one who must declare war; after all, he was the one who slew Apsu.

The tablet becomes fragmentary at this point, but it is clear that Anu, Ea's father, attempts to confront Tiamat but makes a hasty retreat after seeing the extent of her preparations. Ea, too, suffers a failure of nerve. Their last resort is to choose the youngest among them, Marduk, Ea's son. Anshar warns Marduk, "My son, (don't you realize that) it is Tiamat, of womankind, who will advance against you with arms?" Possessing the brash courage of youth yet also a shrewdness beyond his years, Marduk accepts with one condition: if he is the one to defeat Tiamat, he must be granted absolute sovereignty so that any decree of his will automatically find fulfillment. With this offer, Anshar appeals to have a council convened to hear Marduk's case. Amid a lavish banquet and drunken revelry, the "great gods who fix the fates" confer upon Marduk the right of sovereign reign. Also included in the deal is a "princely shrine." "May your utterance be law, your word never be falsified," the gods proclaim. By way of confirmation, Marduk is invited to command the destruction of a constellation and then speak it back into existence. When Marduk does just that, proving his word to be efficacious, the gods proclaim him king, bestowing upon him the enduring symbols of royalty. By unanimous consent, Marduk is made the one and only sovereign deity.

Within the history of religious thought, Marduk's assumption of supremacy marks a significant step toward monotheism, which would be fully reached with the elimination of all other deities. But not here; that is another story (cf. Ps 82). As for this story, the gods remain with Marduk. He receives from them "an unfaceable weapon to crush the foe," a "flood-weapon," and Marduk sets forth "on the path of obedience and peace" with the gods' command ringing in his ears: "Go, and cut off the life of Tiamat! Let the winds bear her blood to us as good news!"

The battle begins. At the start, Tiamat feigns friendliness. But Marduk sees through the ruse and accuses her of forsaking her maternal compassion by carrying through on her dead husband's plan, now with Qingu in charge. Amid the clash of arms, Marduk dispatches a fierce wind to force open Tiamat's mouth, into which he shoots an arrow "which pierced her belly." Marduk vanquishes the deity of watery chaos, and the North Wind carries her blood off as "good news" to Marduk's "fathers." Singing in the (bloody) rain, the council's gods celebrate victory. Having conquered chaos, Marduk rests.

Conflict, war, and victory, the essential ingredients of any ancient epic, prepare for the following scene: creation. The conflictive narrative of theogony, specifically theomachy, leads to cosmogony. As Marduk rests, a brilliant idea comes to him. He dismembers Tiamat's corpse:

> (Marduk) divided the monstrous shape and created marvels (from it);
>> he sliced her in half like a fish for drying:
> Half of her he put up to roof the sky;
>> he drew a bolt across and made a guard hold it.
> Her waters he arranged so that they could not escape.[6]

From death comes life. In addition to fashioning the universe from Tiamat's body, Marduk establishes his own estate, Esharra, a great shrine constructed in the "image of Apsu," and sanctuaries for other gods, including his forebears Anu and Ea. Marduk fixes the constellations, including the planets and the moon, and with them the divisions of the calendar year. He pierces Tiamat's eyes to form the sources of the Euphrates and the Tigris rivers. Her tail is bent up into the sky to fashion the Milky Way, and her crotch is used to support the sky. As for Qingu, Tiamat's lover, Marduk divests him of the "tablet of destinies" and takes possession of it. Marduk then declares his intention to establish his own house in front of Esharra as his "cult center," whereby the gods can find rest whenever they want. This is Babylon, the guest house of the gods and, thus, "the center of religion."

But there is one other matter, one more ingenious idea to implement. Marduk consults Ea, his father, to create humankind. Such a plan is not without cause: imposed upon this earthly creature will be the "work of the gods," thus making possible the "leisure of the gods." Ea suggests that Qingu's blood, the blood of the rebel, provide the raw material for humankind's creation. And so it happens.

Grateful, the gods offer to make Marduk the sanctuary of his dreams, their night's resting place, Babylon. In two years they build a magnificent ziggurat for the Apsu, complete with individual shrines for themselves. Merrymaking accompanies the completion by which the destinies are established, assignments made, and decrees fixed. The order of the universe has found its completion. In the final act, Marduk receives from Anu a miraculous bow, a symbol of Ishtar, the goddess of love and war. In the final act of obeisance, the gods swear an oath of absolute fealty to Marduk. The epic concludes with the conferral of fifty throne names, drawn from the fifty gods in attendance, highlighting different aspects of Marduk's rule over heaven and earth. Marduk, in essence, absorbs the various powers of the gods.

In this epic account, creation begins when conflict ends. Creation is the corollary of conquest, the outcome of victory. A carcass is required for the cosmos, and the bloodshed of a rebel is deemed necessary for human creation. Apsu and Tiamat, the elemental powers of creation, are divested of their personas, including their divinity. They are demythologized with a vengeance, as it were, to serve as creation's raw material. Construction follows destruction. The act of creation, moreover, is vivid testimony to the exercise of imperial power. Marduk's exaltation and the world's creation are bound together. The victor's prowess on the battlefield is matched by his ingenuity on the construction site. Marduk's credentials include conquest in addition to creativity. And so they must in this imperial drama. But munificence is also included. The creation of humankind in particular is testimony to Marduk's sovereign generosity, which grants a measure of freedom to the gods. Their freedom from servitude, however, is the burden humankind must shoulder. Humankind, made in the image of gore, from the blood of a rebel god, is the quintessential servant of the gods. Humanity's creation is humanity's bondage.

Atraḥasīs

Humankind's genesis is further detailed in the related Babylonian account of the flood, known from its central character, Atraḥasīs ("Extra-wise"), and dates to at least the early seventeenth century BCE, if not earlier.[7] The Old Babylonian

Version opens with the gods, "instead of man," laboring over the building of irrigation canals (an endemic feature of Mesopotamian agriculture). The severity of such servitude prompts the gods to revolt against their taskmaster, the high god Enlil. The ensuing rabble rouses Enlil from his blissful sleep. Violent conflict seems eminent. But a cooler head prevails, namely that of Ea, the god of wisdom. Consequently, the normally irascible Enlil listens intently to the rebels' complaint. A plan is devised: instead of outright war, one of the gods, a leader of the revolt, will be selected "for destruction." From his blood the womb-goddess Belet-ili, or Mami, will create "a mortal so that he may bear the yoke," the "load of the gods." A "god who had intelligence" is slaughtered, and the midwife of the gods mixes his blood with clay so that "a ghost [comes] into existence from the god's flesh," that is, something of the god's spirit becomes preserved in humankind.[8] Fourteen pieces of clay are used to fashion seven males and seven females; "two by two" they are created. Mami jubilantly proclaims that her creation experiment has succeeded: the gods are now relieved of their excessive work. As in *Enūma elish*, divine liberation is brought about by human bondage.

But there is more. Upon completing her genetics experiment, Mami makes an ominous pronouncement:

> I have imposed your load on mortals,
> You have bestowed noise on humankind.[9]

Along with the work that human beings must shoulder is "noise." And it is such noise that causes trouble in the divine realm, the noise of proliferation. The narrative then proceeds with a series of dire episodes in which human beings multiply like rabbits, and their collective "noise" reaches such a volume that it is likened to a "bellowing bull." Not surprisingly, Enlil loses sleep, and with each bout of insomnia he pronounces a disease or famine to decimate the population. Each time, his plan is ameliorated, if not outright thwarted, by Ea, who divulges it to one Atrahasīs. The narrative reaches its climax in the divinely agreed-upon plan of a global flood to wipeout the human population altogether. Like the biblical character of Noah, Atrahasīs is instructed to build a boat to ensure the survival of the human race. (Animals are not mentioned.) The flood comes, roaring like a bull with the storm-god Adad bellowing from the clouds, surpassing all other "noise," including human. But Atrahasīs and his family escape. When the waters recede, he offers sacrifice to the gods, who gather "like flies over the offering." While Enlil is enraged that the human race has survived, the gods are remorseful over the swath of destruction they have wrought. The epic concludes with a resolution between Enlil and the gods by which the growth of the human race is checked by certain measures set in place, including infant mortality inflicted by a demon that snatches babies during birth. This ancient epic would have made Thomas Malthus proud.[10]

Ba'al

Somewhat closer to Israel's cultural home is the so-called Ba'al Epic discovered in the ancient coastal city of Ugarit, now known by its Arabic name as Ras-Shamra. Located beside the Syrian coast of the Mediterranean Sea at roughly the northern latitude of Cyprus, Ugarit was a thriving city of Canaanite culture during the Late Bronze Age (1500–1200 BCE) until it was destroyed by the so-called Sea Peoples, invaders from Crete and other Greek islands. From the city's royal archives comes a six-tablets-long epic about its chief god Ba'lu/Haddu, a storm-god whose dual name apparently means "Lord/Thunderer." Ba'al is widely known in biblical tradition as well.[11] The Ugaritic narrative recounts the deity's various struggles to establish his cosmic kingship. Although creation per se is not a theme of this epic, striking parallels and differences can be observed between Ba'al's story and the creation stories of the ancient Near East, including Israel's. Unlike Marduk, Ba'al is a much more vulnerable deity. His struggles are fraught with greater suspense and mixed outcomes. Whereas Tiamat's fate is a foregone conclusion before Marduk's unflappable confidence, not to mention sovereign power, Ba'al's nemesis Mot, the god of the underworld, is a more even match. Ba'al's temporary defeat by Mot accounts for the seasonal cycle of fructifying rain and withering drought that is standard along the Mediterranean Seaboard, a perennial meteorological "struggle."

But the episode that is most pertinent for our purposes is the one recorded on the first two tablets, which present Ba'al's struggle against Yamm or "Sea." As in *Enūma elish*, the conflict between a warrior deity and a god of watery chaos is central. Here, Yamm, known also as "Judge River," poses no cosmic threat to the pantheon of deities; in fact, he is supported by the high god El, who pronounces Ba'al to be Yamm's slave. But this will not do for a god with royal aspirations. Ba'al, the "cloudrider," commissions the construction of two weapons from the divine craftsman Kothar wa-Hasis and with them engages in battle against Yamm. The first attack does not phase Yamm, but the second causes him to collapse. "Ba'al drags and dismembers (?) Yamm; he destroys Judge River."[12] A victory feast ensues. Ba'al's victory paves the way for the construction of his longed-for residence, a palace for his enthronement. The construction, not coincidentally, takes seven days to complete (cf. Gen 1:1–2:3).[13]

Ba'al's victory over watery chaos, however, is not limited to Yamm. A related creature is also defeated in the battle. Mot, Ba'al's greater nemesis, addresses Ba'al prior to their fateful engagement:

> When you killed Litan, the Fleeing Serpent,
> Annihilated the Twisting Serpent,
> The Potentate with Seven Heas,
> The Heavens grew hot, they withered. (*KTU* 1.5 I 1–4)[14]

Ba'al is credited by Mot for having defeated, along with Yamm, the sea monster Litan, whose biblical equivalent is Leviathan, as found in an apocalyptic text from Isaiah:

> On that day the LORD with his cruel and great and strong sword will
> punish Leviathan the fleeing serpent, Leviathan the twisting serpent,
> and he will kill the dragon that is in the sea. (Isa 27:1)

The "twisting serpent" can also be found in a biblical psalm that looks back to the primordial past in order to motivate God to action in the present:

> Yet God my King is from of old,
>> working salvation in the earth.
> You divided the sea by your might;
>> you broke the heads of the dragons in the waters.
> You crushed the heads of Leviathan;
>> you gave him as food for the creatures of the wilderness.
> You cut openings for springs and torrents;
>> You dried up ever-flowing streams. (Ps 74:12–15)

Whereas the sea dragon in the Ugaritic narrative and in the psalm is DOA (see also Isa 51:9–10), Leviathan in Isaiah 27:1 remains alive and twisting, though slated for destruction at the end times. Etymologically related, Leviathan and Litan are virtually identical. By whatever name, this monster of the deep must be vanquished before something new can be created. For Marduk, that "something new" was all of creation, including his own residence. For Ba'al, it was the construction of his multiwindowed palace, all in seven days.

Creation's Self-Evolution: Egypt

On the other side of the Levant comes a series of creation accounts that differ significantly from the construction-from-conflict stories of Mesopotamia and Ugarit, and not just in terms of content. The genre, too, is different. No gripping epics of creation have been found in Egypt. In the land of the Nile, the origins of creation are found mostly in funerary compositions, such as the Pyramid Texts, Coffin Texts, and the Book of the Dead, as well as in magical spells and hymns for certain temple rituals. For Egyptians, death prompted near mystical reflection on a cosmic scale. The most significant creation accounts come from three separate religious centers, each of which was home to a different patron deity and to a distinctive creation account: Atum in Heliopolis, Ptah in Memphis, and Amun in Thebes.[15]

Heliopolis

Egyptians believed that death could be overcome by identifying with the source of all matter. According to the priests of Heliopolis, that source was Atum, and the way creation came about was through Atum's own "evolution,"[16] a process of self-differentiation. Atum "evolved" into manifold creation from a primordial watery state: "I am the one who made me. I built myself as I wished, according to my heart."[17] In a Pyramid Texts spell, Atum produces the atmospheres above and below the earth (Shu and Tefnut, respectively) through masturbation, one of the common metaphors used to explain the origin of the world from a single material source.[18]

In another text, Atum's son and daughter, Shu and Tefnut, are sneezed, spat out,[19] or begotten.[20] Shu is first "created" in Atum's heart before being "exhaled" from his "nose."[21] He is identified with Life. The identity of Tefnut, Atum's "daughter," is Order (Ma'at).[22] Unlike the theogonic pairs in Mesopotamian creation, Atum is a single parent, like Israel's God YHWH. He is the self-evolver,[23] the god who differentiates himself into the physical world, developing from the one to the many, from simplicity to complexity. As Shu, his first progeny, claims, "I am one who is millions."[24] The sun is identified with Atum's Eye, whose dawning from the Lotus flower on the primeval hill marks the first sunrise, thereby establishing the very order of creation. Divinity, thus, adheres to creation in all of its diversity. Divine immanence pervades the elements of nature as order characterizes the cosmos. Indeed, creation's order is itself divine (Ma'at). The ordered cosmos is the differentiated body of divinity.

Memphis

An alternative account of creation comes from a dedicatory inscription composed for the temple of Ptah in Memphis, set up during the Twenty-fifth Dynasty under the reign of the Nubian pharaoh Shabaqo (or Shabaka, 716–702 BCE). The inscription purports to be a copy of an older document going back at least to the Nineteenth Dynasty (1293–1185 BCE). Here, the creator is Ptah, the "lord of order," originally a local craftsman deity. Instead of masturbating, Ptah employs speech. The process is precise: creation begins in the "heart" and is fulfilled in the "tongue" or command.[25] The "heart" is associated with (and evolves into) the scribal god Thoth; the "tongue" becomes developed into Horus, the god of kingship. It is in the conjunction of thought and word, of intentionality and instrumentality, that creation comes about. This is "intelligent design" at its most precise in the ancient world.

Ptah's commands activate (literally "excite") order in the cosmos for "life and dominion."[26] As "He who made totality and caused the gods to evolve," Ptah is also identified with Ta-tenen, "the land that becomes distinct" or the primeval hill.[27] The prototype of the pyramid, Ta-tenen designates the arable land that emerges from the waters when the Nile recedes.

Thebes

From Thebes, the god Amun is given the role of primordial creator. His description represents something of a synthesis of Atum and Ptah. Amun's role in creation is delineated most clearly on a papyrus dated at the end of Ramesses II's reign (1279–1212 BCE). Like Atum, Amun is a self-evolver: "You began evolution with nothing."[28] But unlike Atum, Amun is "concealed from the gods, and his aspect is unknown."[29] He remains separate from the created order, transcendent vis-à-vis the gods, "farther than the sky" and "deeper than the Duat" (the realm of watery darkness in the underworld). Light is "his evolution on the first occasion."[30] Known also as the "Great Honker," Amun "honked by voice" the first dawn.[31] Creation began by Amun's "speaking in the midst of stillness, opening every eye and causing them to look."[32]

Despite the variety of views represented in these traditions, Egyptians saw the world as governed by an inviolable rhythm of eternal repetition.[33] During its daily "cycle," the sun makes its circuit across the sky to then traverse the Duat below, where it joins with the body of Osiris and receives the power to quell the chaos of night and be reborn for a new day. Order was considered a daily, recurring phenomenon. The sunrise reestablishes the cosmic order through the act of "separating," itself an act of justice.[34]

Conclusions

Though sharing some degree of commonality, Egyptian and Semitic versions of cosmogony essentially offer competing visions. Chaos plays a role in both, but to a much greater degree in Mesopotamia and Ugarit than in the land of the

Nile. The "evolutionary" model of Egyptian cosmology, particularly in its Heliopolian and Thebian forms, contrasts sharply with the highly charged combat motif developed in Mesopotamia and Canaan. In both, however, the spoken word is crucial. As Ptah creates by word, so Marduk proves his power by speaking a constellation into existence. But it is in Egyptian, specifically Memphite, cosmology that creation by word takes center stage. For Marduk, the word is a matter of divine display. When it comes to creation, the victorious Marduk is more a "hands-on" deity. On the other hand, the self-differentiating Atum produces the most sacramentalized version of creation: creation is his own body. In all the accounts, however, both Egyptian and Mesopotamian, creation constitutes a divine stroke of genius, the product of careful thought, planning, and execution, whether by construction or by evolution.

Ancient Israel inherited the creation stories of its mighty and not-so-mighty neighbors and, consequently, mediated and transformed them. The biblical accounts were to some degree ancient Israel's responses to the mythic accounts and traditions of its neighbors. In comparison to the elaborate epics and mystical poems of the ancient Near East, they seem straightforwardly simple. Nevertheless, the biblical accounts of creation, like the Egyptian cosmogonies, share a remarkable diversity of views, which taken together point to a complexity about the world unmatched by any one account. Beginning with the seven days of creation, the Bible speaks about creation in seven discrete ways.

3

The Cosmic Temple

Cosmogony According to Genesis 1:1–2:3

Some say that the origin of life brings order out of chaos—but I say, "order out of order out of order!"

—Günter Wächtershäuser[1]

We are as gods and might as well get good at it.

—Stewart Brand[2]

Compared to the imperial drama of Mesopotamian cosmogony, the first creation account of the Bible seems positively ethereal. Set beside the mystical poetry of the Egyptian traditions, Genesis 1:1–2:3 (henceforth "Genesis 1") reads like a dispassionate treatise. By its own measure, Genesis 1 resembles more an itemized list than a flowing narrative,[3] more a report than a story. It reflects a literary austerity, an "eerie abstractness,"[4] that avoids the fray of epic conflict and eschews the pathos of ancient poetry. Genesis 1, known among scholars as the "Priestly" account of creation, acknowledges the magisterial character of its primary character, God, while describing the non-imperious ways God goes about fashioning a world of ordered complexity. To press a metaphor, creation in Genesis 1 is a construction zone in which various building blocks are joined together to build a cosmic edifice replete with order and variation. Call it "construction with modification."[5]

Translation

Preface
(1:1) When God began[6] to create the heavens and the earth, (1:2) the earth was void and vacuum,[7] and darkness was upon the surface of the deep while the breath[8] of God was hovering[9] over the surface of the waters.[10]

Day 1
(1:3) Then God said, "Let there be light." And there was light.[11] (1:4) God saw that the light was good,[12] and God divided between the light and the darkness. (1:5) God named the light "Day," and the darkness he named "Night." Evening came and then morning, day one.

Day 2
(1:6) Then God said, "Let there be a firmament[13] amid the waters to serve as a divider between the waters." (1:7) So God made the firmament, and it divided the waters below the firmament and the waters above the firmament. And it was so. (1:8) God named the firmament "Heavens." Evening came and then morning, second day.

Day 3
(1:9) Then God said, "Let the waters under the heavens gather themselves[14] to one place to let the dry land appear." And it was so. (1:10) God named the dry land "Earth" and the gathering of the waters he named "Seas." God saw that it was good.

(1:11) Then God said, "Let the earth sprout forth vegetation:[15] plants bearing seed, fruit-trees producing fruit containing seed,[16] each according to its kind on the earth." And it was so. (1:12) So the earth brought forth vegetation: plants bearing seed, each according to its kind, and trees producing fruit containing seed, each according to its kind. And God saw that it was good. (1:13) Evening came and then morning, third day.

Day 4
(1:14) Then God said, "Let there be lights in the heavenly firmament to divide between the day and the night, and let them be signs for determining[17] seasons, days, and years. (1:15) Let them be lights in the

heavenly firmament to illumine the earth." And it was so. (1:16) So God made the two great lights, the greater light to rule the day and the lesser light to rule the night, as well as the stars. (1:17) And God set them in the heavenly firmament to illumine the earth, (1:18) to rule the day and the night, and to divide between the light and the darkness. God saw that it was good. (1:19) Evening came and then morning, fourth day.

Day 5
(1:20) Then God said, "Let the waters produce swarms of living beings, and let the winged creatures fly about on the earth, across the surface of the heavenly firmament." (1:21) So God created the great sea monsters and every living being that moves, of which the waters produced swarms, according to their kinds, and every winged creature, each according to its kind. God saw that it was good. (1:22) God blessed them, saying, "Be fruitful, multiply, and fill the waters in the seas, and let the winged creatures increase on the earth." (1:23) Evening came and then morning, fifth day.

Day 6
(1:24) Then God said, "Let the earth bring forth living beings, each according to its kind: domestic animals, crawlers, and wild animals, each according to its kind." And it was so. (1:25) So God made the wild animals, each according to its kind, and domestic animals, each according to its kind, and everything that crawls on the ground, each according to its kind. God saw that it was good.(1:26) Then God said, "Let us make humanity in our image, after our likeness, so that they may rule over the fish of the sea, and over the winged creatures in the heavens, over the domestic animals, over all the land,[18] and over everything that crawls on the land." (1:27) So God created the human being[19] in his image, in the image of God he created it, male and female he created them. (1:28) God blessed them, and God said to them, "Be fruitful and multiply; fill the earth and subdue it and rule over the fish of the sea and over the winged creatures in the heavens and over every creature[20] that crawls on the ground."

(1:29) God said, "*Voila!* I hereby give to you every seed-bearing plant that is upon the surface of all the land and every tree whose fruit bears seed. To you it shall be for food. (1:30) And to every wild animal, to every winged creature in the heavens, to every creature that crawls on the ground in which there is a living self, (I hereby give) every green plant for food." And it was so. (1:31) God saw everything that he had made, and, *voila*, it was extremely good. Evening came and then morning, the sixth day.[21]

Day 7
(2:1) Thus the heavens and the earth and all their hosts were completed. (2:2) On the seventh day God finished the work that he had done; he ceased[22] on the seventh day from all the work that he had done. (2:3) God blessed the seventh day and made it holy, because on it he ceased from all the work that God had done in creation.[23]

Character of Creation

As the Bible's first creation account, Genesis 1 enjoys pride of place. Positioned as the cosmogony of cosmogonies, the Priestly account is also the most carefully structured text in all of Scripture. Its intricate arrangement reflects something of creation's own integrity, as we shall see. But first, back to the "beginning."

A Relative Beginning

As the opening verses make clear, creation's integrity comes not from nowhere. Creation began in a primordial state of "void and vacuum" (*tōhû wābōhû*). But this initial condition is not nothingness, for God's first act takes place in a given setting, a dark, watery mess.[24] But not to worry: the "deep" (*tĕhôm*) and the "waters" (*mayim*) lack the combative chaos that raged in *Enūma elish*, the Ba'al Epic, and for that matter Psalm 74:12–17.[25] Neither Tiamat nor Leviathan is lurking under the surface.[26] Rather, the curtain of creation rises to reveal a benign primordial soup. Although there is nothing salutary about impenetrable darkness (*ḥōšek*),[27] nothing about it is deemed inimical to God. Indeed, God's breath suspended, "hovering," over the dark waters, like a mother eagle over her nestlings (Deut 32:11), could suggest a relationship of intimacy as opposed to enmity. At the very least, it establishes a positive relationship with the deep, as borne out by the waters' constructive role demonstrated later in creation. God's division of the waters above and below is no dismemberment of conquered chaos but a constructive separation that establishes form out of formlessness.

Powers of Seven

The account of Genesis 1 is carefully structured around seven days within which eight acts of creation and ten commands are listed.[28] The number seven is no random counting.[29] God "saw" and pronounced creation "good" seven times; "earth" or "land" (same word in Hebrew) appears twenty-one times; "God" is repeated thirty-five times. The number seven, or multiples thereof, also crops up within certain discrete passages: Genesis 1:1 consists of seven words; 1:2 features fourteen words; Genesis 2:1–3 renders a word count of thirty five. In fact, the total word count of the narrative proper (1:1–2:3) is 469 in Hebrew (7 × 67).[30] In light of such numerical virtuosity, the great nineteenth-century biblical scholar Julius Wellhausen asked, "What sort of creative power is that which brings forth nothing but numbers and names?"[31] A very impressive power indeed! Although underappreciated by early German scholarship, the order inscribed in this account imparts a remarkable mathematical aesthetic, the quantifiable order of a fully stable, life-sustaining, differentiated world.

Why seven? The number is a perfectly odd integer. It is also oddly powerful. Mathematically, the number seven is unique in that it, along with its multiples, divides any non-divisible number with the same repetitive pattern: $1/7 = .142857142\ldots$; $2/7 = .2857142857142\ldots$; $3/7 = .42857142\ldots$, with the seventh decimal position marking the return of the repeated series! The number seven bears a distinctly varied but repeated order. Repeatable variation, in fact, characterizes the text's structure. Moreover, the number seven may derive from the seven cosmic non-stellar entities visible to the ancients,[32] consonant with the fact that Genesis 1 is the most cosmically oriented creation text in the Hebrew Scriptures. But most crucial for the Priestly account, the number seven connotes a ritual sense of completion or fulfillment.[33] It is no coincidence, then, that creation's completion is affirmed on the seventh day, rather than on the sixth (Gen 2:2).[34] Marking the formal completion of all work, the seventh day is the culmination and conclusion of the Genesis account. It is not an afterthought.

Cosmic Temple

The end result of this sevenfold scheme is a fully differentiated creation. Throughout Genesis 1 God goes about the work of "separating out" creation: light from darkness (1:4), waters above from waters below (vv. 6–7), and day from night (vv. 14, 18). As a result, discrete domains are established: light, water, sky, and land, each of which accommodates various entities, living and otherwise. In the course of the Genesis narration, both the domains and the members of these domains reveal an overarching symmetry. Call it the "Genesis Code."

Genesis 1:2	
Void and Vacuum	

Day 1 (1:3–5)	**Day 4** (1:14–19)
Light	Lights
Day 2 (1:6–8)	**Day 5** (1:20–23)
Firmament	Aviary life
Waters below	Marine life
Day 3 (1:9–13)	**Day 6** (1:24–31)
Land	Land animals
	Humans
Vegetation	Food

Genesis Code (1).

According to their thematic correspondences, the six days of creation line up to form two parallel columns.[35] Their chronological ordering, in other words, gives rise to a thematic symmetry. Days 1–3 in the left column establish the cosmic domains, which are then filled or populated in the right column with various entities (Days 4–6). Read vertically, these two columns address the two abject conditions of lack described in Genesis 1:2, "void and vacuum." The left column (Days 1–3) gives *form* to creation through the establishment of discrete domains, including light on Day 1,[36] with Day 3 climactically depicting the growth of vegetation on the land. This concluding act vividly changes the earth's primordial condition from its formless state of barrenness: the earth is no longer a "void" (*tōhû*) but a fructified land, providing the means for sustaining animal life. The right column, Days 4–6, fills these domains with their respective inhabitants, from celestial bodies that "rule" both day and night to human beings, who exercise "dominion." The creative acts on Days 5 and 6 specifically change creation's primordial condition from "vacuum" or emptiness (*bōhû*) to fullness.[37] Genesis 1, in short, describes the systematic differentiation of the cosmos that allows for and sustains the plethora of life.

The six-day schema exhibits symmetrical correspondence, but the symmetry is not perfect. Within its literary patterning, Genesis 1 features a number of what J. Richard Middleton aptly calls "nonpredictable variations"[38] or literary imbalances. Vegetation, for example, occurs on the third day, concluding the left column, even though plants, like the animals, fill the land.[39] The sixth day would have been a better literary fit for the creation of plants. Days 5 and 6, moreover, are one-sidedly weighted with the language of blessing, which bears no correspondence to Days 2 and 3. Structurally, certain literary building blocks such as the fulfillment report and transition formula ("and it was so") either do not appear in a consistent order or, in certain cases, are entirely absent.[40] Finally, of all the days enumerated in the account, only "the sixth day" and "the seventh day" bear definite articles. The text, in short, manifests a symmetry supple enough to allow for variation and surprise.

These small variations, however, pale in comparison to the most significant case of dissymmetry in the text, namely, Day 7. Having no corresponding partner, the seventh day is unique. By its presence, the tight six-day symmetry of the Genesis account is broken. Nevertheless, this distinctly odd day does establish a *vertical* correspondence to creation's initial, pre-creative condition, as described in 1:2, which one could call paradoxically "Day 0."

Day 0 (1:2)
Void and Vacuum

Day 1 (1:3–5)	**Day 4** (1:14–19)
Light	Lights
Day 2 (1:6–8)	**Day 5** (1:20–23)
Firmament	Aviary life
Waters below	Marine life
Day 3 (1:9–13)	**Day 6** (1:24–31)
Land	Land animals
	Humans
Vegetation	Food

Day 7 (2:1–3)
Creation completed
Holy

Genesis Code (2).

Together these two "days" form a surprising correspondence, the static "day" of non-creation and the "theo-static" seventh day. The time*less* character of "Day 0" shares a subtle affinity with Day 7: the final day lacks the temporal formula "evening came and then morning." It, too, is a day suspended above temporal regularities. Yet these two "days" could not be more different: "Day 0" refers to creation's empty formlessness; Day 7 marks creation formed and filled (2:1).[41] The final day serves as the capstone for the entire structure, for it shares something of God's holiness, set apart from creation yet not set against it.

Without this symmetry-breaking, concluding seventh day, the creation pattern would lose a distinction that remains hidden to modern readers not acquainted with the ancient architecture of sacred space. As we know from their remains, many temples of the ancient Near East, particularly in the Syro-Palestinian region, followed a threefold or tripartite structure, a pattern also found, not coincidentally, in the literary symmetry of the Genesis text.

For example, the temple as described in 1 Kings 6 consisted of an outer vestibule or portico flanked by two pillars, the nave or main room, and an inner sanctuary or holy of holies (*dĕbîr*), as diagramed here.

Portico
Nave
Holy of Holies

This threefold arrangement of sacred space corresponds to the way in which the various days of creation are distributed both chronologically and thematically. The first six days, by virtue of their correspondence, establish the architectural boundaries of sacred space. The last day inhabits, as it were, the most holy space.

As creation unfolds "daily," it becomes constructed in the *imago templi*, in the model of a temple.[42] What took Solomon seven years to complete (1 Kgs 6:38), God took only seven days, and on a cosmic scale no less![43] In the holiest recess of the temple God dwells, and on the holiest day of the week God rests.[44]

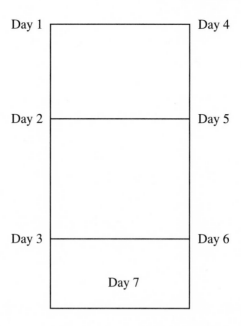

Imago Dei

The cosmic temple code of Genesis 1 also reveals something significant about humanity's role and identity in the creation account. Many an ancient temple contained an image of its resident deity within its inner sanctum. In Jerusalem, however, the physical representation of God was expressly forbidden, at least by the time of the exile (sixth century BCE), as one finds, for example, conveyed in a certain commandment of the Decalogue: "You shall not make for yourself an idol." (Exod 20:4; Deut 5:8). Such is what scholars call Israel's "aniconic" tradition, a categorical prohibition of divine images, which distinguished ancient Israel's worship practices from some of its more religiously elaborate neighbors.[45] Because God was considered to be without form, at least in some religious circles, the deity could not be represented (see Deut 4:12–18). According to biblical lore, the inner sanctum of the Jerusalem temple contained the "ark of the covenant" covered by the outstretched wings of two cherubim (1 Kgs 6:23–28). The ark was regarded as God's footstool or throne and as the container of the stone tablets of the Decalogue.[46] In place of a statue, the statutes of God were housed.

Genesis 1, however, does not jettison the language of divine image but recasts it by identifying the *imago Dei* with human beings, created on the sixth day:

> Then God said, "Let us make humanity (*'ādām*) in our image (*ṣelem*),
> after our likeness (*děmût*)"

> So God created the human being (*hā'ādām*) in his image (*ṣelem*),
>> in the image (*ṣelem*) of God he created them;
>>> male and female he created them. (Gen 1:26a, 27)

The creation of humanity is unique. Nowhere else in Genesis 1 does God command the divine assembly, and nowhere else is the term "image" used. Human beings alone, according to the text, bear an iconic relation to the divine. The Hebrew term for "image" (*ṣelem*) is elsewhere used for idols.[47] By virtue of their unique creation, human beings take on the status of near divinity; they bear a distinctly theophanic presence in creation. While God lacks a blatantly anthropomorphic profile in Genesis 1,[48] humanity is unequivocally "theomorphic" by design. Cast in God's image, women and men reflect and refract God's presence in the world. The only appropriate "image of God," according to Genesis, is one made of flesh and blood, not wood or gold.

There are at least four interrelated ways by which to understand how human beings are linked to God through the language of "image" in Genesis.[49]

ESSENTIALIST LINK. Human beings bear something of God's very nature. In Genesis 5:3, the Priestly narrator provides a genetic analogy by which to understand this connection. Adam becomes "the father of a son in his likeness, according to his image." Seth bears the image of his father as Adam does of God. The son bears an essential resemblance to his father and, therefore, a certain resemblance to God, passed on from one generation to the next. "[A]s long as there are descendents of Adam on earth, the *imago Dei* will not and cannot be lost."[50]

FUNCTIONAL LINK. The *imago* reflects something of humanity's role and place in creation. Scholars have long noted that the language of divine "image" is frequently applied to the king in both Egypt and Mesopotamia. Here, too, the Priestly author of Genesis 1 does something quite innovative, breaking with religious convention. The image of God so typically associated with the king is applied to all humankind, male and female. Some have called this a "democratizing" move on the part of the biblical narrator,[51] though the term smacks of political anachronism. Better is "universalizing": the whole of humanity bears the image. All human beings rightfully exercise "dominion" over the world for their welfare (1:28). Humankind is "humanking." The account of humanity's creation in Genesis 1 involves, in short, the *humanization* of royal status and, in turn, the *universalization* of humanity's "dominion" over creation, the nature of which will be discussed later.

STATUS LINK. Human beings are invested with uniquely royal and priestly dignity. Throughout the ancient Near East, human beings were considered servants of the gods. This necessarily involved taking care of the gods' images
in and outside the temple. Not so from the Priestly perspective in Genesis:
humankind is created not to serve but to embody the *imago*, to reflect God in
the world.

This remarkably elevated view of humanity is also indicated in a subtle yet
dramatic transformation of the divine realm that unfolds in the very act of creating humankind.[52] Only with respect to humankind's creation does God enlist
a plurality ("Let us . . . our image . . . our likeness," 1:26). The references to
"our" and "us" most likely refer to the members of the divine assembly or heavenly council rather than to a private conversation held within God's inner self.[53]
(For a fuller picture of the divine council at work, see 1 Kgs 22:19–23 and Job
1:6–12; 2:1–6.) God speaks to the divine or angelic agents that constitute the
heavenly council and enlists them in a collaborative effort. The creation of
humankind is set apart as a creation by divine committee, albeit one that is
remarkably efficient.

In the act of creation, God comes to take full ownership of humanity's
creation: the reported agency of creation in 1:27 is singular ("he created them").
What have become of the divine agents enlisted by God in v. 26? What role do
they play? The text is tantalizingly silent. It leaves an intentional "gap" that the
Babylonian epic of creation, for example, fills with violence: the slaughter of a
rebel god and the mixture of blood and soil to fashion seven gendered pairs of
humans.[54] Yet the biblical text does leave a subtle clue in the shift from divine
command to fulfillment. As the divine realm proceeds from the many ("our" in
v. 26) to the one ("he" in v. 27), so humankind moves in reverse: from one
("human being" in v. 27a) to many ("them" in v. 27b). The move from the divine
plural to singular suggests that human beings, at the moment of their creation,
have displaced God's divine assembly. They are, after all, the "images of God."
The creation of humanity has, in effect, dissolved the divine community and,
in so doing, created and elevated the human community. Human beings, in
other words, have become God's new pantheon.

GENDER LINK. Humankind created "male and female" is central. Does the gender differentiation of humankind in any way reflect the divine? Again, the text
is suggestively silent. At the very least, the author had reasons to underline this
fundamental, but by no means species-specific, characteristic of humanity. Biologically, most animals share in sexual differentiation, and, textually, they too are
commanded to multiply (Gen 1:22). What sets *Homo sapiens* apart from the other

animals, in the eyes of the Priestly author, is that human beings engage sexually not only for the sake of procreation. "Male and female" constitute the most differentiated, socially intimate, and, yes, generative of relationships, one that, as with all animals, preserves the genealogical survival of the species. But there is more. Genesis 1 seems to suggest that humanity's relational and reproductive capacities reflect in some attenuated way the communal and generative dimensions of the divine. Moreover, along with the so-called democratization of the *imago* there is also its diversification. Created as male and female, God's image is differentiated. And because the separate genders do not reflect God in the same way, God's very self is shown to be differentiated, indeed at the moment of humankind's creation and the divine council's displacement.

Cooperative Process of Creation

Regardless of how the link between God and humanity is to be construed, a more fundamental question must be raised: What is it about God that human beings, as God's "images," are said to reflect? The answer lies in the various ways God goes about creating in Genesis 1. Nowhere does God dominate or conquer. Unlike Marduk, God is no imperious deity. God commands light into being but does not slay the darkness to do so. The sky is created through the commanded separation of the waters, not by their dismemberment. The various cosmic domains are not the bodily remains of cosmic chaos. Neither is humanity fashioned from the blood of conquered deities. God is a differentiator, not a terminator.

Not unlike the Egyptian Ptah of the Memphite cosmogony, God creates by word rather than by sword.[55] God creates by verbal decree that is at once commanding and invitational. For every act of creation, God begins by commanding and then, in certain cases, lets creation happen, as in the case of light (1:3), land (v. 9), and vegetation (v. 12). In other cases, God is directly involved in the act of creating (vv. 7, 21, 25, and 27). In several striking cases, God enlists the elemental forces of creation to further the process. In Days 2 and 3, for example, the waters are commanded to separate themselves, and the earth is commanded to produce plants (vv. 9–11). Far from being inert, passive entities, the waters and the earth act at God's behest. They are bona fide agents of creation, as demonstrated also in the creation of life: the earth is commanded to bring forth land animals, and the waters are beckoned to produce sea life.

> And God said, "Let the waters produce swarms of living creatures,
> and let birds fly above the earth across the dome of the sky." (v. 20)

> And God said, "Let the earth bring forth living creatures of every kind: cattle and creeping things and wild animals of the earth of every kind." (v. 24)

In both instances, creation is accomplished by divine agency:

> So God created the great sea monsters and every living creature that moves, of which the waters produced swarms. (v. 21)

> So God made the wild animals, each according to its kind, and domestic animals, each according to its kind. (v. 25)

Nevertheless, as v. 21 makes clear, divine agency is accompanied by the active participation of the waters and the earth. God works *with* the elements of creation, not over and against them, much less without them, elements enlisted by God as "empowering environments."[56] Creation is a cooperative venture exercised not without a degree of freedom.[57]

But such freedom is not mythologized in Genesis. Lacking in creation are the trappings of divinity. The sun, moon, and stars are not gods but simply lights set in the firmament by God (1:16–17). Nevertheless, they exhibit active agency by serving as "signs" to "rule" the "seasons," including religious festivals. According to Genesis, the celestial bodies are designed not for divination, as one finds in Babylonian horoscopy, but for observation.[58] As the creator of the cosmos, God is neither limited nor localized; God is certainly not the force of good struggling against combative chaos. The alternation between night and day does not replay the ancient script of primal struggle but ensures the rhythmic continuance of a self-sustaining order. Everything is created "good."

In Genesis, "good" is a many-splendored thing. God's approbation evokes a wide range of significance, from the aesthetic to the ethical. "Good" acknowledges creation's ordered integrity and its intrinsic value as beheld by God. "Good," moreover, affirms creation's self-sustainability, its proclivity for fecundity. Creation deemed good by God is creation set toward the furtherance of life. Hence, the emphasis on seeds and sex: plants and fruit trees regenerate through their seeds; vegetation, in turn, sustains all land animals, including humans; and all animals reproduce ("multiply"). Robust, resilient, fecund life is part and parcel of creation's "goodness."

As a whole, creation takes place in Genesis 1 from the top down and from the bottom up. God commands from on high for creation to happen, yet much of the creative process emerges from below. Both the earth and the waters contribute to the emergence of life. God's engagement with creation is thoroughly interactive. The creative process is no singular event; neither is it a unilateral process. The result is a creation that exhibits structure and variety, a cosmic living temple, a creation deemed "extremely good" (1:31).

Character of God

The primordial rhythm of six-day work and seventh-day cessation challenges the Mesopotamian view of the divine and human realms.[59] In the Babylonian creation myths, the gods were not inclined to work, at least not endlessly so. According to *Atrahasīs*, they became fed up with the difficult work of channel construction and imposed their labor upon a lower class of gods, prompting a revolt and setting the stage for the creation of human beings.

The God of Genesis, however, is no irascible Enlil or imperial Marduk.[60] Israel's God does the commanding *and* the heavy lifting, freely taking on the role of creator by word *and* by deed. The work of creation is not beneath the God of Genesis. Whereas Marduk's rest came at the conclusion of conquest, God's "rest" in Genesis concludes creation itself. God is finished establishing the parameters and vitalities that make for a self-sustaining world. God has no need to stand on guard to vanquish any opponent that may rise up to threaten creation. God counts on creation to continue on its own under human "dominion."[61] By ceasing to create, God entrusts to humanity the power and responsibility of managing creation[62] and, as explicated elsewhere in biblical tradition, the privilege and responsibility of resting.[63]

God in Genesis 1 is a composite and complex character. By freely taking on the task of creating a "good" order, God is no begrudging creator. Indeed, God acts in many ways like a priest whose charge is all creation. God blesses[64] and consecrates;[65] God separates and establishes distinctions, something that priests were charged to do (see Ezek 22:26). The priests of ancient Israel were the handlers of holiness and the cultic medium of blessing.[66] So also God, cast in the *imago sacerdotis*, the image of the priest. More elevated still, God is the consummate *royal* priest, or priestly king: God's words are akin to decrees.[67] Though preeminently magisterial, God is no imperial deity in Genesis 1. As generous sovereign, God grants creation its freedom.

Finally, God is the artisan who fashions a self-sustaining creation, replete with order and beauty, and is satisfied with the outcome. The God of Genesis 1 is a temple builder with an artistic bent. In view of the tabernacle's construction recounted in Exodus 25–40, which mirrors the narrative of creation in Genesis 1, this creator God functionally combines the three discrete roles represented by Moses (the instruction-giver), Aaron (the priest), and Bezalel (the artisan).[68] Bezalel, most remarkably, is endowed with God's "spirit" (*rûaḥ*), along with "ability, intelligence, and knowledge" (Exod 31:3), suggesting that God's "breath" in Genesis 1:2 (also *rûaḥ*) signifies "God's creative presence hovering in the wings."[69]

Character of Humanity

As God in Genesis 1 is no imperious warrior, so human beings are not conquerors of creation. The language of dominion lacks all sense of exploitation (1:26, 28). The hoarding of resources is implicitly forbidden in the account: seed-bearing plants and fruit trees are granted to animals and humans alike (1:30). Absent is any hint of the savage competition for resources. God's gift of sustenance is one of abundance, not scarcity, to be shared, not hoarded. Note the repeated use of the word "every/all" (*kōl*) for plants and animals in 1:29–30 (seven times!). These final verses depict a community flourishing in a "common realm of nourishment."[70] Only after the Flood (Gen 9:1–4) is permission granted to human beings to eat flesh, but with certain restraints.[71] Thus, humanity's "regime" over the world in Genesis 1 is constructive, even salutary, consonant with God's life-sustaining creation. The language of ruling and subduing is in the hands of the Priestly narrator transformed.[72] Humanity's "dominion" is unlike any other kind of dominion, certainly far from "modernity's well-known triumphalistic anthropology."[73] It is "filled" with the collaborative, life-sustaining practices set by the creator God. Divide and conquer[74] is banished from the primordial world of the text. Enter separate and flourish.

"Dominion" transformed is also transformative, particularly for an agrarian-based society such as ancient Israel. The cultivation of the land, difficult as it was in ancient times, is deemed in Genesis to be an eminently noble enterprise, raising subsistence agriculture to a regal level. Seedtime and harvest become, in effect, the seasons for royal ventures. Like the humanization of kingship invested in the *imago Dei*, the hierarchical nature of royalty is leveled out and extended to incorporate the most fundamental base of ancient Israel's cultural existence, agriculture. The charge to "subdue" the earth (1:28) acknowledges that cultivating the land involves painful exertion. The terracing of slopes, for example, to prevent land erosion and maximize crop production required great effort. Human life in the land, Genesis 1 acknowledges, cannot prosper effortlessly. The world created "good" is not perfect; the ground can be difficult (cf. Gen 3:17–19). Necessary is the tiller, the farmer who would be, indeed is, king.

Socio-Historical Context

The cooperative, as opposed to conquering, spirit that suffuses the creative process described in Genesis 1 reveals something of the text's socio-historical setting. This cosmogony renders a fully differentiated cosmos made possible by the tacit cooperation of its various elements, all serving at the behest of its beneficent ruler and artisan. Creation is founded not on the conquest of chaos but on its enlistment and transformation. Such an irenic view of creation reflects a profound effort on the part of ancient Israel to put the painful past of conquest and exile behind and to point the way to a new future. Genesis 1 offers a powerfully edifying and inclusive vision of a community poised to reestablish itself in the land.

The Babylonian exile of 587 BCE had left the land of Judah more than decimated. From the perspective of those most affected, imperial conquest and deportation rendered the land "void and vacuum." The survivors experienced such national trauma as nothing less than a resurgence of cosmic chaos, leaving the land "empty," stripping the community of its national identity, and leaving the temple in ruins.[75] The good news of Genesis 1 is that God can work with such chaos to bring forth new creation. Heard in the time of exile, the message of *imago Dei* in Genesis would have been a "clarion call to the people of God to stand tall again with dignity and to take seriously their royal-priestly vocation as God's authorized agents and representatives in the world."[76]

For an exiled people, this "clarion call" also had a more specific function. With the edict of release issued by Cyrus II in 538 BCE, an unprecedented opportunity availed itself for the exiles: they could return and rebuild their homeland. In its historical context, Genesis 1 offered a hopeful vision of how to proceed, namely, by appealing to all sectors of the dispersed community, by enlisting them in the great collaborative task of reconstruction, a monumental, if not cosmic, task. Genesis 1 provided a programmatic, cosmic vision for communal restoration, one that did not require a monarchy (forbidden under Persian hegemony) but instead affirmed the kingship of all. At the same time, the account reminded the returning Jews that God's temple was far more than anything they could ever rebuild on site; it was creation itself. For them, God's cosmic temple was an edifice of hope.

Creation and Science: Connections and Collisions

Genesis 1:1–2:3 is perhaps the Bible's closest thing to a natural account of crea-tion. "If ancient people had consciously set out to articulate a worldview con-genial to science, it is hard to imagine how, in terms available to them, they could have done better."[77] Compared to the rough-and-tumble, divinely micro-managed, theogonic world of Mesopotamian creation, Genesis 1 is an exercise in mythological reduction, on the one hand, and an acknowledgement of crea-tion's freedom and integrity, on the other. Creation in Genesis is replete with dynamic order and structure, cosmic qualities readily discerned by science. The growth and character of cosmic structure is a fruitful starting point of dia-logue between science and Genesis 1.

Cosmic Stuff

But in order for the dialogue to proceed, an obvious collision between the ancient text and scientific account must be addressed. The cosmos as por-trayed in Genesis 1 bears little resemblance to its physical character as observed by scientists today. Cosmologists have discovered an expanding universe populated by roughly a hundred billion galaxies, each of which con-tains a hundred billion stars, many presumably hosting planetary systems of their own. By contrast, the world according to Genesis is a three-tiered uni-verse or "Astrodome":[78] the waters above are contained and held aloft by a firmament or dome, constituting the sky, and the waters below are bounded by land. The stars interspersed across the celestial vault on the fourth day act as conduits for the primordial light that established the first day. In between the firmament and the waters below lies a narrow band of space within which terrestrial life flourishes.

We know better from science, which has much to say about what is out there "above" us. The vastness of the cosmos is not filled to the brim with water poised to inundate us from above. On the other hand, the celestial heavens are not composed entirely of some alien fifth element, such as Aristotle's "ether."[79] What, then, is the universe out there filled with? Actually, water would not be a bad guess. There is, in fact, abundant water in the universe, as found, for exam-ple, in comets and on other planets.[80] The Oort Cloud, a spherical cloud of comets that surrounds the solar system, consists mostly of frozen water, ammonia, and methane. In addition, interstellar clouds, wherein stars are born, contain vast amounts of carbon, hydrogen, and oxygen in the form of CO_2 and, yes, H_2O.[81]

The universe, however, is filled with much more than the simple molecules with which we are so familiar. Physicists speak of "dark matter," the dominant form of mass in the universe. Even though it interacts gravitationally with ordinary (baryonic) matter, dark matter remains unidentified at this point.[82] Responsible for the growth of structure in the universe, this invisible matter helps to keep galaxies intact amid accelerating cosmic expansion. Proposed first in 1933 by Fritz Zwicky, dark matter is commonly described as an invisible "halo" that surrounds ordinary matter. It constitutes up to 85 percent of all the matter in the universe, outweighing ordinary matter by a factor of ten.[83] It neither emits nor reflects nor absorbs light.

Though invisible to the eyes, dark matter can be mapped. By following the curvature of space in relation to dark matter's gravitational effect on light (known as gravitational lensing), scientists from the Cosmic Evolution Survey (COSMOS) have produced a crude three-dimensional map of the large-scale structure of dark matter, "a network of filaments, sheets, and knots,"[84] whose intersections are found at the location of galactic clusters. Simply put, the universe is a "cosmic web" that sharpens in focus as the universe becomes more mature.[85] The formation of dark matter is one of increasing "clumpiness" over time, forming "dark matter congregations."[86] Dark matter has played an indispensable role in the evolution of the cosmos, beginning with the Big Bang, by converting the relatively even distribution of matter of the early universe into something quite "lumpy," resulting in the formation of stars and galaxies. In the words of astrophysicist Heidi Newberg, throughout the cosmos "there are big lumps, little lumps, smeared-across-the-sky lumps. It's just lumps everywhere."[87] By its lumpiness, dark matter provides a "gravitational scaffold into which gas can accumulate, and stars can be built."[88] Without it, there would have been insufficient mass in the universe for galaxies to form.

Filled as it is with exotic forms of matter and energy, the universe has taken its share of lumps.[89] But therein lies its "virtual parallel" with Genesis 1. For the ancients also, the cosmos "above" the earth was composed of cosmic material, dark "waters." They, too, recognized that the observable world was only a small part of what was "out there" beyond the firmament or sky. For them, it was the realm of impenetrable transcendence. For modern cosmologists, all that can be seen is but a small slice of cosmic stuff. As Peter Coles puts it, "Ordinary matter . . . may be but a small contaminating stain on the vast bulk of cosmic material whose nature is yet to be determined."[90] Or put more simply: "Luminous matter is like the foam on the ocean."[91] Not unlike what modern cosmologists have theorized about the origins of the universe and all that lies beyond human visibility, the ancient cosmogonists of Genesis imagined God creating out of a primordial soup, with creation proceeding from formlessness to form. Like the waters gathering into "one place" in Genesis, dark matter clumps together into "congregations," thereby making possible cosmic structure.

In addition to dark matter, the dark "waters" of Genesis can be considered parallel (virtually) to the interstellar dust and gas responsible for stars and planets. This diffuse mixture fills what astronomers call the "interstellar medium," the region between stars, which even in its greatest density is a "much better vacuum than can be created on Earth in a laboratory."[92] But this cosmic vacuum is not entirely empty; physical and chemical processes occur therein that give birth to celestial bodies and ultimately to the basic molecules of life. Such is the generative role of interstellar medium, a vacuum that contains, paradoxically, the elements necessary for life. Whether understood as the ancient counterpart to dark matter or interstellar medium, the dark "waters" of Genesis 1 are by no means inert or inimical to continuing creation. They play a formative role in the creative process. As the waters "below" contributed to the genesis of life (Gen 1:20), so what about the waters "above"? The ancient text is silent, but pregnant.

More down to earth, the ancients acknowledged the thin transparent expanse or firmament (*rāqîaʻ*) that holds up the "waters above," preserving the earth from an inundation of destruction and ensuring the flourishing of life (cf. Gen 7:11). That expanse is what we now call the atmosphere, the blanket of air that shields the earth's surface from the sun's ultraviolet radiation (by means of the ozone layer) and from the onslaught of extraterrestrial material such as comets and meteors.[93] It is the crown of the biosphere. Created, in part, by the "miracle" of photosynthesis,[94] the oxygenated atmosphere is thinner vis-à-vis the earth than the skin of an apple relative to the apple.[95] The Bible's hardened firmament and the earth's protective atmosphere, though conceived quite differently, share the same function for life on Earth: life-sustaining protection.

Cosmic Structure and Symmetry

More broadly, the depiction of creation in Genesis 1 highlights two defining characteristics that science has also observed but in far greater detail: structure and variety. Structure in Genesis is found in demarcated domains and thematic correspondences, while variety is indicated by broken symmetries and life's diversity, "each according to its kind." According to gravitational theorist Lee Smolin, structure and variety are the hallmarks of a world that is "hospitable to life."[96] Each of the forty-one orders of magnitude that constitute the physical universe as we know it, from half a billion light years to one hundredth of the diameter of a proton, exhibits structure. Even galaxies are found to be concentrated in great cosmic clusters and "sheets," such as the "Sloan Great Wall" discovered in 2003, an enormous collection of galaxies 1.37 billion light-years long. It wins the prize for being the largest structure in the universe. Closer to home, the earth gives evidence of at least eight living levels of organized life: from the organelles of cells to the biosphere as a whole.[97]

Both the ancient cosmogonist and the modern cosmologist also agree that the universe bears a history, a history that begins in symmetry. As Brian Greene notes, *"everything* we've ever encountered is a tangible remnant of an earlier, more symmetric cosmic epoch."[98] Cosmic symmetry comes in many forms and degrees: there is "translational symmetry" or "invariance," which assures that the laws of physics apply everywhere throughout the universe. There is also "rotational symmetry," which grants every spatial direction an equal footing. Rotate the universe on any axis, and it will look nearly the same. On the 300 million light-year level (or 10^{24} meters according to others), the universe exhibits near perfect homogeneity, the result of an overall uniform distribution of matter and energy from the Big Bang. The near uniformity of microwave radiation throughout the universe is another sign of early cosmic homogeneity. Uniformity is how the universe began, and it is still evident at the largest scale of perception.

But zoom in more closely and you find countless irregularities. As the universe expanded and cooled, it underwent a series of "phase transitions," which accounts for its clumpiness. When water, for example, undergoes its phase transition from liquid to solid, cracks, bubbles, and crystals spontaneously appear.[99] Analogously, the galaxies, stars, and planetary systems that populate the cosmic expanse are the result of gradual "clumping" or stellar condensation, thanks to gravity and a gradual drop in temperature. Such developing variations can be traced back to quantum-level fluctuations at the time the universe was infinitesimally small, unimaginably hot, and poised to expand. Cosmic history is littered with broken symmetries, those anomalies and variations that have propelled the cosmos toward new levels of self-differentiation and complexity.[100] As physicist David Gross observes, "The secret of nature is symmetry, but much of the texture of the world is due to mechanisms of symmetry breaking."[101] Perfect symmetry makes for a lifeless, useless universe.[102] So does the uniform state of "chaos" described in Genesis 1:2.

One case of symmetry breaking that made, literally, a world of difference was the production of matter and antimatter soon after the Big Bang (10^{-34} seconds). When the universe cooled sufficiently for quarks and antiquarks to "condense out," each pair resulted in total annihilation, releasing a photon of energy. Fortunately for us and the cosmos, the symmetry between matter and antimatter was skewed ever so slightly: for about every billion pairs of quarks and antiquarks engaged in mutually shared destruction, there was an extra quark, a positive oddball upon which the entire future of the universe hung. If perfect symmetry had ruled the "day" (or better millisecond), the universe would have quickly evolved into "a random collection of gamma rays."[103] Owing to this slight imbalance, cosmic evolution was weighted toward matter, resulting in life as "specified complexity."[104]

Though entirely ignorant of cosmic background radiation, rotational invariance, and cosmic clumpiness, the author of Genesis 1 observed that creation is indelibly imprinted with a skewed symmetry. There are, as we have seen, cases of broken symmetries or "nonpredictable variations" littered throughout the ancient account. Most significant is Day 7, which presents the greatest "anomaly" in the creation schema, yet provides the key to the entire structure. Again, perfect symmetry makes for a lifeless universe. Variation, by contrast, constitutes the story of cosmic evolution as it does the Genesis story of creation.

Chaos Redefined

The uniform "chaos" of Genesis 1:2—the hubbub of *tōhû wābōhû*—shares virtual semblance with the uniformity that scientists posit about the primordial, "soupy" state of the universe at its inception. As noted above, the "chaos" of pre-creation in Genesis 1 is anything but chaos, at least in the inimical, combative sense one finds elsewhere in the biblical and ancient Near Eastern tradition.[105] Darkness, water, and emptiness do not make a monster, at least in Genesis 1. Such pre-creation "stuff" is neither Tiamat *redivivus* nor inert matter. Nor is it "nothing" (*nihil*). Undifferentiated "chaos" designates the initial state and stuff of creation, and its containment is essential for life to flourish.

But there is more to watery "chaos" in Genesis than simply its confinement that sets the conditions for creation. Put positively, creation gives the primordial waters its form *for* life, and the waters, in turn, contribute to the formation of marine life (1:20). Watery "chaos" persists in creation, but it does so constructively. No conquering is required. Put ironically, Genesis 1 defies any kind of "chaos theory" in the "biblical" (i.e., conflictive) sense, but it does share some semblance with the chaos theory of science.

Science has revealed the constructive side of chaos. Henri Poincaré, the father of chaos theory, demonstrated in 1889 that despite the best of Newtonian physics, the gravitational interactions of three or more objects (the so-called three-body orbital problem), even in simple isolated systems, defy prediction. [106] The motion of balls on a frictionless pool table, for instance, cannot be predicted beyond eleven collisions.[107] Proving Poincaré's conjecture mathematically with the help of a primitive computer in 1961, Edward N. Lorenz of

MIT developed the modern theory of chaos, thereby ushering the "third scientific revolution of the twentieth century."[108] Complex systems such as the weather exhibit such variability that they are unpredictable beyond a certain finite mathematical horizon.

Another prime example of "chaos" is the turbulent flow of fluids, a highly complex and unpredictable system, something the ancients themselves recognized in their association of chaos with water. Whether rushing torrents, sweeping hurricanes, bouncing billiards, or water droplets skimming a convex surface, such situations are considered "bounded but unstable."[109] Mathematical equations can determine the boundaries of these interactive events, such as the probable "chaos curves" charted on a graph, within whose boundaries the objects in question are subject to internal fluctuations that defy prediction.[110] Randomness reigns, but not absolutely so. Chaos in the scientific sense is not utter chaos. Order is evident: "underlying chaotic behavior . . . are elegant geometric forms that create randomness in the same way as a card dealer shuffles a deck of cards or a blender mixes cake batter."[111]

Chaos can, in fact, be a precondition for order. As the Nobel Prize–winner Ilya Prigogine has shown, "disorder at one level leads to *order at a higher level*."[112] Stuart Kauffman refers to a computer model that begins with a random configuration of activity that results in identifiable patterns, such as an array of 100,000 light bulbs each randomly turning on and off in relation to input from its nearest neighbors. The eventual picture is that of a limited number of identifiable on/off patterns, 317 states in contrast to a near infinite number of possible states.[113] Order out of chaos? Kauffman calls it "order for free."[114] Chemical and biological systems in a state of disequilibrium can give rise to self-organized complexity.[115] Such evolving systems tend toward ordered states that flourish near "the edge of chaos," balanced delicately between stability and change. On the one hand, something that exhibits perfect internal order, such as a crystal, cannot experience change. On the other hand, something utterly chaotic, such as boiling water, exhibits insufficient order to permit change. The system that is most evolvable, exhibiting the capacity to innovate as well as to integrate, falls in between order and chaos.

Such is confirmed on the subatomic level. Particle physics, according to Gary Zukav, "is a picture of *chaos beneath order*," a "confusion of continual creation, annihilation and transformation" out of which certain "permissive laws" emerge.[116] Alan Cook, drawing from Heinz Georg Schuster, prefers the label "deterministic chaos."[117] The deterministic element in chaos is reflected in how random interactions exhibit predictable patterns according to the limits of possible outcomes. Such interactions can be captured in mathematical equations, yielding a picture of randomness constrained,[118] or of chaos confined. "Chaos gives way to order, which in turns gives rise to new forms of chaos. But on this swing of the pendulum, we seek not to destroy chaos but to tame it," says mathematician Ian Stewart.[119] By giving chaos a constructive role in creation, the ancient narrator of Genesis has not only "tamed" chaos, stripping it of its mythological antecedents and threatening character, but also complexified it. Chaos theory, thus, finds an ancient precedent in Genesis as an essential, constructive part of creation's evolving order. The "chaos" depicted in Genesis 1:2 can be read as the disorder that sets the precondition for order. It is the kind of disorder that God can work *with*, not against.

Space and Time

Out of chaos cometh space and time. Both Genesis and science regard space and time as fundamentally related. With the establishment of light *as a domain* in Genesis 1:3,[120] space and time emerge together. From Einstein's theory of special relativity, modern science discerns an even more intimate connection. Space and time constitute an interdependent whole. Physicists talk of "space-time" as a single entity. Brian Greene likens this framework to a loaf of bread, which can be sliced in different ways depending upon one's frame of reference.[121] But the overall integrity of the loaf remains fixed. Space and time constitute an encompassing framework of four dimensions (three spatial and one temporal). But just as when a balloon is squeezed, making one part larger and another smaller, so when objects and events are examined from different frames of reference, a certain amount of space is exchanged for time and time for space. For example, as an object approaches the speed of light (186,000 miles per second) from the perspective of an observer at rest, its dimensions change significantly: length contracts, mass increases, and time dilates. A moving clock runs more slowly as its speed increases relative to one that is at rest. What may take one hundred years for an astronaut to reach a particular star relative to earth time may take only five for the astronaut whose ship is able to approach the speed of light.

　　Neither space nor time is the empty arena in which events simply occur. Bending and curving, spacetime suffuses all with which we are familiar.[122] But at the most minute level, even it breaks down. From the quantum perspective, "space must only emerge as a kind of statistical or averaged description, like temperature."[123] On the smallest, most fundamental level, distance bears no relevance; space is an "illusion" that arises from a system of networks, "bundles of string" knotted together at the Planck level,[124] or a "foam-like topology of bubbles connected by tunnels."[125] Regardless of the metaphor, only at a larger scale does space seem "smooth and featureless."[126] Similar for time: if time is fundamentally grounded in change and change is meaningless both at the quantum level, where correlated moments do not exist, and at the largest scale, where uniformity reigns, then time too is an illusion. Like space, "time is an approximate concept."[127] The non-fundamental nature of time and space would suggest that "creation" itself marks a radical shift in scale whereby space and time became dynamically interrelated as they emerged together. The universe literally took time and space to develop.

Big Bang

As for the "moment" when time and space had their genesis, the Big Bang model is frequently invoked as a scientific parallel to Genesis 1:1–3. When it was first proposed, however, the Big Bang model met stiff resistance from the famous astronomer Fred Hoyle, who actually coined the phrase in a disparaging remark. Hoyle, instead, championed the "steady-state" theory in part because the Big Bang, in his opinion, found too great a "conformity with Judeo-Christian theologians."[128] Regardless of any alleged "conformity," the Big Bang won. Nevertheless, caution is in order. First, some common misconceptions require correction. The theory of the Big Bang does not explain the *origin* of the cosmos but accounts for its *evolution* a fraction of a second *after* time zero (10^{-43} seconds), prior to which the laws of general relativity breakdown. Similarly, the first verse of the Bible begins not at the absolute beginning of creation.

"In the beginning, at the Big Bang singularity, everywhere and everything was in the same place."[129] This statement by cosmologist Peter Coles counters another widely held misconception, namely, that the universe began as a point located in space from which everything exploded. But, contrary to conventional opinion, there was no point *in* space, no center of the universe, for space itself was contained, reduced, as it were, almost infinitesimally at the moment of the "bang." Hence, there is no place to which one can point that marks the spatial center of the universe, just as there is no center described in Genesis 1:2 with the omnipresent waters. Space itself inflated, and it continues to expand. In our runaway universe, all galaxies are moving away, each from the other, as if every galaxy were a center from its own relative position.[130]

Both the ancient narrator and the modern cosmologist find a sliver of common ground: creation commences in extremis, in a condition of incredible (some would say "infinite") density, uniformity, and intensity. Physicists describe this initial condition as an inconceivable situation of extreme intensity in which the temperature remained above a million billion degrees Kelvin (10^{15})

with wild fluctuations of energy so severe that all known particles assumed zero mass, a state of perfect symmetry . . . and chaos! Call it energized *tōhû wābōhû*. Although the universe may have existed in a state of spacetime "singularity" or point of infinite compression, as some physicists describe it,[131] that state was also thoroughly "jittery and turbulent" on the quantum level.[132] A singularity is by definition "a pathological property" in which a quantifiable variable comes to have infinite value in the course of calculation, thus breaking down all equations.[133]

But primordial pathology was not all bad in the course of creation. It reflected, according to the Grand Unification Theory (GUT), a condition in which all the known forces were unified: gravity, electromagnetic force, strong nuclear force,[134] and weak nuclear force.[135] It was a "perfect unity"[136] until the Bang—an "explosion" occasioned by a false vacuum state or condition of gravitational repulsion.[137] This state of negative pressure expanded space by a factor of at least 10^{30}, lasting for the briefest of moments.[138] Such was the universe's most dramatic expansion: the smallest suddenly writ large, setting in less than the twinkling of an eye the cosmic pattern for variety amid uniformity.

So marked the earliest stage of cosmic evolution, one that began in "perfect unity" and was shattered by a false vacuum, putting the "bang" in the Big Bang. Genesis 1:2 analogously describes a state of colossal potential, of cosmic readiness for creation. God's breath suspended above the dark waters; pent-up quantum energy poised to burst forth.[139] Both indicate a generative fluctuation of astronomical or (I can't resist) biblical proportions. As one theologian notes, "The 'Spirit' that is moving over the waters of chaos expresses . . . God's readiness to speak and to act 'from eternity,'"[140] or, as one could restate, from the initial state of "chaos" within which time and space had no bearing, the state in which God's breath was held, ready to explode and unleash creation.

Big Flash

"[T]he most obvious and fundamental medium of our connection to the universe is light," according to Lee Smolin.[141] Such was God's first act of creation in Genesis.[142] But what the biblical author did not know was that along with the creation of light as a domain there was also the cosmic expansion that, according to astrophysicist Adam Frank, "flooded the infant universe" with "light brilliant beyond description."[143] Characterized by a diffuse sea of photons, the first epoch of light commenced. In this "smooth, hot, dense soup of exotic high-energy particles," matter and radiation were nearly indistinguishable ("coupled"), and any photon could travel only for an infinitesimally small distance before being absorbed and re-emitted.[144] The visual result was a glowing fog in all directions as bright as the noonday sun.[145]

During this time (including the first second), fundamental atomic particles began to emerge, assuming form from formless plasma, the primordial soup of quarks. Within the first three minutes, the nuclei of helium and lithium appeared. When the temperature dropped below 3000 degrees Kelvin after 380,000 years, electrons and protons combined to form neutral hydrogen, the most abundant element in the universe. Matter was no longer composed of highly charged plasma but of neutral atoms. Consequently, matter and light parted ways: photons could now travel unimpeded and straight, their wavelengths stretching along with the expanding universe. Born was the cosmic microwave background (CMB), which remains today as the Big Bang's cosmic fingerprint. And so the "fog" lifted and suddenly became "transparent."[146] "Let light be unleashed!" Still the "first day."

But then darkness returned less than one million years after the Big Bang. Astronomers refer to this new era as the Dark Ages, which lasted far longer than the more familiar kind associated with Western civilization. Its cosmic precursor persisted close to a billion years, occasioned by the growing dominance of neutral hydrogen gas throughout the universe. Although neutral hydrogen gas could not absorb cosmic background photons, it efficiently absorbed visual and ultraviolet (UV) light, thereby trapping it close to any source producing it.[147] Visible light became "reshackled." Hence, the universe came to appear dark and "extraordinarily smooth." "The Dark Ages mark[ed] an era of transitions, not only from blackness to light, but also from formlessness to form."[148] And then there was "evening."

Under the cover of darkness, gravity began to pull matter, however diffuse, together. What began as infinitesimal perturbations or "inhomogeneities" from the Big Bang grew larger and denser, and in those contracting regions cosmic expansion was halted even as the space between such regions kept stretching. The first objects to become gravitationally effected were "clumps" of cold dark matter.[149] Dark-matter halos had to grow 10,000 times our sun's mass before hydrogen and helium could "clump" at their center, eventually forming stars and galaxies. The first stars began to appear 400 to 600 million years after the Big Bang and were massive (from fifty to a hundred times greater than our sun), producing a torrent of ultraviolet photons that ionized the surrounding hydrogen gas to well beyond 1,000 light-years by stripping the neutral hydrogen atoms of their electrons. That was not enough, however, to illuminate the universe. The Dark Ages finally came to an end as more and more early stars and galaxies began to ionize most of the universe. Thanks to stellar "re-ionization," a new era of light commenced, the "fourth day" according to Genesis. Enabling telescopes to peer far into the cosmos, ionized hydrogen remains dominant to this day. The first billion years of the universe, in short, proceeded from diffused light to unfathomable darkness to focused light.

Usually considered a textual conundrum, the creation of light in Genesis on "Day 1" and that of the celestial bodies on "Day 4" find a clear, albeit virtual, parallel in science. Genesis acknowledges that the creation of primordial light is temporally distinguishable from the light emitted by the stars. From the biblical perspective, the realm of primordial light was created first, and it lies inaccessibly beyond the firmament fashioned on the second day. Such transcendent light is transmitted, the ancients concluded, by the celestial bodies that God set in the firmament but not until the fourth day. "There was light," but only thereafter—three days or half a billion years—came the stars.

Eye of the Beholder

For the Priestly author of Genesis 1, the primordial or pre-celestial light of Day 1 made possible perception. To state the obvious, light is necessary for eyesight; the sensitivity of eyes can actually be measured in terms of photons. Everything we see is the result of light either directed or reflected at us. According to Genesis, God's sight plays a constructive role in creation; it is integral to the creative process. Seven times the text mentions God "seeing" creation and declaring it "good" (vv. 3, 10, 12, 18, 21, 25, 31).

The creative role of perception has a startling parallel at the quantum level. In the smallest of worlds, the act of scientific observation is "deeply enmeshed in creating the very reality" that is being measured.[150] This holds true, for example, in determining an electron's position. Prior to measurement, an electron's location remains merely a probability to be realized only when measured. According to the Heisenberg Principle, uncertainty is woven into the ultramicroscopic level of reality. Without observation, subatomic particles maintain a "nebulous, fuzzy existence characterized solely by a probability that one or another potentiality might be realized."[151] The act of observation, in other words, compels all possible potentialities to "yield a single outcome."[152] What remains in the fuzziness of quantum limbo comes into sharp focus through the act of observation. "Reality remains ambiguous until perceived."[153]

From Cosmic to Biological Evolution

At the broadest level, both ancient and modern cosmologists affirm that the inception of the universe as we know it was not a one-time (or "one-day") event. It was a process that developed from simplicity to complexity, from uniformity to diversity, from "singularity" to manifold abundance. The four fundamental forces that give the universe its shape are each dramatically different in terms of range and interactive strength. The Grand Unification Theory (GUT) marks a valiant attempt to unify them. But whether entirely unified or not "in the beginning," these fundamental forces have fashioned the world as we know it. "Eliminate any one, or change its range or strength, and the universe around us will evaporate instantly."[154]

Creation is as much historical as it is history in the making: 13.7 billion years and counting of interstellar expansion and formation, of stars living and dying, setting the stage for new cosmic life. More down to earth, scientists have reconstructed nearly three billion years of biological evolution, from unicellular to human life. The ancient sages of Scripture, of course, had no real inkling of such cosmic and biological agedness. But what they share with modern science is a fundamental awareness that the universe took time to develop to what it is now, that it began with an initial defining event characterized by a cosmic effusion of light, the emergence of time, the structuring of space, and eventually the formation of life.

Set within the grand evolution of the cosmos is the smaller-scale evolution of life on planet Earth. For the first 500 million years of existence, Earth was too inhospitable for life. Once the atmosphere formed, thereby shielding the planet from a constant barrage of devastating meteors and asteroids, microbial life burst upon the scene within a mere 150 million years. Primordial cells at first existed independently but soon began to interact with each other and exchange DNA. When one cell developed a protein giving it an advantage, its neighbors quickly acquired the capability. Driven by "horizontal gene transfer," a kind of genetic tool swapping that offered a remarkably efficient way of "keeping up with the Joneses," new properties rapidly spread throughout the microbial kingdom.[155] Thus began the community of life.

The emphasis in Genesis 1 on life "filling" every domain offers ancient testimony to creation's robust fecundity and resilience. Such also is the testimony of evolution. "Life has been expansionist from the beginning."[156] The ancients knew full well the power of reproduction. Genesis 1 acknowledges the regenerative power of life in trees and animals alike. To the evolutionary biologist, the "goal" of every living species is the transmission of genetic identity that enables the fitness of both organism and species to the environment. There is no fittest kind of organism, only an organism fit for its niche.[157] To fill is to fit.

In the divine ordering of things, according to Genesis, all species have their "fit" in certain domains by virtue of their natural constitutions. Fitness in the biological sense affirms the adaptability of every species vis-à-vis its environment by virtue of its evolved genetic makeup. "Biological capacity evolves until it maximizes the fitness of organisms for the niches they fill, and not a squiggle more."[158] And so Genesis places marine life in the seas, birds in the sky, and land animals on land, each created according to its domain. The "great sea dragons" are fit for the sea, where they thrive among the swarming fish. The land, too, is the natural domain of various kinds of animals, from wild to domestic, including mammals and reptiles ("crawlers"). Such is the ancients' rudimentary awareness of biological fitness. Fins and wings, gills and lungs, legs and tails: all have their functional place within their respective domains. Life in all its variety corresponds to its varied domains.

But as better adapted species emerge, others are forced into a corner. Death follows life. Unacknowledged in Genesis 1 is the dark side of evolution's creative force. "Life today was earned at the cost of the death of almost all that went before," states geneticist Steve Jones. Extinction "is a crucial part of the evolutionary machine and is as inevitable as is the origin of species." [159] In addition, natural history testifies to several cataclysmic extinctions, the most recent one caused by a meteorite's collision with the earth across the northern Yucatan Peninsula 65 million years ago, wiping out nearly every land animal bigger than a bread box and destroying almost all surface marine life. Thus began a new stage in evolution, the Cenozoic Era or the Age of Mammals. The dinosaurs' extinction left an ecological window that allowed mammals to evolve from former "dino-hor d'oevres,"[160] such as tree shrews, to the great variety of mammal forms found today, including human beings.

For some biologists, the pervasive feature of death and extinction inherent in evolution's "progress," if one can call it that, "casts doubt on the perfection of God's plan."[161] It must be repeated, however, that Genesis 1 does not depict a *perfect*, let alone perfectly micromanaged, creation, only a "good" one. The distinction is crucial.[162] God enlists the natural elements of creation such as the earth and the waters to participate in the creative process, a conceivably messy, somewhat unregulated affair. The creation narrative, moreover, acknowledges the continuing struggle of life (1:26, 28). To "fill" the earth involves, from a Darwinian perspective, competition. "Subduing" the earth, from the perspective of Genesis, is required for the flourishing of human life (1:28). Biologically speaking, the commission to "subdue" extends to all species for their survival. But "man" does not live by competition alone, and the same goes for any other species. The flourishing of life also depends upon sustaining various kinds of interdependent relations through cooperation, collectivity, and balance.

Genesis 1 depicts in its own way something of an overarching "progression" of life from simple to complex forms: vegetation sprouts forth on Day 3, aquatic and aviary life emerges on Day 5, and on the sixth day land animals and human beings are created. On the other hand, one cannot say that insects and reptiles, created on the sixth day, are more "advanced" than the "great sea monsters" of Day 5. The biological "progression" of life does not fully apply in Genesis 1, only partially at best.[163] The chronological order, moreover, is not biologically accurate: life originated in the sea, not from plants on land. More accurately, life emerged from the fertile interaction of water, air, and land.[164] That all life, from bacteria to human beings, is composed mostly of water is itself sufficient testimony of our primarily watery origins. But to the Priestly narrator's credit, sea life is presented the "day" before terrestrial life. Plants, however, only began to fill the earth millions of years *after* life in the sea had its start. Day 6 would have been the natural "period" for the sprouting of vegetation, not Day 3. Nevertheless, the ancients discerned a developing hierarchy of life, with humanity placed at the top as creation's latecomer.

From a biological standpoint, the development of life is no Jacob's ladder: "Evolution has no escalator of increase," according to Jones. "The idea of evolution as a ladder is (or ought to be) dead, but life has certainly got more complicated."[165] Complexity, in fact, does not necessarily reflect improvement or progress. "Some lineages get more complicated, some simpler, and much of life has to struggle to stay in the same place."[166] Nevertheless, complexity remains a useful way of charting the broad arc of evolution's work. Complexity does not arise solely from competition. Natural selection cannot by itself generate complexity and innovation.[167] Complexity emerges, rather, from the dynamic interaction between mutation and necessity.[168] The fact that genetic copies are never perfect is key. Because something is always "lost in translation,"[169] variation is a given. "The capacity to blunder slightly is the real marvel of DNA. Without this special attribute, we would still be anaerobic bacteria and there would be no music."[170] Error is the basis of innovation. Also key to the emergence of complexity are cooperation and specialization, beginning even on the cellular level.[171] Lynn Margulis and Dorion Sagan add synergy: "In synergy two distinct forms come together to make a surprising new third one," one that is invariably more complex than its precursors.[172] Self-organization and synergetic interaction are just as critical to evolution as natural selection and adaptation.

With regards to the notion of "progress" in evolution, it is important to distinguish between "net progress" and "uniform progress."[173] The former denotes the "tendency for life to expand, to fill in all the available spaces in the livable environments."[174] Thomas H. Huxley (Darwin's "bulldog") likened such expansion to the filling of a barrel: filled first with apples, then with pebbles in between the apples, then sand, and finally overflowing with water. But as Francisco Ayala points out, Huxley's analogy fails to take into account that as "the more species appear, the more environments are created for new species to exploit."[175] In other words, the barrel itself should be expanding. But there are limits. Living space and resources are not inexhaustible. The barrel can expand only so far before bursting.

Through the rhetoric of blessing, Genesis mandates the expansion of life to "fill" the waters and the land (1:22, 28). And the evolutionary drive to fill these domains gives rise to biodiversity: an environment can be filled more effectively with a variety of organisms than with only one kind. To fill an environment also presumes the ability to perceive it, to "gather and process information about the environment," and to act upon it.[176] That, too, can serve as a criterion of evolutionary "progress," and by most criteria *Homo sapiens* is the "most progressive organism on the planet."[177]

So opens the latest (but by no means last) chapter of evolution's encompassing epic: the creation of humanity, infinitely more complex than the first (and current) bacterium. Both the biologist and the ancient sage agree that humanity is a newcomer, an endnote to life's developed diversity. If the evolution of the cosmos could be reduced to within the span of a year, then the appearance of hominids would take place "within three hours of the stroke of midnight on New Year's Eve" and *Homo sapiens* a mere twenty seconds before the hour.[178] Humanity's tardiness on the evolutionary scene accounts for its uniformity: *Homo sapiens* is a solitary species within a single genus that has not had sufficient time (only about 200,000 years) to build up the kind of diversity found in other primates.[179] Chimpanzees, humanity's nearest cousins, have two species. "The [human] world is divided by politics, but it is united by genes."[180]

All members of the human species are 99.5 percent identical at the DNA level. Of course, there is much more to the human species than its uniformity. Aardvarks share the same characteristic of low genetic diversity. From a biological standpoint, it is the brain's size relative to the body that makes human beings distinctive. The hominid brain has doubled in size the past two million years. Everything from complex social life to problem solving abilities requires grey matter. "Human progress had made a simple but crucial move, from body to mind. That mind is built from genes, but what it can do has long transcended DNA."[181]

Human beings exhibit critical acumen far beyond the capacities of their primate ancestors and "cousins." Although our protein molecules are only 3 percent different from those in chimpanzees, our cranial cortex is three times larger. For all that is similar, if not nearly indistinguishable, between humans and chimps on the genetic, chromosomal, cellular, and neuro-anatomical levels, "the developmental modifications" that gave rise to human cognition and culture "culminated in a massive singularity," observes J. Craig Venter.[182]

The human brain is of such complexity that quantifying it reaches astronomical proportions. One estimate is 10^{13} neurons, each equipped with several thousand synapses. Such a lively network of informational exchange results, in principle, in an explosion of "more possible thoughts than there are atoms in the universe"[183]—not that this is ever achieved on a regular basis.

Because the human mind is qualitatively different, "to speculate about its evolution is largely futile," cautions Jones.[184] Nevertheless, the mind did not suddenly appear fully formed ex nihilo. It emerged within the last six million years of evolutionary history, when the lineages that led to modern humans and chimpanzees diverged. The human brain seems to have reached its current development within the last 50,000 years, as evidenced in the spectacular cave art of Western Europe, dated to the Upper Paleolithic.[185] During this time, several lines of cognitive and cultural development quickly converged, including spoken language, symbolic thinking, and religious practice, all made possible by what archaeologist Steven Mithen calls "cognitive fluidity."[186] According to Mithen, the mind developed initially through the specialization of discrete "cognitive domains" or "intelligences," each advancing a specific type of behavior such as linguistic ability, natural history intelligence, and tool-making intelligence.[187] Mithen employs the image of a Swiss army knife to describe the mind at this stage, with each blade or tool representing a specialized cognitive domain or "module."[188]

Mithen describes the latest phase in cognitive development as integrative, flexible, and fluid. The reflective power of consciousness took hold as the brain's various cognitive operations began to interact and work together, making possible the use of symbols and metaphors. Specialized modules no longer had to work independently. The result was an unprecedented shift from an earlier compartmentalized way of thinking to genuinely creative activity. It set off what Mithen calls the "big bang" of human culture,[189] matched by what Holmes Rolston calls a neurological "combinatorial explosion."[190]

This cognitive shift prompts a new metaphor. No longer a Swiss army knife, the mind is now a "cathedral" in construction.[191] Its development began with a central "nave" of generalized intelligence, then multiple "chapels" of specialized intelligences were added, and finally these domains or intelligences became connected when "doors and windows were inserted in the chapel walls,"[192] allowing for the free flow of thoughts and information across the cognitive domains. Once separate "chapels" came to form a "superchapel," a cathedral more Gothic than Romanesque.[193] Or one could call the fully evolved mind a "temple," complete with various side rooms all connected to a central nave, including an inner sanctum constituting a religious frame of mind. But whether as cathedral or temple, the fully evolved mind is evidenced in the *collaboration* of various cognitive "modules." The mind's interconnectivity is what makes possible creative reflection, the exercise of full consciousness.[194]

In the light of evolutionary science, the *imago* dimension of human creation in Genesis 1 encompasses all the cognitive, emotive, and cultural aspects that constitute a distinctly human identity. "No single trait or capacity like intelligence or rationality should ever be taken as the definitive word on human uniqueness."[195] One should not even identify the *imago* as an individual matter at all. Human beings were created, according to Genesis 1, as a plurality, and out of this plurality arose culture. The individual human brain with all of its networking neurons cannot hold a candle to the "collective of human brains and their psychological processes that make up human culture."[196] However it is to be defined, culture is something more complex than the simple sum of human individuals. It "seems to involve the creation of something whole, something cohesive, and possibly something which seems to be greater than its constituent parts."[197] Similarly, the *imago* is much bigger than the human brain and mind; it encompasses humanity collectively, culturally, and, according to Genesis, theologically.

It is no coincidence, then, that central to humanity's cultural complexity is religion or the capacity for worship. Religion, along with cognitive fluidity and the development of the arts, is an evolved phenomenon, one that rests on symbolic thinking, ritual practice, and communal collaboration, evident particularly among Cro-Magnons or early *Homo sapiens*.[198] As a development of symbolic intelligence, religion was a key factor in human evolution. Indeed, the rise of religion may have been instrumental in the development of agriculture around 10,000 years ago. Currently identified as the earliest "man-made holy place," the monumental "megatemple" at Göbekli Tepe (ca. 11,000 years old) in southeastern Turkey, with its decorated pillars depicting a frightful array of wild animals, suggests that large-scale worship may have prompted the initial steps toward agriculture among hunter-gatherers, the key to the Neolithic revolution.[199] Contra Marx, religion was not the opiate of the masses, but it may very well have provided food for the masses.

The various capacities that constitute human uniqueness, individual and cultural, artistic and religious, scientific and poetic, all converge for a specific purpose in Genesis: human beings reflect the manifold character of divinity required for managing creation. In Genesis 1, the functional side of the *imago Dei* is expressed in humanity's ascendancy to power *for* creation. Only humans, as God's images, are given the responsibility of dominion. In the words of biologist Edward O. Wilson, "[W]e are the first species to become a geophysical force, altering Earth's climate, a role previously reserved for tectonics, sun flares, and glacial cycles."[200] On the evolutionary scale, humanity has clearly achieved such dominance, changing the face of the earth. The world has become humanity's niche, for better and for worse, a mixed blessing at best. It is deeply ironic that the most intelligent, gifted, powerful creature of the planet, God's very image no less, is now endangering much of life, including its own. Humanity's rise to power is by itself no legacy of the *imago Dei*, though it is evolution's legacy. Dominion exercised on behalf of creation most fully embodies the image of the God who created without subjugation to ensure creation's "goodness." In the end, the *imago Dei* entails the *imitatio Dei*, with human dominion reflecting divine dominion as modeled in Genesis 1.

No matter how the *imago Dei* is to be interpreted, humanity's uniqueness is firmly embedded in evolution's panoramic story. Even Darwin acknowledged the "god-like intellect" of the human being, along with the other "exalted powers" of "sympathy" and "benevolence," while in the same breath affirming the "indelible stamp of [humanity's] lowly origin."[201] Evolution, indeed, has its own ethical payoff: it "releases us from the narcissism of a creature that is one of a kind. It shows that humans are part of creation, because we are evolved."[202] Genesis, likewise, shows that humans are very much part of creation, sharing resources with all life. We are distinguished not apart from but in relation to all other life forms, with whom we share the biosphere. The *imago Dei*, thus, does not champion our superior uniqueness over other animals. Rather, it lifts up our special purpose to bear God's presence in the world for creation's sake and for God's.

Rereading Genesis 1

What are the hermeneutical payoffs of bringing the ancient text and contemporary science into dialogue? Our discussion has already begun to address these questions by noting possible points of connections ("virtual parallels"), while acknowledging the collisions. But we have only scratched the surface. As we explore further, only preliminary observations are possible. Nevertheless, I have found that such observations, tentative as they are, help sustain the kind of wonder that is at the root of both science and faith.

Genesis 1 brings into conversation a number of scientific disciplines. Its coverage of creation is vast, and for good reason. This ancient report provides the most comprehensive account of creation anywhere in the Bible (which also explains this chapter's interminable length). Genesis 1 attempts to chart the rise of form out of formlessness, the emergence of structure and life from "chaos," no mean feat! And it does so with remarkable brevity. What Genesis touches upon ever so briefly, science has explored with great depth, detail, and ongoing investigation. Like Genesis 1, science charts the course of the cosmos from the initially amorphous character of the universe to the wealth of structure we see today. To explain cosmic evolution today, one must plunge into quantum physics, astronomy, and biology, and that's only the tip of the iceberg. In all of its specialized disciplines, science puts flesh on the skeletal account of Genesis, but not without some surgical rearrangement. The following discussion takes up particular passages from Genesis and re-reads them in dialogue with science.

"When God Began to Create"

We return, thus, to "the beginning," or more accurately to *a* beginning in the broadest cosmic measure. Such was perhaps the Big Bang, one among possibly many "bangs" preceded by cosmic "crunches" within a "cyclic universe."[203] Or, as others speculate, our universe is one among many that are continually erupting, a mere bubble in a vast quantum sea of universes, old and new, that lie beyond the scope of our perception.[204] The jury is still out; extra dimensions and parallel universes have yet to be demonstrated, much less observed, if they ever can. Even the nature of time remains in dispute among cosmologists.

Nevertheless, the beginning of creation in Genesis remains tantalizingly open. Even though time and space, in their structured form, are interlocked in the creation of light (Gen 1:3–5), the biblical author posits something "before," namely, the pre-creative condition of "Day 0." Is the scene described in v. 2 set beyond time or in "imaginary time"? Is it cast in quantum superspace? So what was God doing back "then"? Augustine is well known for having ridiculed such pondering, claiming that time came into being at the moment of creation,[205] consonant with much modern cosmology.[206] But is it time *as we know it* (i.e., "phenomenological time") that makes its dramatic appearance at creation? Can the hovering *rûaḥ* of God suspended over the restless waters reflect a jittery, chaotic world of the quantum realm set for just the right fluctuation to begin its fated expansion? Perhaps. But whether as a singularity or as a derived event within a larger cosmic cycle, what is described in Genesis 1:1–3 is no less dramatic and no less ambiguous than what science speculates regarding the explosive origins of the universe. Both give pause to ponder and marvel.

As for dark "chaos" described in Genesis 1:2, science sheds a more positive light. Watery chaos has its creative side, as evidenced in God's enlistment of the "waters below" to produce marine life. Nothing in the text, however, is said of the generative capacities of the "waters above." The author of Genesis fully acknowledges the limits of human perception. Science has succeeded in pushing those limits by discerning the generative nature of interstellar medium, the dilute "chaos" of gas and dust that by dint of gravity leads to the formation of stars and galaxies. Creation unfolds on the "edge" between chaos and order, not polar opposites but polar partners engaged in the ongoing process of creation. Chaos is essential to an order that generates structure and complexity. Though immensely vast, the dark waters of Genesis are not so ominous after all.

"And God Said"

As for the divine word and its defining impact on the creative process, science offers rich analogies. We have already seen how the theory of cosmic inflation (the "Big Bang") lends an "explosive" nuance to the word that generates the "Big Flash" (1:3). Thereafter, God's word enlists and unleashes the elemental powers of creation. One could think of God's words as including the laws of physics, which direct the evolution of cosmic structure and, eventually, life. To take another analogy, the divine word acts as a "voice activation" that triggers creation.[207] As God's "word" is crucial in the Genesis story, "sound" is instrumental in the cosmological history, so much so that astronomer Peter Coles refers to the ripples present in the early universe as the "sound of creation."[208] These primordial sound waves, whose traces are evident from "the hot and cold spots on the microwave sky," were generated by the Big Bang, traveling approximately at the speed of light for 380,000 years. Within this primordial sound spectrum is stored a wealth of information about the universe—information yet to be fully deciphered. In view of cosmic evolution, God's word is a "resounding cosmic fanfare."[209]

God's word in view of Genesis, however, does more than resound. It commands and, thus, informs. The divine word in Genesis is not only formative in the process of creation, it is also informative such that it defines the roles and activities of other agencies. Information, not coincidentally, is increasingly recognized for its formative role in biology. Coded in DNA, information exhibits a particularly active agency. Like language, DNA transmits such information, specifically a message containing four different modules or nucleotides (A, C, T, G).[210] Like letters in a written language, different sequences of DNA modules transmit different information. The interpretation or decoding of such information lies in certain biochemical mechanisms that translate the sequence of DNA nucleotides into amino acids to form proteins. It is these proteins that ultimately determine what an organism is and does. As Holmes Rolston points out, the genetic code is more than causal; it is purposive.[211] The process of transmission leads to "the achievement of increasing order, maintained out of the disorder."[212] So also, analogously, is the aim of the divine commands in Genesis.

Information is communication: it is coded, transmitted, and then decoded.[213] Perhaps it is no coincidence that the divine commands in Genesis 1 exhibit their own rhetorical shape in distinction from the surrounding narrative material.[214] Carried forth by the divine word, creation is ushered into the information age! God's commands deliver an input of information,[215] which, once interpreted, activates, directs, and sustains creation. Indeed, the transition formula ("and it was so"), usually sandwiched between God's command and creation's fulfillment,[216] could indicate a crucial intermediate stage in the creative process, one that involves the decoding and interpretation of the command.

But more than interpreted information is needed to initiate creation. Cooperation is required in decoding and responding appropriately. The value of cooperation within the natural world is fully evidenced in evolutionary development, so much so that Martin Nowak, director of the Program for Evolutionary Dynamics at Harvard, claims cooperation as "one of the three basic principles of evolution."[217] Whereas natural selection and mutation lay the groundwork for the emergence of new traits for a species, cooperation is "essential for life to evolve to a new level of organization."[218] Cooperation, for example, can be found even among bacteria, the most successful life form on Earth. The notorious *Escherichia coli* (*E. coli*), credited with recent outbreaks of food contamination, exhibits a complex social life marked by cooperation and strife within the microbial world.[219] Certain species of bacteria produce enzymes that break down food, which all bacteria can then consume. Energy is expended in the production. But if a bacterium refuses to cooperate by ceasing to produce enzymes, it can still enjoy the meal, thereby gaining a potential reproductive edge over bacteria that cooperate.[220] According to Eshel Ben Jacob of Tel Aviv University, when certain "opportunistic bacteria" operate independently, these "'cheaters' are shut out of the community" by the cooperative bacteria, which alter "the expression of their genes, shifting into a dialect that becomes unintelligible to the cheater."[221] Even bacterial "betrayal" carries its own just deserts!

Cooperation in nature, even at the microbial level, gives reason to pause over those verses in Genesis 1 that imply collective, cooperative activity in the ongoing work of creation. The waters, for example, are enlisted by God. Specifically, they are *co-opted* in the process of creation. Their cost? Energy. Their gain? Being "filled" with life. Possibility of defection? Yes, although such an option is not chosen in Genesis 1. Admittedly, the waters do bring to Genesis 1 a rather dubious reputation, for elsewhere in ancient Near Eastern tradition, including biblical tradition, they are well known for their unruly, threatening behavior.[222] In Genesis, however, their reputation is transformed by God's co-optive will. In the natural world, co-option plays a definitive role in generating novel possibilities. Lens crystallins in eyes, for example, once exhibited a function altogether different from their current function. They were originally heat stress proteins that happened to be transparent, and because of this incidental quality, they got "co-opted" to make eye lenses, something entirely new.[223] The chemical compound acetylcholine, the first neurotransmitter to be identified, was for millennia operative in sustaining botanical and microbial life. But when nerves made their appearance on the evolutionary stage, the compound got co-opted for use in synaptic transmission, enabling neurons to communicate with each other.[224] In each case, something new emerges from co-opting the old.

In general, the chemical and biological constituents of life can be co-opted to perform new roles as evolution makes its meandering, incremental "progress." Using the biological analogy, the waters and the land in Genesis get co-opted for an altogether new role, namely, that of producing life. Life is the result of co-opted emergence. "Life did not take over the globe by combat," note Lynn Margulis and Dorion Sagan, "but by networking."[225] Analogously, by speaking, God also networks, co-opts, and delivers critical information. Come to think of it, so does an effective priest.

"And God Separated"

Much of creation in Genesis occurs through separation (1:6, 9–10; 2:3). Separation is also key to the physical formation of the universe. According to GUT, the four fundamental forces, originally constituting a single "superforce," separated off in rapid succession within the first millisecond of the Big Bang. Separation and gathering were critical to the formation of atoms. As photons were separated off, atoms formed through the attractive nuclear force of elementary particles. Had the early universe not cooled sufficiently for atoms to form, everything would have remained as undifferentiated plasma.

Heavier atoms formed through the violent deaths of massive stars or supernovae explosions: separation and gathering occurring, one might say, with a vengeance. Increasingly complex molecular structures formed from combinations of various atoms. Finally, jumping over the yawning gap between molecules and cells, life itself came about through separation. The translucent, semi-permeable membrane formed by ancient lipid molecules was critical for the formation of the first cell.[226] "Life is recognizable by its partial separation from the environment by way of a membrane."[227] Indeed, the isolation of inside from outside is "one of life's most basic requirements."[228] This fundamental boundary allows for the containment of energy and temporarily thwarts the entropic tendency toward equilibrium. According to both Genesis and science, separation by boundaries was necessary for a fully differentiated, dynamic, living cosmos. The primacy of boundary in Genesis is indicated even at the very outset with references to the "surface of the deep" and "the surface of the waters" in v. 2. Not coincidentally, science has determined that "a surface is a logical site for life's origins," where two different materials meet, such as water and air, or rock and water. It is where molecules congregate.[229] In the murky state described in Genesis 1:2, water and wind meet, setting the stage for creation.

"No separation, no order" could be a motto for Genesis. Yet order through separation is only one side of the creational coin. The other is interdependence. While boundaries separate, they also bind together. For every separation established, for every boundary formed, an integral connection is forged. Sea and land are bound together for the flourishing of life. As life is established through separation, it is sustained by connection, by "cross-kingdom alliances,"[230] by evolving interdependent relations between organisms, between species, even between predator and prey. Space and time are experienced as separate, yet they are interdependently related. The human brain is a composite of separate modules, each with its own specialization, and an interdependent whole. Our very bodies are "joint property" that provide a wholesome environment for innumerable bacteria, fungi, roundworms, mites, and only your resident biologist knows what else.[231] "Differentiated relationality," in short, is the key to creation.[232]

"And God Saw"

Another pause for wonder is the role of sight in the creative process. The determinative influence of scientific observation, particularly on the quantum level, offers a potent interpretation of God's beholding creation seven times. As with the act of scientific measurement of an electron, divine sight is no passive matter in Genesis 1. Literarily, it sets the occasion for God's approbation. Hermeneutically, however, there is more. Literally, the biblical text reads:

1:3 God saw the light, that it was good (see vv. 10, 12, 18, 21, 25).

1:31 God saw everything that he had made, and *voila!* it was extremely good.

The syntax is peculiar to Hebrew and is usually not reflected in English translations, including my own.[233] But there is an unmistakable syntactical pause between the clauses in Hebrew: "God saw X, *that* it was good," quite different from the seamless movement featured in the typical English translation: "God saw that X was good." The pause between the two clauses is key, marked by the Hebrew *kî* ("that"). However one translates the sentence, the most obvious sense is that of an artisan evaluating her work—in this case a cosmic edifice— and finding it "good." As William Foxwell Albright noted, the sense is "And God saw how good it was."[234]

But more can be inferred from these verses than simply "evaluative perception."[235] The analogy of scientific observation, particularly on the quantum level, sets in sharp relief the active, defining role that divine perception has in the process of creation. The text comes close to suggesting a causal relationship between divine sight and creation's goodness. God's "seeing" does more than set the stage for God's approval. Sight occasions a causal pause, as if to say that divine perception is itself instrumental, indeed necessary, for the realization of creation's goodness. God's sight carries creation to its completion. It helps *make* creation "good." The repeated role of divine sight in the Priestly cosmogony highlights all the more the fundamental character of light created on the first day. Its opposite, the unrelenting darkness of Genesis 1:2, precludes observation, and is therefore by itself not "good."

God's sight completes each discrete act of creation, with the final observation concluding the six days of creation (Gen 1:31). Without divine sight, creation would remain incomplete at some level: light not fully effulgent against the darkness, the land still partially submerged within the nebulous waters, life not fully formed and filling its domains, creation not having reached its full integrity. In short, creation is not fully "good" until it is "sighted" by God at every step in the creative process. Divine sight, thus, can be placed on a nearly equal footing with divine command and act in Genesis 1. In quantum physics, observation is fully interactive: the collapse of a wave function marks, among other things, the transference of information.[236] Observation is itself informative, indeed creative. As God's word commences each creative step, so God's perception concludes it. God "sees" creation to its orderly end. Creation, by analogy with quantum mechanics, requires an Observer.[237] As some cosmologists are fond of saying, the Big Bang was "the greatest ever particle physics experiment."[238]

"That it was Good"

Following each and every act of sight, God expresses approval over the outcome of every step of creation, whether executed by God or fulfilled apart from God. As already noted, God's pronouncement of "good" in Genesis 1 conveys a wide range of meaning. Science, however, introduces another nuance, one that opens wide the question: Is the creational product strictly determined by God? Are the various steps of creation that meet God's approval simply a matter of fulfilled design? The theory of evolution, for example, generally avoids the language of purpose or teleology. Evolution does not drive toward perfection or toward anything else beyond the organism's fitness for its environment. Species are the products or outcomes of evolution, not fulfillments. Evolution does not operate with preset goals; it is open-ended, ongoing, and flexible. "Evolution is pure improvisation."[239]

To be sure, the "plot" of Genesis 1 is governed by the teleological link between God's command and creation's fulfillment. Each command sets a goal that is fulfilled. Evolutionary science, however, suggests a more dynamic and flexible reading of the creative process, namely, that creation is as much a product or outcome as a realization. In the case of the waters and the earth, one can imagine God issuing a command and then stepping back to see what happens: vegetation and life, for example, emerging in manifold forms, the product of unleashed creative energies. God's declaration of "good," thus, could convey a complete sense of fulfillment on God's part that also acknowledges the freedom and flexibility of the "co-opted" elements of creation in the ongoing process. That is to say, God's approval may be open to a range of possible fulfillments, species included, but without lowering the high bar that it sets.

Biologically speaking, species are neither absolute nor static. They are always accumulating genetic and trait-bearing differences that can eventually lead to new species ("speciation"). Species are dynamic, ever shifting and mutating in order to "fit" better within their respective environs. That, too, is part of creation's glorious "goodness." Biology, thus, inserts a healthy dose of improvisation into the creational process as outlined in Genesis 1. When God proclaims the whole of creation to be "extremely good" (1:31), it is creation's big picture that is most highly prized, the vast dynamic, interrelated sum.

"On the Seventh Day God Finished"

The story of creation in Genesis 1 is rhythmically structured according to the regularized passage of day and night. Such is not the way science demarcates cosmic evolution. Consider the milestones identified by science regarding cosmic, geological, and biological evolution (with approximate numerical figures):

1. Big Bang	0 seconds (13.7 billion years ago)
2. Gravity force separates out; inflation begins	10^{-43} seconds
3. Inflation ends	10^{-34} seconds
4. Electromagnetic and weak forces separate	10^{-10} seconds

5. Matter dominates antimatter 10^{-3} seconds

6. Fusion stops; nuclei form 3 minutes

7. Formation of atoms and CMB 380,000 years

8. Onset of darkness 1 million years

9. First stars 400–600 million years

10. Sun 8.7 billion years (5 billion years ago)

11. Earth 4.55 billion years ago

12. Beginning of life on Earth 4 billion years ago

13. Microbial life on Earth 3.85 billion years ago

14. Photosynthesis 2.7 billion years ago

15. Multicellular organisms 1.7 billion years ago

16. Cambrian explosion 540 million years ago
    (invertebrates)

17. First (jawless) fish 510 million years ago

18. Land colonized by algae 500 million years ago
    and insects

19. Plants with seeds, first forests 408 million years ago

20. Animals on land 370 million years ago

21. Trees, ferns, reptiles 345 million years ago

22. Dinosaurs, mammals 230 million years ago

23. Extinction of dinosaurs 65 million years ago

24. *Homo sapiens* 200,000 years ago

The twenty-four-hour day is nowhere a factor in the astronomical or biological scale of evolution. Rather, the temporal scale of cosmic evolution swings wildly from minute fractions of a second to millions and billions of years. In addition, the order of cosmic evolution is different from what one finds in Genesis 1, as already noted.

What's in a day in Genesis 1? The temptation is to make Genesis more compatible with science by counting the biblically enumerated "days" as symbolic of much longer stretches of time. One could cite 2 Peter 3:8 as a precedent, which likens one day "with the Lord" to a thousand years.[240] But by harmonizing the text with cosmic evolution, one would quickly find the chronological value for each "day" in Genesis to vary widely scientifically, some measured in billions of years, others in millions, others in milliseconds.

Symbolic days with wildly divergent time spans are clearly not what the text has in mind. The alternation between night and day, darkness and light, evening and morning, is fully regularized in Genesis 1. A hermeneutic of harmonization, moreover, cannot resolve the conflict between Genesis and science over the chronological ordering of cosmic and biological events. The scientific witness demands a wholly different understanding of the six-day order in Genesis. From a scientific standpoint, one can credit the biblical author for imagining a distinction between primordial light and stellar light, between diffused photons and radiant stars, as well as discerning something of a movement from biological simplicity to complexity. But the *scientific* appreciation stops there. It is precisely the incompatibility between science and Genesis that forces an alternative approach to the Priestly delineation of seven days of creation.

The theological question generated by this collision is: "Why is creation ordered according to seven days?" So that it models the weekly rhythm of work and rest. As the Sabbath commandment in Exodus states:

> Remember the sabbath day, and keep it holy. Six days you shall labor
> and do all your work. But the seventh day is a sabbath to YHWH your
> God; you shall not do any work—you, your son or your daughter,
> your male or female slave, your livestock, or the alien resident in your
> towns. For in six days YHWH made heaven and earth, the sea, and
> all that is in them, but rested the seventh day; therefore YHWH
> blessed the sabbath day and consecrated it. (Exod 20:8–11)

In contrast to the same commandment in Deuteronomy 5:12–15, the Exodus version grounds Sabbath in the ordering of creation. As God created in six days and ceased on the seventh, so should the Israelites from week to week. The orderly unfolding of creation shapes the orderly conduct of human activity. God's blessing for humanity is a sabbath-tempered "dominion." The magisterial outworking of God's creation in Genesis ensures that Sabbath is more than an afterthought. It is a destination. Though radically distinct from all other days, Sabbath remains an essential part of the creative process (cf. 2:2). Sabbath, the symmetry-breaker, ensures the integrity of creation.

Another way to interpret Sabbath is to see it as creation's attainment of a dynamic steady state characterized by stability and self-sustainability. A dynamic form of stasis is maintained by various "feedback loops" or perpetual cycles of circulating material and energy.[241] The dynamic equilibrium that is achieved between the conversion of carbon dioxide to oxygen by plants and the conversion of oxygen to carbon dioxide by animals is but one example. Far from being in thermodynamic equilibrium (the highest and final state of entropy in which everything dissipates into a "featureless chaos"),[242] a dynamic steady state is the "goal" (if I may use that term) of consciousness itself: "Consciousness acts and reacts to achieve a dynamic steady state," notes E. O. Wilson.[243] Is Sabbath, then, the goal of divine consciousness? Can rest be the crowning achievement of human consciousness? Is Sabbath the goal to which all creation yearns? Such ponderings highlight, in any case, the centrality of Sabbath for God and for creation.

Materially, Sabbath confirms the sufficiency of cosmic and biological organization achieved during the first "six days." In Sabbath, God releases creation to thrive on its own. Sabbath confirms the sufficiency of information for the world to flourish, to develop and change yet remain stable and secure. God's rest and release of creation parallels what Lynn Margulis and Dorion Sagan call life's "autopoietic" nature, its "continuous production of itself."[244] God pauses, and creation sustains itself. Ethically, Sabbath directs the human will toward acknowledging creation's self-sustaining grace and humanity's central, responsible place within it. The seven-day scheme of creation, thus, serves to regularize human activity, in imitation of God, in a way that ensures creation's flourishing.

"Let us Make Humanity in Our Image"

Francis Collins, head of the Human Genome Project, describes what he considers the "special qualities" of *Homo sapiens*: an awareness of right and wrong, a developed use of language, an awareness of self, and the ability to imagine the future.[245] Although such distinctive human qualities, when compared to non-human animals, may be more a difference in degree than in kind,[246] God's "image" in Genesis also encompasses various aspects of human identity and function. Our species-specificity operates on a number of different levels, so also God's specificity. Thus, it is best to think of the *imago Dei* not as something that reflects a singular aspect of the divine off a singular aspect of the human but as a prism refracting the various ways human beings, beginning with their gendered diversity, are capable of conveying the manifold character of God in the world.

To be sure, *Homo sapiens*, as the taxonomic classification implies, is distinctly "wise." The proof is in the results: "We know time and space across 40 orders of magnitude, and we build cultures that . . . exceed anything known in animal cultures by 40 orders of magnitude."[247] But we are also the only species capable of destroying the earth forty times over. The proof of our *sapientia* is ambiguous. Ultimately, the "image of God" comes down to human agency, which is most fully evidenced in the transmission, evaluation, and appropriation of ideas through the use of symbolic logic. In this respect we are unique: "Animals have communication systems that are sometimes impressively sophisticated, but they neither invent them nor teach them to others."[248] Apes possess high levels of intelligence, enabling them to use language when taught, but they lack the distinctly human capacity to invent symbolic language.[249] Humans are, as concisely put by Wilson, the "babbling ape."[250] They are also the praying ape.

If pressed to identify the most centrally distinctive human trait, we must consider language, whose origin remains perhaps the "hardest problem in science."[251] The evolutionary origins of human language are still very much open to debate.[252] But by any accounting, language constitutes the ultimate biological innovation of human development, the emergent genius of the human genus. In such a light, humanity's "refracting" of God takes on added hue. What makes language unique is its remarkable creativity. As God began creation with a word, so human beings can sustain or destroy creation by word, for on language hangs all of human agency, from "playing God" in exercising power willy-nilly to "imitating God" in care-filled stewardship. God's initiating work of creation in Genesis 1 is the work of word—the imparting of information, the harnessing and co-opting of natural forces, and the invitation to collaborative work, not to mention the blessing. The same goes for God's talking images.

4

The Ground of Being

The Drama of Dirt in Genesis 2:4b–3:24

My skin is happy on the black dirt, which speaks a language my bones understand.

—Barbara Brown Taylor[1]

The body of a soil is a sky where seeds and worms and ions fly.

—William Bryant Logan[2]

God writes such short stories about humankind.

—Francis S. Collins[3]

Creation comes crashing down to Earth in Genesis 2:4b–3:24. God exchanges the royal decree for a garden spade. The God from on high becomes the God on the ground, a down-and-dirty deity. Known as the Yahwist account (J) for its prominent use of the divine name YHWH,[4] this second creation story is altogether different in tone, content, and scope from the first. Compared to the lofty liturgical cadences of its canonical predecessor, this account reads more a Greek tragedy. Methodical progression gives way to narrative bumps and twists. While the Priestly account teeters on the edge of abstraction, the Yahwist story, with its focus on the family, revels in messy drama, the drama of dirt.

The opening scenes of these two accounts could not be more different: whereas the Priestly account begins with dark, watery "chaos," the Yahwist account opens with a dry, barren terrain—a land of lack. Yet both scenes emphasize an initial setting of emptiness and desolation. In their canonical ordering, the reader exits, as it were, the cosmic temple of Genesis 1 and enters "an inhospitable field of clay."[5] Here the land takes center stage, itself a major character in the narrative. Fruit and soil replace distant stars and ominous sea-monsters. The grand cosmogony of Genesis 1 is followed, in short, by a clod-laden anthropogony, a tale of God's "dirty" deeds.

Text and Narrative

Just as the Priestly account follows a discernible pattern, the Yahwist account bears its own literary logic. But as drama, the garden story exhibits greater flexibility in its presentation. It unfolds in four scenes, each containing parallel elements. Every scene, except for the last, identifies a deficiency followed by God's response. The plot is driven by the fits and starts of creative activity, more improvisational than meticulously executed. The "creating God is not only the acting God, but also the reacting God, the God who responds to what has been created."[6] The Yahwist tale is the story of God's responsiveness to certain variables, the most unpredictable being humankind. In contrast to Genesis 1, creation according to the Yahwist is a series of "not-goods" made good.

Scene 1 (2:4b-17): The Groundling in the Garden

(2:4b) On the day YHWH Elohim[7] made Earth and Heaven—(2:5) before any pasturage[8] was on the earth and before any field crops[9] had sprouted, for YHWH Elohim had not sent rain upon the earth, and there was yet no groundling ('ādām) to serve[10] the ground ('ădāmāh), (2:6) though a spring would emerge from the land and saturate the ground's entire surface—(2:7) YHWH Elohim formed the groundling out of the dust[11] of the ground and blew into his nostrils the breath of life, and the groundling became a living being.[12]

(2:8) YHWH Elohim planted a garden in Eden, in the east, and placed there the groundling whom he had formed. (2:9) YHWH Elohim caused to sprout from the ground every tree, pleasant to the sight and good for food, including the tree of life in the middle of the garden and the tree of the knowledge of good and bad.[13]

(2:10) Now a river flows from Eden to water the garden, and from there it splits and becomes four branches. (2:11) The name of the first is Pishon, which encircles all the land of Havilah, where there is gold. (2:12) The gold of that land is fine; bdellium and onyx stone are (also) there. (2:13) The name of the second river is Gihon, which encircles the whole land of Cush. (2:14) The name of the third river is Tigris, which flows toward Assyria, and the fourth river is the Euphrates.

(2:15) YHWH Elohim took the groundling and placed him in the garden of Eden to serve[14] it and preserve[15] it. (2:16) And YHWH Elohim commanded the groundling, saying, "From every tree of the garden you may most certainly eat, (2:17) but from the tree of the knowledge of good and bad you shall not eat, for on the day you eat from it you will surely die."

Similar to the opening lines of *Enūma elish*, the Yahwist account opens with a stark description of lack. This first, rather extended, sentence catches creation in midstream: Earth and Heaven have been differentiated, but the former remains a barren wasteland, devoid of vegetation. This lack stems from two other deficits, namely, rain and a cultivator (v. 5b). But not for long: a spring emerges that moistens the ground, thereby setting the conditions for an *'ādām* or "groundling."[16] Like Mami, God works the "dust of the ground," the topsoil of the arable land,[17] as a potter.[18] However, in distinction from the mother goddess of Akkadian lore, the God of the Yahwist animates this dirt-laden creature, this muddy hominid, with divine breath. Flesh is akin to ground, and the *'ādām*'s enlivening happens not from a discreet divine touch—as one finds in Michelangelo's depiction featured on the ceiling of the Sistine Chapel—but from something akin to CPR, specifically mouth-to-nose resuscitation. Human life, according to this ancient tale, begins with God's intimate, fleshy exhalation.

In contrast to creation by magisterial word in Genesis 1, this story charts a distinctly dirty creation. The *'ādām* is made not in the *imago Dei* but in the *imago terrae*, in the image of Earth. Humanity's identity is bound to the ground by a remarkable Hebrew wordplay: the *'ādām* from the *'ădāmāh*. "Humanity was thus born of Earth," to quote Edward O. Wilson.[19] According to the ancient story, humanity is literally a "groundling" or earthling. But humanity's tie to the arable land is more than an etymological accident.

Contrary to Genesis 1, the *'ādām* is the first rather than the last of God's creatures to be made. Yet, as will soon become clear, this hominid of the humus remains incomplete and requires further work. For now God places the deficient *'ādām* in a garden, a horticultural feast for the eyes and the appetite (2:9). As both gardener and potter, God works naturally with creation. As "the divine farmer,"[20] God grubs about in the soil, planting trees and fingering clay. God's hands are dirty! If in the Priestly account God is King of the cosmos, in the Yahwist account God is King of the compost. Highlighting God's organic work with the earth, the Yahwist depicts the garden as the quintessence of creation.

The garden, indeed, provides creation's sustenance. Out of it flow four major rivers to water the world: the Pishon, the Gihon, the Tigris, and the Euphrates. The first two extend to unknown lands, marking Eden as half real and half imaginative. In fact, the etymological root of the word "Eden" has to do more with condition than with geography; it designates a state of delight and plenty (*'dn*), the sheer opposite of lack. "Garden of Plenty" is the *'ādām*'s home.[21] But Eden is no pristine paradise of leisure: it must be tended. The human must "serve" and "preserve" the fruit-bearing trees.[22] The *'ādām* is tied to the *'ădāmāh* in service, as the *'ădāmāh* yields to the *'ādām* its productivity. One is bound to the other.

Unlike some creation accounts, humans are not created to perform menial labor for the gods, who themselves loathe such toil. The God of Genesis 2 creates the *'ādām* and bestows upon him the commission to preserve the garden that nourishes him. The garden exists for the groundling and the groundling for the garden. The *'ādām*, moreover, is given wide-ranging freedom with only one specific restriction: he is granted access to "every tree of the garden" except one, the "tree of the knowledge of good and bad" (2:16–17). The capacity for autonomous judgment is not the groundling's prerogative. It is, for the time being, reserved for the gods (see 3:22).

Scene II (2:18–25): Creation of a Companion

(2:18) YHWH Elohim said, "It is not good that the groundling be alone; I shall make for him a co-helper."[23] (2:19) So YHWH Elohim formed from the ground every wild animal and every winged creature of the heavens and brought (each one) to the groundling to see what he would call it, and whatever the groundling called each living creature,[24] that was its name. (2:20) The groundling gave names to every domestic animal and winged creature of the heavens and to every wild animal, but for the groundling no co-helper was to be found.

(2:21) So YHWH Elohim caused a deep sleep to fall upon the groundling, and he slept. He took one of (the groundling's) sides[25] and closed up flesh in its place. (2:22) YHWH Elohim then fashioned the side that he had taken from the groundling into a woman and brought her to the groundling. (2:23) The groundling proclaimed,

> This now is bone of my bones
> and flesh of my flesh;
> this one shall be called Woman,
> for from Man this one was taken.

(2:24) This is why a man leaves his father and his mother and clings[26] to his wife,[27] so that they become one flesh. (2:25) And the two of them, the groundling and his wife,[28] were naked and not ashamed.

For the first time in Genesis, God declares creation "not good."[29] While it provides sufficient support for the 'ādām's physical welfare, the garden is deemed deficient for his well-being. The groundling lacks a companion, a corresponding other. As a remedy, God creates animals of various species also "out of the ground" and brings them to the 'ādām to be named. The groundling of the garden is the first taxonomist. By naming the animals, the primal human forms a community whose members all share common ground, including the 'ādām. But he has yet to find his match and mate; he sees nothing of his own "flesh" and "bone" in the animals brought before him (see v. 23). God's initial "experiment" has failed, but its results are not discarded. The animals are left to flourish along with the 'ādām. The garden has become a community, but one that remains deficient from the 'ādām's perspective.

God, thus, resorts to Plan B, a more invasive procedure. God fashions another creature, but this time from the 'ādām's own flesh and bone, not from the dust of the ground. This new creation requires, however, a new lack: a part of the 'ādām must be removed. This new being is no one-way derivation, however, for from the creation of the woman ('iššāh) the 'ādām becomes fully a "man" ('îš). The narrative acknowledges that the man and the woman are of common stock and are mutually engendered. By virtue of her genesis, the groundling finds himself "mutated," as it were, into a man. With the creation of the woman, humanity is now "genderly" separated. In Genesis 1, separation plays a key role in creation. In Genesis 2, the separation of flesh makes possible gendered differentiation and, in turn, sexual union. Call it splitting the 'ādām.

Through the creation of the woman, the groundling has become a man. In his genetic transformation, nothing is subtracted. He remains a groundling. Put paradoxically: through the act of "surgical" removal, the 'ādām is not neutered but rather made a man. The woman bears a functional and physical correspondence to this new man, and his response is one of utter (and uttered) jubilation, which acknowledges the woman's formal, intimate affinity to him. His cry of joy acknowledges their shared identity. Their affinity is more than biological.[30] The woman and the man are made in the image of each other, physically, socially, and covenantally.[31] Having been fashioned in the *imago terrae*, the *'ādām* discovers *himself* to be fashioned also in the *imago feminae*, and the woman in the *imago viri*. Together, they form a community of correspondence, enjoying mutual companionship and help. A new social world unfolds before them, but one that remains firmly grounded in the garden.

With the creation of the woman, the man now embodies a dual identity: he remains kin to the ground in his humanity as he has become kin to the woman in his gendered identity. As the ground is receptive to the *'ādām*'s labors, the *'ādām* receives the fruits of the ground. And so, analogously, the man and the woman are receptive to each other. No subordination pertains in the garden. The *'ādām*'s service to the garden is rooted in his kinship with the ground. Marriage, according to the Yahwist, is founded on the kinship intimacy of partnership and companionship (2:24). Life in the garden is one of fruitful work, abundance, and intimate companionship. In the garden there is neither fear nor shame, even before God. These are "lacks" that are meant to endure. But, alas, they do not.

Scene 3 (3:1–7): Temptation and Disobedience

(3:1) Now the serpent was craftier[32] than any other wild animal that YHWH Elohim had made. It said to the woman, "So it's really the case that God said, 'Do not eat[33] from any tree in the garden.'"[34] (3:2) But the woman said to the serpent, "From the fruit of [every] tree of the garden we may eat. (3:3) But as for the fruit of the tree that is in the middle of the garden, God said, 'You shall not eat from it, nor shall you touch it, lest you die.'" (3:4) But the serpent said to the woman, "Surely, you will not die, (3:5) for God knows that on the day both of you eat from it, your eyes will open and you shall be like gods knowing good and bad."

(3:6) The woman saw that the tree was good for food and that it was a delight to the eyes and that the tree was desirable for acquiring wisdom. So she took from its fruit and ate and gave it also to her husband, who was with her, and he ate. (3:7) Then the eyes of both of them opened, and they realized they were naked, so they sewed fig leaves together and made loincloths for themselves.

The scene of disobedience begins with a new lack introduced by a new character. The serpent is no garden-variety animal, but neither is it a satanic figure. The snake distinguishes itself by its "craftiness," which explains its ability not only to speak but also to manipulate. It is, moreover, a creature of the wild, an outsider. Dietrich Bonhoeffer aptly observed that the serpent initiates the first conversation about God,[35] which would qualify it as the Bible's first theologian! By engaging the woman in dialogue about God's intentions, the serpent aims to generate a *perceived* lack, one that prompts the primal couple, not God, to action. The serpent draws the woman into conversation with an absurd claim that, if true, would have resulted in the couple's starvation. Correcting the serpent's feigned ignorance, the woman responds by faithfully recalling God's prohibition given to the *'ādām* in 2:17, recasting it even more stringently: not only are they to avoid eating from the fruit tree growing in the middle of the garden, they are not to touch it (3:4). But the serpent is interested not in the precise form of God's prohibition but in its consequences. This sly creature elicits doubt about the consequences of disobedience by claiming that wisdom and power will follow: their eyes "will open," and they will thereby assume divine status.

Prompted by the serpent's "clarification," the woman considers the tree. She "perceives" that its fruit is edible and desirable for wisdom. Such wisdom equips the self for autonomous judgment and agency, for determining what is beneficial and what is detrimental. Such wisdom, the serpent promised, will result in self-enhancement of the highest order, an apotheosis no less. Plucking the fruit, the serpent claims, is tantamount to grasping and appropriating divine power.

Yes, the woman partakes first, but she does so with the full complicity of her partner, "who was with her" (v. 6b). This all important phrase is conveyed by only one word in Hebrew, a suffixed preposition (*'immāh*). With it, the narrator makes clear that the woman does not act alone, even though the focus is on her initiative and on the attraction the tree holds for her. Was the woman deceived by the serpent? Contrary to the weight of early Jewish and Christian interpretation, much of it influenced by the Greek myth of Pandora's Box,[36] the answer given by the narrator is a resounding "No!" It is significant that the narrator lingers over the woman's lingering over the tree. Rather than duped by the serpent, she is tempted by the tree itself and makes her own judgment (despite her defense in 3:13).

The results prove disappointing. The couple's eyes open wide enough to recognize their nakedness, but the desired trappings of divinity, such as power and immortality, they do not receive. Such would have been the benefits offered by the *other* tree of the garden, the "tree of life." But as God later confirms, eating from the "tree of knowledge" did bring the couple a step closer to divinity (3:22). And so the serpent was technically, if not ironically, correct: their eyes were opened and, moreover, they did not die upon eating the fruit. Yet naked vulnerability and resulting shame were neither anticipated nor welcomed by the primal couple. Through their disobedience, they become only pale, impotent images of the divine.

Their newly attained condition, however, does not leave them paralyzed. By gaining the fruit, they are thrust into a conventional world order in which their nakedness, once the occasion for delight and intimacy, now signals deficiency and defensiveness. Their world is now governed by power and pride, honor and shame, a world in which they find themselves woefully ill equipped. So, with the meager resources that they have, they must now "grow up." They take action to address their self-inflicted condition of deficiency, the vulnerability that their nakedness now signifies. They clothe themselves, a move, if not toward divinity, at least toward humanity.[37] But they can only do so pathetically. Fig leaves quickly shrivel up after being cut.

The perceived lack of divine status and self-enhancing sagacity prompted the couple's grasping for knowledge and power, but as the story recounts, gaining wisdom leads only to more "lacks." In addition to the lack of adequate clothing, the couple now lacks the confidence and comfort of continued life in the garden, as well as the joy of mutual interaction with each other and with God. The curtain falls, only to rise on judgment day.

Scene 4 (3:8–24): Curse and Expulsion

(3:8) They heard the sound of YHWH Elohim walking around in the garden during the breeze of the day, and the groundling hid himself along with his wife from YHWH Elohim's presence amid the garden's trees.[38] (3:9) YHWH Elohim called out to the groundling and said to him, "Where are you?" (3:10) And he said, "I heard your sound[39] in the garden, and I was afraid because I was naked,[40] so I hid." (3:11) And (YHWH Elohim) said, "Who told you that you were naked? Did you eat from the tree that I commanded you not to eat?" (3:12) And the groundling said, "The woman whom you gave to be with me, she gave[41] to me from the tree, and so I ate." (3:13) So YHWH Elohim said to the woman, "What is this that you have done?" And the woman said, "The serpent misled me, and so I ate."

(3:14) So YHWH Elohim said to the serpent,
> "Because you did this, you are cursed among every
>   domestic animal
>   and among every wild animal.
> Upon your belly you shall crawl,
>   and you shall eat dust all the days of your life.
(3:15) Moreover, enmity I will put between you and
>   the woman,
>   and between your progeny and her progeny.
>   (Her offspring) will strike you at the head,
>   and you will strike (her offspring) at the heel."

(3:16) To the woman he said,

> "I will greatly increase your pain in pregnancy;
>> in pain you shall bear children.
> Even though your husband remains the object of
>> your desire,[42]
>>> he shall rule over you."[43]

(3:17) And to *'ādām* he said,

> "Because you heeded your wife,
>> and ate from the tree about which I commanded you,
>>> 'Do not eat from it,'
> the ground is cursed on your account;
>> in pain you shall eat from it all the days of your life.
(3:18) Thorns and thistles it shall produce for you,
>> and you will eat the field crops.
(3:19) By the sweat of your nostrils you shall get bread to eat,
>> until you return to the ground,
>>> for from it you were taken.
> For you are dust,
>> and to dust you shall return."

(3:20) Then the groundling named his wife "Eve," for she was the mother of all living.

(3:21) YHWH Elohim made for the groundling and his wife garments of hide and clothed them.

(3:22) YHWH Elohim said, "See, the groundling has become like one of us, knowing good and bad. And now, lest he stretch out his hand and take also from the tree of life and eat and live forever," (3:23) YHWH Elohim expelled him from the garden of Eden to cultivate[44] the ground from which he was taken. (3:24) And he expelled the groundling and stationed east of the garden of Eden the cherubim and the flame of a sword turning every which way to guard[45] the way to the tree of life.

This concluding scene charts the consequences of the couple's disobedience. It opens with God's entrance, remarkable for its informality. God casually strolls through the garden during a comfortable time of day, as has been routine, to enjoy the company of the garden's creatures. But this time, instead of greeting God, the man and the woman hide in fear. God discerns that they are not immediately present and so inquires as to their whereabouts.

The man's response is steeped in irony (3:10). The Hebrew idiom "to hear the voice (or sound) of God" can also mean "to obey God," precisely what the couple did not do! The language of willing obedience is twisted to indicate fear and self-loathing.

God immediately knows what has transpired and demands an explanation. The man blames the woman ("whom you gave to be with me"), who blames the serpent (it "misled me"). This blame game, however, does not come full circle, at least within the narrative's scope, for God chooses not to interrogate the serpent. Had God done so, the serpent could have easily responded with "I only told the truth," putting God on the defensive,[46] the God who created the tree and the serpent in the first place.[47]

The curses that cycle through the various perpetrators are not punishments, however. According to Claus Westermann, they are "states that reflect the condition of separation from God,"[48] and I would add: states that reflect the condition of life as was known by the author. The curses serve to explicate certain well-known painful aspects of life. They bring the primordial world of the text into the author's contemporary world of pain and conflict. The curse delivers, in short, an etiology, an account that explains the human (and serpentine) condition as a primordial event. The serpent is condemned to a life of slithering and suffering human hostility. The woman faces the pangs of childbirth and subordination. The most extensive curse is reserved for the man: he too will face "labor pains," the pangs of hard labor on the ground. His hardship involves the backbreaking work of cultivating crops on a resistant soil. "No pain, no grain" is the curse's motto. The groundling is no longer the ground's kin but slave. Henceforth, the man shall lead a life of enslavement that is fully consummated in his death, when the 'ādām returns to the 'ădāmāh, when the human and the humus become one. His genesis was an act of separation; his death marks a reunion.

The 'ādām and the 'iššāh, the groundling and the woman, are each fashioned from their respective sources: the groundling from the ground, the woman from the man. The curse twists these life-sustaining agents into oppressive agents. The woman becomes subordinated to the man; the groundling becomes painfully bound to the ground. The woman's return to the man results in painfully borne life amid the ever-present threat of death in childbirth. The man's painful work with the soil inexorably leads to his return to the ground. Only at the expense of their lives can the man and the woman generate and sustain life outside the garden.

In this final scene, there is no lack that invites God's response in the furtherance of creation. Instead of lack there is only loss. The consequences of the couple's disobedience, of their grasping to *gain* power and wisdom, entailed the *loss* of gendered mutuality and harmony in the garden. Instead of intimacy, there is alienation; instead of joyful reciprocity, painful subordination. Fear displaces freedom and communion. The final loss involves the couple's banishment: they have lost the garden for the cursed ground.

Contrary, however, to the weight of interpretive tradition, the primal couple did not *lose* immortality by their disobedience, for it was not theirs to begin with. God acknowledges in 3:22 that the groundling "has become like one of us," fulfilling the serpent's promise. To guard against full divination, including the attainment of immortality, God exiles the couple from the garden to prevent them from partaking of the tree of life. Thus, the loss is not immortality per se but the chance to gain it.[49] As Jon Levenson suggests, God's curse of the *'ādām* serves as an etiology not of death but of burial: as the groundling emerged from the ground, so he will return to it, all the while remaining a slave to it.[50]

Guarded by formidable gatekeepers, the garden is now a barred temple,[51] and the man and the woman must eke out their existence outside it. God casually strolled inside a sacred garden, in an inner sanctum now cut off from human entry, a secret garden. But the primal couple is not cut off from God. Before their expulsion, they are clothed by God (3:21). Fig leaves will not do. As seamstress, God equips the couple for a harsher life, a life of culture and convention. God also helps Eve in procreation (4:1) and even protects Cain (4:15). Outside the garden, God remains their sustainer, reacting to new situations brought on by humanity's exercise and abuse of freedom.

The Tree of Life

The tree of life, the other tree in the narrative, offers the fruit of life everlasting.[52] It makes a brief appearance in 2:9, paired with the tree of knowledge, but reappears not until 3:22 as the tree barred from human contact. Its arboreal partner takes center stage for most of the narrative, supplying the narrative's turning point. Located on the periphery of the narrative but stationed nonetheless in the middle of the garden, the tree of life outside of the book of Genesis is intimately associated with wisdom.

> [Wisdom] is a tree of life to those who lay hold of her;
> those who hold her fast are deemed happy. (Prov 3:18)

Elsewhere, the "tree of life" is identified with the "fruit of the righteous" (Prov 11:30), "desire fulfilled" (Prov 13:12), and a "gentle tongue" (Prov 15:4), all marks of wisdom. In Proverbs, wisdom provides moral instruction *and* abundant life. In wisdom, knowledge and power, blessing and insight are united. In the garden, however, they are severed, represented by two separate trees, as if the Yahwist himself had split arboreal wisdom in two.[53]

Having two trees in the garden serves the Yahwist's narrative logic, for one leads to only a partial and, thus, pathetic apotheosis: knowledge is gained, but the chance to possess divine power and everlasting life is lost. In addition, the two trees correspond to the gendered separation of humanity into male and female. It is the woman who first partakes of the fruit from the tree of knowledge, and it is specifically the man who is barred from partaking from the tree of life. Such paired associations may not be coincidental from the Yahwist's perspective. Knowledge and the feminine are intimately connected in the personified figure of Wisdom.[54] Men seek godlike power and immortality but are ever destined to lose it, as in the case of Gilgamesh, the legendary king of Uruk.[55] And so it should be, for otherwise the human attainment of divine power would jeopardize all that God has created, garden included.

The Seeds of Paradox

By explicating the similarities and differences that mark the human family, this thoroughly "organic" etiology sows the seeds of paradox. This messy narrative elucidates how men and women are so deeply connected to each other yet so profoundly alienated from each other.[56] They are of one flesh, but each is subordinated to their originating sources: one is bound to sustain the soil's productivity; the other is bound to procreate flesh and blood.

Another paradox is planted in the very act of the couple's disobedience. In retrospect, the primal couple's free act of disobedience smacks of inevitability. The story is as natural and expected as a child's defiance of a parent's command. The man and the woman made a necessary choice, a willing but inexorable one. The willful grasping of knowledge marks humanity's coming of age, of humanity's entrance into humanity, into human culture. In the grand scheme of things, the fall is a fall forward, or better a *fail* forward to complexified, ambivalent personhood, that is, a fully human self.

Another paradox: Whereas the curse acknowledges the reality of subordination, it is not an entirely necessary reality. The curse is explicative but not prescriptive. It is not a mandate. Subordinate status is not mandated for wives any more than crop failure is commanded for farmers and the pain of childbirth for women. Cultivating weeds is not a moral obligation; neither is imposing subordination. The consequent state of affairs in Genesis 3 is deemed tragic but not morally binding. The curse reflects the consequences of the failure to live out the mutuality and responsibility for which human beings were created. The world of curse is neither what God intends nor what human beings are to strive for. The curse acknowledges that mutuality is harder to embody. Outside the garden, mutuality loses its facileness, its naturalness, but not its possibility.[57] The way of mutuality does not lose its appeal anymore than one's appetite for fresh fruits and vegetables diminishes. As the land is cultivable only through painful effort, so mutuality remains possible, indeed necessary, requiring greater resolve. For the Yahwist, it is the life of blessing *within* the garden, not the life of curse, that sustains meaningful existence, however much life outside the garden falls short. The *need* for mutuality is by no means mitigated by the reality of curse but in fact made all the more urgent.

Kingship and Kinship: Socio-Historical Context

The historical background of this ancient tale draws from ancient Israel's experiment with monarchy. Every king had his garden, and Jerusalem's king was no exception. On the west bank of the Kidron valley, east of the fortified city, was the "king's garden,"[58] watered by the Gihon spring.[59] The royal garden of Jerusalem, the city of God (see Pss 46:4; 87:3), was in some sense a replication of, or perhaps the basis for, the primordial garden of Eden in Genesis. Assyrian annals indicate that kings were as proud of their horticultural expertise as they were of their prowess on the battlefield. They frequently transplanted the exotic botanical species of conquered territories, boasting that they thrived better under their green thumb than in their natural habitats.[60] Many Akkadian and Sumerian rulers assumed the epithets "gardener" and "farmer,"[61] for their kingdom was their garden, and their ordained task was to cultivate it. The royal garden was the kingdom in miniature.

But underneath the garden story's surface lie the seeds of critique. The garden story is a cautionary tale about the human surge to consolidate power to the point of godlike status. No ascendancy to power takes place in the garden, only a failed coup with debilitating consequences. Primal man is, at most, a stripped king.[62] The Yahwist is painfully aware that imperial policy and agricultural livelihood do not mix well. An agrarian-based society cannot flourish if its citizens are routinely drafted for royal, urban-centered projects.[63] Solomon, for example, enslaved his own people to build his temple and palace (1 Kgs 5:13–15). Conscription interrupted the vital work of farming, compromising humanity's God-given identity and task. Though the *'ādām* is "taken out" of the ground

(Gen 3:23), he cannot be divorced from it. The groundling who would be king remains a struggling farmer.

Science and the Ground of Life

Despite its historical context, the garden narrative, like its canonical predecessor, widens its scope of meaning as it engages the reader's contemporary context. The Yahwist tale distinguishes itself from the Priestly report in Genesis 1 by its more "bottom-up" perspective on creation. Bottom-up accounts tend to stress continuity between past and present, between human and animal.[64] So it is with the Yahwist account of creation: in place of the magisterial God issuing commands from on high we find in Genesis 2 the intimate God experimenting with dirt and flesh. Though lacking an evolutionary sense of natural history, this second account offers a grounded vision of common origination and open-ended improvisation.[65]

Unlike Genesis 1, this earthy story of creation proceeds not in any linear, incremental fashion. At best, creation happens fitfully, replete with setbacks and "false" steps, as well as modest advances, not unlike Darwin's great idea: "life [is] a series of successful mistakes."[66] For the Yahwist, the "successful mistakes" include the experimental and the tragic—God's initial failure to create a fulfilling social context for the groundling and the couple's "failing forward" in their disobedience. For the biologist, the "mistakes" are the "errors of descent" that make up "the stuff of evolution."[67] For the Yahwist, the false steps and setbacks constitute the very stuff of human development.

Recent evolutionary study suggests that the development of life has over time proceeded not only in small, graduated steps (known as "phyletic gradualism"), but also in prolonged periods of "stasis" or morphological stability lasting millions of years, occasionally "punctuated" by relatively short bursts of rapid change. Some call them evolutionary "explosions," such as the Cambrian "explosion" of 590–525 million years ago. The proposal of "punctuated equilibrium," first made in 1972, remains controversial, but regardless of whether such "explosions" are as dramatic as Niles Eldredge and Stephen J. Gould have argued,[68] it is clear that evolution proceeds in a somewhat irregular rhythm of intensive periods of species generation and diversification, on the one hand, and conservation and stability, on the other, much of which is dependent upon environmental pressures and conditions.[69] The story of evolution, in short, proceeds with its own narrative bumps and twists.

The garden story likewise proceeds with its own fits and starts: as "lacks" are filled, new ones emerge. The scenes are set in stages punctuated by experimentation, from the hand of God no less, but not exclusively so. God's creatures are also at work, for good and for bad. The process of creation is "full of compromise."[70] Otherwise, there wouldn't be much of a story to tell. The same goes for evolution. But first, back to the story's beginning.

A Dirty Beginning

The curtain of the ancient drama rises to reveal a dry, inhospitable land characterized by deficiency. Geologically, the earth formed about 4.55 billion years ago from the gravitational condensation or "accretion" of dust. Planets developed as the result of dust and gas "pushing each other around while constantly being stirred and jolted by magnetic fields and gravitational torques."[71] Once formed, Earth was a sterile mass, and for about a half a billion years it was too hot for life. With a thin crust forming on the surface, Earth gained its "skin," a membrane separating its hot core from cold outer space.[72] Volcanic eruptions ensued as the earth was pummeled by comets and meteorites. Indeed, the most spectacular collision ever in geological history was one that, as hypothesized by scientists, happened about 4.4 billion years ago: a Mars-sized body colliding with Earth that knocked off what eventually became the moon.

Geologically speaking, water—the *sine qua non* of life as we know it—was an acquired trait. Today, oceans and seas span nearly 70 percent of the earth's surface. About 200 million years after our planet's accretion, liquid water appeared.[73] Through successive cycles of heating and cooling, elemental oxygen produced from stellar fusion reactions came to bond with hydrogen to form water.[74] Rain poured down nonstop on the land, perhaps for as long as twelve thousand years at one time.[75] Some of Earth's water also arrived via water-laden meteorites and comets.

In addition to relentless bombardments from above, volcanic eruptions, fissures, and fumaroles were releasing gases trapped within the earth's hot interior (called "outgassing"), spewing out mostly water vapor and carbon dioxide. In the early years, the sun's powerful ultraviolet rays split water vapor into molecules of hydrogen and oxygen ("photodissociation"). By 2.2 billion years ago, a new atmosphere was in the making, one in which oxygen (O_2) made at first only a modest appearance, constituting no more than 1 percent of the atmospheric level, but later dramatically increased in concentration, beginning 750 million years ago when phytoplankton blanketed the ocean's surface.[76] At the same time, the terrestrial surface was undergoing significant change. Through the processes of sedimentation, accretion, and subduction, much organic matter[77] came to be buried deep within the earth's mantle, thereby increasing continental mass and permitting the net buildup of oxygen in the atmosphere.[78] Thus, in its tectonic movements, the "ground" played an indispensable role in making possible the oxygenated "breath of life."

Another primordial contributor to the atmosphere were, of course, land plants, which began to make their dramatic appearance some 360 million years ago. The terrestrial invasion of plants, trees in particular, eventually doubled the planet's oxygen production.[79] Photosynthesis, consequently, came to exceed aerobic respiration, making possible the air that we enjoy today, which consists roughly of 21 percent oxygen. And so from above and below, the oceans and the atmosphere were formed. The "spring" from the ground and the "rain" from above (Gen 2:5–6), were the natural results of great geological upheavals and biological transformations.

Life began more than 3.8 billion years ago, before oxygen was present in any significant amount. Some scientists theorize that the organic material necessary for life was also introduced from "above" in the form of volatile rich comets and meteors. For example, the meteorite that fell near Murchison, Australia, in 1969 contained seventy-four different amino acids, eight of which were protein amino acids.[80] From comets and their gently descending dust, "prebiotic organic compounds came from space, and were brought again and again for aeons until they found the right 'little pond' to get life started."[81] Perhaps that "warm little pond" imagined by Darwin was not too far off the mark.[82] Essential for the origins of life were the interactions of air, rock, and water.[83] Life began on the edges, banks, and intersections of these elements. In the Yahwist's world, "dust" and water set the stage for life. So also on Earth.

Children of Dust

The Yahwist's account of human genesis claims that life descended from "dust." This is virtually true in ways that the ancient author could not have imagined. From interstellar dust and gas, stars and planets are formed. Dust in the universe, in fact, consists of many atoms that form complex molecules, including silicates, carbon grains, and iron oxides.[84] Cosmic gas clouds are composed primarily of hydrogen and helium. By dint of gravity, dust and gas together become the necessary ingredients of cosmic "clumpiness." A star is an interstellar cloud that has collapsed under its own weight, generating enough heat (10 million degrees Kelvin) to produce thermonuclear fusion for billions of years.[85]

The higher-mass stars manufacture dozens of elements in their cores, beginning with the conversion of hydrogen into helium and then to carbon, nitrogen, and oxygen, all the way to iron. In their final phase, many of the larger stars explode into supernovas, effectively spilling "their chemically enriched guts throughout the galaxy."[86] Elements heavier than iron are forged in these interstellar death throes. It is precisely these "guts" that provide the elements necessary for the formation of everything from planets to people. The origin of life itself is indebted to the violent deaths of these high-mass stars. It is, in the words of one astrophysicist, a "messy business."[87] Human beings are thus constituted by dust of the most elemental kind, stardust.

Power of the Ground

Closer to home, dust is the primary source of fertility. Windblown dust, called "loess," has provided the deepest, richest soils in the world. Northern China's yellow loams, some hundreds of feet deep, were formed by dust blown in from the Gobi desert.[88] Just as astounding, the Brazilian rainforest is actually "an artifact of the Sahara."[89] Silt from Lake Chad is borne aloft by the wind, carried across the Atlantic to be picked up by storms that eventually pass over the Amazon Basin, dropping rich nutrients, including phosphate. It is estimated that up to 12 million tons of Sahara dust drop on Amazonia each year, maintaining its fertility.[90]

Watered "dust" produces potter's clay, which according to Genesis God used to create the 'ādām. From a mineralogical standpoint, there is also the kind of clay that provides a rich repository for biosynthesis, the beginning of organic life. Clay minerals are based on hexagonal rings composed of silicon and oxygen, which could have served as templates for organizing the ring structures of many carbon compounds.[91] Amino acids, for example, tend to concentrate on clays to form small, proteinlike molecules.[92] Having a large reactive surface and being rich in iron and potassium, clay was likely a "necessary partner for the birth of the organic realm,"[93] acting as "scaffolding" for the formation of RNA, the polymer that enables protein synthesis.[94]

In accounting for the development of organic life on the ground, one cannot forget humus. Composed of partially decayed organic material and holding the most amount of water of any topsoil, humus is a habitat for microbial life and a repository for mineral nutrients.[95] Given the fertile power of humus, and of soil in general,[96] no wonder the ancient storyteller chose the "dust of the ground" as the raw and rich material for all terrestrial life. Both the Yahwist and the scientist acknowledge the ground as Earth's life-support system. The return to dust is flesh returning to Earth's flesh, thereby sustaining Earth's ecosystem. Maggots, mites, fly larvae, beetles, and worms are Earth's lowly custodians, which usher the body's return to its generative womb, to "dust."

'Ādām *and the Animals*

One unavoidable collision between the biological account of humanity's gene-sis and the Yahwist's anthropogony is the order of appearance: in the biblical account the *'ādām* is created before the animals (2:7, 19). From an evolutionary perspective, humanity is, so far at least, the endnote to the sweeping saga of life's development, beginning with the microbial. Yet credit is due this ancient narrator for recognizing the *common* ground of life.

The fact that all organisms we know share the same kind of genetic cod-ing (DNA), with only slight variation, is itself testimony that life on Earth descended from the same group of primitive bacterium-like cells. These rudi-mentary cells eventually evolved from simple prokaryotic cells to the more complex eukaryotic variety, which features a tightly organized nucleus con-tained within a porous membrane. The next major evolutionary advance was the emergence of multicellular life, manifest in such forms as crustaceans and mollusks, each bearing sense organs and a central nervous system. And, finally, "to the grief of most preexisting life forms, came humanity."[97] One could say that the Yahwist conflates in one fell swoop the sweeping saga of evolution by claiming that humans, with their unmatched complexity, emerged from the ground up, whether one calls such "ground" primordial stardust, organically rich soil, microbial material, or simply "slime."[98] By any name, the "ground" constitutes our humble beginnings, whether told by a Darwinian or by a Yahwist.

By claiming such a simple, bottom-up beginning, both the ancient narra-tor and the evolutionary biologist acknowledge the linkage of all life. The basic biochemical and genetic unity of life suggests a single biological (spe-cifically "monophyletic") origin for all known living beings. Gene counts between human beings and much simpler organisms such as "worms, flies, and simple plants" all fall in the same range, "around 20,000."[99] Among primates, humans (*Homo sapiens*) and chimpanzees (*Pan troglodytes*) are 96 percent identical at the DNA level,[100] making chimps humanity's closest non-human relatives. While the human has twenty-three pairs of chromosomes, the chimpanzee (along with the gorilla and the orangutan) has twenty-four. The difference lies in the fusion of two ancestral chromosomes shared by chimpanzees resulting in Chromosome 2 of *Homo sapiens*.[101] Among the primates, the human is the genetic result of a simple fusion of two short chromosomes.

"Fusion" also pertains to humanity's evolution in another way. Recent DNA research conducted at the Broad Institute of Harvard and the Massachu-setts Institute of Technology suggest a picture of human origins far more detailed than what the fossil record reveals. When the ancestors of human beings and those of chimpanzees parted ways some 6.3 million years ago, it was by no means a clean break. There was extensive interbreeding for more than a million years before going their separate ways for good. As geneticist James Mallet comments: "We probably had a bit of a messy origin."[102]

The messiness of genetic kinship between humans and other primates extends into the social and perhaps even the ethical realm. Chimpanzees, for example, exhibit a remarkable range of behavior and skills. They employ and even build tools, hunt in groups, engage in violence (including a primitive form of warfare), form alliances, and reconcile after quarrels.[103] They are by nature social creatures and appear to exhibit empathy, self-awareness, cooperation, planning, and learning. The linkage between humans and chimps includes far more than just expressive faces and opposable thumbs.

Behavioral similarities, however, are not limited to chimps. Rhesus macaques exhibit what primatologist Dario Maestripieri playfully describes as "Macachiavellian" behavior, the primatological counterpart to Machiavellian conduct: everything from nepotism to competitive politics.[104] "For most of our evolutionary history we probably acted a lot like rhesus macaques, and we still do in our everyday lives," Maestripieri observes.[105] Frans de Waal of the Yerkes Primate Research Center, however, sees more than just self-centered social maneuvering among primates. The antecedents of human morality, he claims, can be found in nonhuman primate behavior.[106] Consolation, for example, is universal among the great apes.[107]

De Waal has observed several common forms of ethical behavior among certain primates: cognitive empathy (empathy combined with appraisal of the other's situation), reciprocity, and fairness.[108] They are, in his words, "moral sentiments."[109] With regards to empathy, the bonobo exhibits more affinity to humans than the chimp.[110] De Waal is convinced that the evolutionary origin of the ape's ability to take another's perspective is to be sought not in social competition but in the need for cooperation and community concern, the results of group living and social pressure.[111] To be sure, the capacity for moral judgment applies only to humans, but as de Waal rightly notes, such abstract reasoning is not all that definitive for *Homo sapiens* in practice.[112]

Recent experiments have shown that when faced with a dilemma requiring a moral decision, we tend to act situationally or emotionally rather than logically. The rational mind is used sparingly in situations that call for a quick decision. Reasoning typically comes *after* the decision is made, "as the brain seeks a rational explanation for an automatic reaction it has no clue about."[113] In situations of argumentation, the brain is like a lawyer: it "wants victory, not truth; and, like a lawyer, it is sometimes more admirable for skill than for virtue."[114]

To sum up: "While it is true that animals are not humans, it is equally true that humans are animals."[115] To deny this is to commit "anthropodenial," de Waal's term for a species-centric hermeneutic that is equally careless as unchecked anthropomorphism. "Even if human morality represents a significant step forward, it hardly breaks with the past."[116] For the Yahwist, the past points to the common ground of all life.[117]

Emergent Human

How does the study of evolution account for this common ground of life, including human life? The evolution of hominids continues to be fleshed out by paleoanthropologists and geneticists. We know much more now than what was known even forty years ago. Nevertheless, central questions persist. The period between 2 million and 1 million years ago, for example, remains a "dark age" for paleontologists, with various hominid species possibly overlapping, suggesting a greater diversity during this time period than currently thought.[118] The story of our evolutionary heritage is still being written.

But some things we do know. The roots of the human tree of life go back at least 50 million years to the common archaic primates. Before that, about 80 million years ago, shrew-like animals began to climb trees and shrubs for edible fruits, leaves, and insects. Developing dexterous fingers and binocular vision, our arboreal ancestors began to diversify into early monkeys and apes.[119] But for many, life in the trees did not last. Because of a housing shortage brought about by climate change, an environmental eviction notice was served. Between 8 and 5 million years ago, the earth suffered an extended drying out period, owing to moisture trapped in ice sheets spreading from the North and South Poles. Dense forests in Africa were replaced with open woodlands or savannas. While our gorilla ancestors remained in the forest, the ancestors of chimps and hominids ventured forth into the open country, adapting to life on the ground. The ecological shift from forest to savanna, from life in the trees to life under the sun, eventually led us to walk upright.[120]

Since the time of Darwin's publication of *On the Origin of Species* in 1859, many missing links have been discovered that fill out humanity's ancestral lineage, which now resembles more a sprawling bush than a towering tree. It began with the discovery of the fossilized remains of a Neanderthal[121] in 1856 and Eugène Duboi's discovery of the "ape-man" (*Pithecanthropus*) in 1891. Other major finds include Raymond Dart's discovery in 1924 of a baboon-sized fossil skull at a quarry near Taung, north of Johannesburg, which he named *Australopithecus africanus* ("the southern ape from Africa" of 3.2 to 2.5 million years ago) but is known colloquially as the "Taung child." Five years later, fossil remains were found in China's Zhoukoudian Cave, eventually classified as *Homo erectus* (770,000 years ago).[122] Perhaps most famous is the discovery of "Lucy" in 1974 at Hadar, Ethiopia, by Donald Johanson, nicknamed after the Beatles song but classified as *Australopithecus afarensis* and dated to about 3.5 million years ago.[123]

Lucy's discovery shocked the anthropological world because she was bipedal yet lacked a brain size anywhere near that of a human. It had been thought that bipedalism was based on large brain size, which we now know developed much later. Many discoveries since then have increased the number of branches on the phylogenetic tree. One of the most recent findings is the discovery of the so-called "hobbit" species, formally named *Homo floresiensis*, discovered in a cave on Flores, an island east of Bali. The skeletal remains indicate a human species of 18,000 years ago that grew no larger than a three-year-old modern child and belonged to the ancestral line of *Homo erectus*.[124]

If we place the modern *Homo sapiens* at the top branch of this tree, then "standing" at the trunk is the partially bipedal primate, which made its entrance into the savanna as early as 4.4 million years ago. As the "prime diagnostic feature of hominids,"[125] the ability to assume a consistently upright posture set a new course in hominid evolution that eventually gave rise to distinctly human capabilities, including music and language.[126] Physical uprightness required the lowering of the larynx, thereby broadening the voice's pitch and enhancing its diversity of sounds. According to paleoarchaeologist Steve Mithen, "bipedalism may have initiated a musical revolution in human society"[127] and, along with it, language. Along with physical uprightness came a leveling of size difference between genders (dimorphism) and a marked increase in sexual communication or "signaling power."[128] In such a light, the biblical tale of the woman's creation takes on additional nuance. As the groundling recognized the woman as his equal, in size and in flesh, as they faced each other in genuine wonder, the man, it could be said, experienced a "double erection."[129] Moreover, his poetic words of jubilation, from an evolutionary perspective at least, were part of his song and dance.

Among the various hominids, the late comer *Homo ergaster* (dated to 1.6 million years ago) was likely the first of our ancestors to have lost most of its body fur, an evolutionary advantage under the savanna's big sky. Hence, this hominid was the "first truly 'naked ape.'"[130] It was also the first to emigrate out of Africa and, consequently, the first to wear clothing,[131] presumably from animal skins, for protection from the cold.[132] *Homo ergaster* may have also been the first to cook food on a routine basis.[133] Its emigration and evolutionary development led to *Homo erectus*, which migrated out of Africa sometime around 2 million years ago and settled parts of Eurasia, including China and Indonesia.[134]

The question of whether *Homo sapiens* is a direct descendant of *Homo erectus* remains debated. Recent findings in Java suggest one separate line of *Homo erectus* that became an evolutionary dead end.[135]

By analyzing mitochondrial DNA, which is passed only from mother to offspring, molecular biologists have discovered that the earliest version of human beings ("archaic *Homo sapiens*") evolved in Africa at least 600,000 years ago and that *Homo neanderthalensis* emerged as an entirely separate species. Compared to modern *Homo sapiens*, Neanderthals carried a larger but lower skull.[136] As such, they lacked the frontal lobes of the neocortex, where highly cognitive thinking is lodged. Nevertheless, this species launched a new technological era marked by a greater refinement of stone tools. In addition, Neanderthals were the first to bury their dead, perhaps with accompanying ceremony.[137] Use of language and art remains a matter of debate. Steve Mithen argues that Neanderthals performed music and dance, but did not exhibit linguistic abilities comparable to those of *Homo sapiens*.[138] They were "linguistically challenged but musically advanced."[139] Their extinction about 28,000 years ago remains a mystery. Although more robust physically, they were evidently no match for modern *Homo sapiens* in the competition for resources when the latter settled Europe around 40,000 years ago.

Mitochondrial DNA analysis traces the ancestry of most human beings or modern *Homo sapiens* back to a single African population that lived around 170,000 years ago, to what was earlier called the "mitochondrial Eve" or African Eve—allegedly one woman from whom all modern humans are descended.[140] Recent analyses, however, point to a discrete group of *Homo sapiens* in East Africa. Other scientists take a more "multiregional" approach by arguing for both African and "local" ancestry regarding our evolutionary heritage.[141] Regardless, for the past two million years Africa seems "to have been the source of 'pulses' of hominin evolutionary novelty."[142] In other words, Africa is Eve.

As with previous hominids millions of years before, our species did not linger at home: the emergence of *Homo sapiens* out of East Africa around 50,000 years ago prompted another, more extensive spread throughout Africa and Eurasia, rapidly replacing their predecessors, including Neanderthals in the west and *Homo erectus* in the east. By 40,000 years ago, Australia and New Guinea were colonized. It was also during this time that Upper Paleolithic technology appeared, prompting the "most rapid and radical cultural change ever recorded in the hominid line."[143] This "Great Leap Forward," as coined by Jared Diamond,[144] or "Creative Explosion," according to David Lewis-Williams,[145] marked an unprecedented leap into religion, new technology, and population growth. Sewing implements were invented, and

full-blown art suddenly flourished, yielding magnificent cave paintings, stat-
uettes, and jewelry.[146] Burials were accompanied by elaborate artifacts, indicating religious ritual and belief in an afterlife. While theories range from
greater competition for resources to neurological transformations of the
brain, the specific reasons behind this cultural explosion presently remain
unexplained.[147]

In any case, hominid evolution passed through 700,000 years of environmental change, "one of the most turbulent periods of environmental instability
in the earth's history,"[148] to bring itself to this benchmark. Such instability
forced greater mobility and versatility upon early humans as they adapted to a
wide range of environmental conditions. According to Richard Potts, versatility
was modern humanity's hallmark, "the capacity . . . to buffer survival risks and
resource uncertainty."[149] From 700,000 to 50,000 years ago, brain size
increased, stone tools diversified, and social interactions intensified. Between
400,000 and 300,000 years ago, hearths and shelters proliferated. Humanity
was constructing its niche in the world.

Up until 10,000 years ago, prehistoric humans remained hunter-gatherers,
the basic way of life for 5 million years, and throughout most of our evolutionary development, gathering took precedence over hunting. But as skills
and organization improved, along with a new range of weapons, prehistoric
humans became so efficient at hunting that they likely wiped out much of the
Pleistocene megafauna, a population of large animals ranging from the giant
ground sloth to the wooly mammoth. By killing off "large mammalian genera
between 10,000 and 50,000 years ago," early humans greatly altered the flora
and fauna of North America,[150] ushering an age of human-induced extinctions
that continue unabated to this day. In addition to pushing many species to
extinction, human predation has also altered the evolution of various species
by causing them to reproduce at younger ages.[151] While this shift improves
the chances of reproducing before being eaten, the change is harmful on the
species level: the spawn of younger fish, for example, are not as robust as
the spawn of older fish. With the exception of domestication, humans have
been reshaping the evolution of many species neither to their advantage nor
to ours.

Mind-full Evolution

As cognitive capacities increased among hominids over time, enter the mind
onto the evolutionary stage. But first the brain: its expansion is "the reason why
human beings instead of baboons or chimpanzees burst out of Africa, occupied
the entire planet, and shaped Earth to their own uses."[152] Most distinctive about

the brain is the large cerebral cortex or neocortex, the brain's outer shell, "perhaps the most complex entity known to science."[153] It is there where most "higher" cognitive functions occur, including the development of language, tool manufacture and use, social skills, and memory capacity.

However it is to be defined, the mind is intimately associated with consciousness—the reflective awareness of one's own self. Paul Ehrlich helpfully distinguishes between "intense consciousness"—the awareness of a continuous sense of self—and ordinary consciousness—"the capacity . . . to have, when awake, mental representations of real-time events."[154] According to psychologist Nicholas Humphrey, consciousness has deep ties to feelings. *Cogito ergo sum* ("I think, therefore I am") is intimately linked to *Sentio ergo sum* ("I feel, therefore I am"), suggesting a continuum between nonhuman animals capable of intentional behavior and human beings.[155] At the whole brain level, consciousness marks what Steven Mithen calls "cognitive fluidity."[156] At the neurological level, "a sense of self arises out of *distributed* networks in both hemispheres" of the brain.[157]

Science treats the mind as an emergent, evolved property. The mind manifests itself in the brain's neurological complexity, which sets the conditions for cognitive and behavioral complexity, some of which can be associated with particular parts of the brain.[158] But not all. Some mental activities require the exercise of the whole brain as an *integrated* network. Hence, one could cite another revision of Descartes' epistemological motto: "I link, therefore I am."[159] Such mind-full activities include the exercise of moral judgment and creative imagination, the work of "evaluating that which moves one to action."[160] Although the mind is no metaphysical entity in its own cognitive right, it does indicate the capacity for what I would call "meta-self engagement": the ability to step outside ourselves and critically examine our presuppositions, practices, and motives, and to act accordingly. This capacity makes us cognitively different in the animal world. Whatever it is precisely, the mind is a function of the highest cultural order that enables self-transcendence. Emerging from its evolutionary roots, the human mind has the power to redirect human evolution culturally, for woe or for weal.

For all that must be kept in mind regarding the distinctions between human and nonhuman animals, the cognitive difference remains one of degree rather than in kind. It is the *degree* to which human beings are self-conscious that makes us unique and marks a "phase shift" in hominid evolution.[161] Acknowledging this biological, specifically neurological, common ground yields a profoundly ethical payoff: "Assuming various degrees of animal consciousness spares us the hubris of seeing ourselves as the sole possessors of consciousness,"[162] as well as, one should add, the sole possessors of dignity.

"Our dignity arises *within* nature, not against it," so observes Mary Midgley.[163] Our dignity as human beings should not feel threatened by "our continuity with the animal world."[164] The Yahwist would agree.

The evolutionary impetus for the brain's enlargement, along with the development of its intellectual faculties, stems as much from the neurological complexity of the human brain as from the growing complexity of social structure with which the brain/self has interacted. Consciousness is no island; its emergence was socially elicited. As witnessed within the last 700,000 years, social and environmental challenges helped to stimulate the brain's evolution.[165] Not coincidentally, the Yahwist identifies certain challenges as formative in the development of human consciousness: the theological dilemma of disobedience posed by the snake, the temptation posed by the tree of knowledge, and the challenge of adapting to the great environmental transition to life outside the garden. The bumps and twists in the Yahwist's tale are formative ones as the human characters undergo their own narrative "evolution."

Gender and Sexuality

According to the Yahwist, gender differentiation originated out of the need for companionship, thereby setting the occasion for sexual intimacy between woman and man and defining the context for marriage. Within the (word)playful logic of the ancient narrative, the "splitting of the Adam" is an act that involves mutual gain. While the man provides the woman her flesh and blood, so the woman contributes to the man's maleness. The ancient tale acknowledges a mutual contribution in the determination of gender. So does genetics.

GENETICS. In reproduction, both the male and the female provide their unique genetic contributions. The female provides mitochondrial DNA (mtDNA), which, unlike nuclear DNA, does not get reshuffled between chromosomes when cells divide. Hence, mtDNA is a powerful tool for tracking maternal lineage, indeed, all the way back to the so-called Mitochondrial Eve. On the other side, the male contributes the Y chromosome. Because it has no female counterpart, the DNA of the Y chromosome also does not get reshuffled.[166] Passed on from one generation to the next, both mtDNA and the Y chromosomal DNA constitute "non-recombining" regions of the genome.[167] While the Y chromosome determines the male gender of the offspring at the embryonic stage, mtDNA transmits certain physical inheritances from the mother. On the genetic level, both the male and the female offer their distinctive contributions to their progeny.

SEXUALITY. While the genetic determinants of gender may seem straightforward, the evolution of sexual behavior is anything but simple. Polygyny—the practice of one male bonding with more than one female—is considered the "basic" mating system of mammals. The reason likely stems from the fact that the male's biological investment in nurturing and raising his offspring is disproportionately lower than that of the female, beginning with gestation.[168] A number of mammal species are known to form harems, each dominated by a single male, and the size of the harem frequently correlates with sexual dimorphism, the size difference between genders. Male gorillas, for example, average three to six mates and weigh nearly twice as much as the females. Male southern elephant seals weigh more than eight times as much as the females and average forty-eight mates.[169] Male gibbons, by contrast, do not outweigh their mates and are monogamous. A "low-grade polygyny,"[170] thus, lurks within humanity's evolutionary upbringing, confirmed by the slightly greater size, on average, that males have over females.

What most clearly distinguishes humans sexually from most other mammals is that the latter have a clearly defined period of estrus (heat) when the female is receptive to copulation. Most of the time, no sexual activity occurs among nonhuman mammals, and in all nonhuman animals, except bonobos, copulation is periodic.[171] For humans, however, sexual activity is relatively constant.[172] No consensus has been reached as to why this is the case. Reasons include the need to recruit more paternal investment in child rearing, to reduce the possibility of male infanticide by concealing knowledge of paternity, and to conceal ovulation.[173] The determining factor behind the loss of estrus in hominid evolution may be biological or sociocultural.[174] Probably both.

Not only is human ovulation concealed, but also human copulation. Human beings engage in "cryptic copulation," evidenced also among chimpanzees and gorillas.[175] Its evolutionary genesis, Ehrlich surmises, may have begun with the successful solicitation of sexual engagement by less dominant "sneaky" males within harems: "Cryptic copulation makes sense when pairs are trying to avoid the attention of dominant males."[176] Regardless of its evolutionary origin, "cryptic copulation" is a relatively distinctive feature of human sexuality. It is signaled in the Yahwist narrative by the acquired "shame" of nakedness, which the primal couple sought to conceal from God.

Regarding sexual desire, research suggests that females are just as psychologically desirous as males. Nevertheless, there is a marked tendency for males to seek multiple partners and accept casual sexual encounters and for females to "focus on the quality of relationships."[177] Ehrlich sees this as a genetically based tendency that for women "cultural conditions can clearly overcome."[178]

But the same could be said for men. In modern society, staying married increases men's chances of living past the age of sixty-five from 65 to 90 percent.[179]

MALE DOMINATION. Much in contrast to the parity of sexual desire among human males and females, human history has consistently testified to the disparity of social power between genders. The problem is an inherited one "rooted in genetics that is gradually being solved by cultural revolution."[180] The genetic inheritance is the gender size and strength difference between males and females, the result of "sexual selection caused by competition for mates."[181] Ehrlich points out that sexual size dimorphism among the great apes is least evident among bonobos. "Male bonobos are only about 15 percent larger than females, and females are basically codominant with them."[182] At the other extreme are male gorillas, which are almost twice as large as females and have larger canine teeth, features related to maintaining exclusive access to harems against other males.

With regards to human evolution, the fossil record reveals that size dimorphism between genders was clearly evident among the australopithecines, with males 30 percent larger than females, and more so with *Homo habilis* (up to 60 percent). *Homo ergaster* and *Homo erectus* were the first of our ancestors to exhibit a reduction of size difference between genders comparable to modern *Homo sapiens* (15–20 percent). The difference, however, remains substantial enough to this day to allow males to assert physical dominance over females. The physical difference is genetically determined, but the consequences are socially shaped. At one extreme is the Yanomamö tribe, near the headwaters of the Amazon and Orinoco rivers in Brazil and Venezuela. Male members frequently exert violence against their wives, from beating to burning.[183] At the other extreme are modern Western societies that have established anti-harassment laws to ensure equal treatment of women and men in the workplace. Yet at home violence continues to be the leading cause of injury to women in the United States.

Along with the freedom from reproductive commitment, the superiority of male strength has fostered many a patriarchal society, from ancient to modern. So also among some of our nonhuman ancestors: "In both chimps and people, fathers, brothers, and sons form the core domestic group; it is ordinarily females that change groups."[184] But this is more the exception: among most other primates, such as baboons and rhesus macaques, it is the male who leaves his own group, like the man in Genesis 2:24, to seek a mate, while the female stays at home.[185] Among these primates, females hold the political power as they form and maintain most social alliances.[186] "If human societies had retained the predominant primate pattern of female bondedness and female domination, things would have been different,"[187] comments

Maestripieri. Indeed. It is as if the Yahwist had drawn from humanity's primate roots, rather than from common patriarchal practice, for its etiology of marriage!

The "Fall" and Its Consequences

Evolutionary science has yet to demonstrate anything resembling the "Fall." There is no paleontological evidence to indicate that hominids once existed in blissfully peaceful relations and that snakes, for that matter, once walked upright and talked. However, a recent survey of 350 adult Neolithic skulls from British burial sites reveals evidence of a dramatic rise in violence around 10,000 years ago. A high number of depressed craters on the left side of the skull, indicating a "startling frequency of overall violence," were observed.[188] Injury rates, moreover, were equal for men and women. The results of the study "bring home the idea that these were not just peaceful farmers living in a rural idyll."[189] Owing to the social pressures of food production and living together in greater concentration, our ancestors became more prone to violence.[190] In his own way, the Yahwist acknowledges the increase of violence in the human family as it suffers the painful transition from life inside the garden to life outside, from gathering food to cultivating the "cursed" ground.[191]

There is, however, a real collision between science and the traditional interpretation of the biblical "Fall" that demands a reckoning. By itself, the Yahwist story offers profound reflections concerning the genesis of human sin, about how human violence begets violence leading up to God's grief-stricken resolve to "blot out" all life on the land by flood (Gen 6:7). But the garden story has been traditionally read, particularly by Christians, as an etiology for all pain and suffering experienced on Earth, from animal predation to earthquakes, all the result of the primal couple's disobedience. Humanity's fall from paradise, it is claimed, brought about nature's fall from perfection. But the story itself is much more limited in scope. Its primary focus is on the human family, not the family of life.

Science critiques all interpretations to broaden the purview of the "Fall" beyond the human family. Savage competition, untold suffering, ravaging disease, and extinction—the "war of nature," to quote Darwin[192]—were all endemic to the natural order long before hominids ever arrived. Nature was very much "red in tooth and claw" prior to the advent of human beings.[193] Think of the *Tyrannosaurus rex* rampaging the land some 80 million years ago or the saber-toothed tiger (*Smilodon fatalis*) of 2.5 million years, whose teeth were perfectly adapted to ripping open the throats of their prey.[194] At no time in evolutionary history did the lion ever eat straw "like the ox" or the leopard lie "down with the kid" (cf. Isa 11:6–7). To attribute predation and suffering to human disobedience is scientifically unfounded. There is no evidence whatsoever, and all evidence to the contrary, for nature to have "fallen" from an original harmonious perfection. Instead, the evolution of life has from the very beginning operated under death, predation, extinction, and competition while, at the same time, giving rise to diversity, complexity, beauty, and the self-consciousness to perceive it all.[195]

Nevertheless, nature has suffered considerably as a result of the emergent human. Humanity has been a "plague mammal . . . with a long track record for transforming, and impoverishing, a range of ecosystems."[196] Currently at 6.7 billion and expected to reach 9.1 billion by 2050, humans have become the most invasive species on Earth. But numbers alone do not tell the whole story. Ever since we learned how to hunt, extinction upon extinction of animal species, beginning in the Pleistocene era, has followed, and now at an alarmingly rapid rate. Within the first few thousand years of the arrival of humans into North America, 70 percent of large mammalian species were pushed to extinction.[197] E. O. Wilson estimates that the rate of species extinction is "now 100 times the rate at which new species are being born," an unprecedented figure since the end of the Cretaceous Period 65 million years ago.[198] If left unabated, anthropogenic degradation of the biosphere could destroy "half the species of plants and animals on Earth by the end of the century."[199] Already 25 percent of all nonhuman mammals are in imminent danger of extinction due to human activity, from habitat loss to hunting pressure and accidental death, as well as 36 percent of marine mammals.[200]

So, yes, the ancient tale of human disobedience, of the urge to snatch divinity and transgress limits, beginning with plucking fruit from the forbidden tree, can be read to highlight the irreversible damage to creation wrought by human hands. As a consequence of humanity's rise to power, countless species have suffered their own "fall" into extinction. And as a "plague mammal," humanity continues to be, as it always has been, plagued by sin, which according to the Yahwist is most evident in the human will to power to the point of violence[201] and in the failure to take responsibility to "serve and preserve" the garden.

Rereading Genesis 2:4b–3:24

There are many other aspects of the Yahwist's tale that can profitably engage science. To be sure, the natural sciences place the ancient story in its proper context, one that is more existential than empirical in orientation. The garden story is unapologetically "mythic." But as William Sloane Coffin aptly notes, "The truth of a myth is not literally true, only eternally so."[202] The primal couple's genesis and expulsion highlights certain features of the human condition as understood by ancient, wise minds. Nevertheless, science, particularly biology, invites today's reader to probe more deeply the wisdom of this seemingly simple tale.

"The Groundling Gave Names"

The chronological order of creation according to the Yahwist, as noted earlier, does not correspond to the evolutionary account, in which animals, including primates, evolved before hominids ever made their appearance. Nevertheless, it must be noted that the Yahwist does affirm that the *full* identity of humankind marks the culmination of the creation of life, occurring only *after* the animals are created. Prior to the creation of the animals, "man" is merely a "groundling," fashioned from the same substance as the animals. His distinction becomes evident in his power to name them.

Against the backdrop of hominid evolutionary development, the specific order of humankind's genesis in Genesis 2 finds its significance in the remarkable scene in which the animals are brought to the *'ādām* to be named. Humanity's temporal primacy in this mythic tale attests to humanity's power to domesticate life on the animal planet. When God brings the animals "to see what he would call them," the groundling is granted the privilege of determining the social and functional place of each animal in the garden. The human being is deemed the quintessential "alpha male" or, better, the "*alef 'ādām*" (in keeping with the Hebrew). Consequently, all forms of domestication, from animal breeding to employing "beasts of burden," find their mythic origin in this primal act of naming. Domestication bears both genetic and cultural power.

From an evolutionary perspective, the modern dog bears a particularly telling witness to the genetic power of domestication. The domestication of the wolf, the ancestor of all dogs, began nearly 400,000 years ago, and the spectacular result is today's American Kennel Club, which operates the world's largest registry of over 150 pure breeds in the world.[203] The wolf destined for domestication eked out its living as a scavenger, roaming "from rubbish tip to rubbish tip" at the fringes of human culture, like many feral dogs today.[204] Eventually, the relationship developed into one of cooperation: domesticated wolves became useful because they could chase down wounded prey.[205] Conversely, humans made hunting a lot easier for their canine companions.[206] A domesticated animal is one that knows its place in relation to its owner, who has given it a name, a "pet name." The Yahwist recognized that without humankind's genesis, specifically without the *'ādām*'s act of naming, many animals would have had a markedly different destiny, genetic or otherwise. For some, domestication was a matter of survival: "To be born free was, for wolves or wild sheep, a dead end."[207]

"And the Groundling Became a Living Being"

To put it bluntly, science has not discovered a soul, that is, an immaterial part or attribute of human nature that accounts for thought, consciousness, and will. Indeed, science cannot. Nevertheless, the neurosciences are fruitfully studying the exercise of rational capacities as brain processes.[208] Thus, we need to ask whether the Yahwist actually champions the soul's existence.[209] As discussed in the translation note for 2:7 (p. 264, n. 12), the issue hangs on how one translates the Hebrew *nephesh* (or *nepeš*), rendered "soul" in the King James Version. Throughout the Hebrew Bible, the term conveys a wide range of meaning, from "throat,"[210] "neck,"[211] and "breath,"[212] to "life,"[213] "person(ality)," and, yes, "soul" and "spirit" in certain cases.[214] Our beginning point, however, is grammatical: the verse in Genesis refers not to the human being having a *nephesh* but to the human being *becoming* one. *Nephesh* identifies the whole groundling as alive. It is a life force as much as blood is elsewhere in biblical tradition.[215] But in Genesis 2, *nephesh* is associated with breathing, and like breathing, a *nephesh* can cease.[216] For the Yahwist, *nephesh* bears no hint of an "indestructible core of being, in contradistinction to the physical life."[217]

Such a conclusion is not to strip *nephesh* of its anthropological and theological import. In our narrative, *nephesh* is the definitive sign of life for every breathing creature. Breath has all to do with animal and bacterial life. As scientist Tim Flannery pointedly observes:

> The time-honored custom of slapping newborns on the bottom to elicit a drawing of breath, and the holding of a mirror to the lips of the dying are bookmarks of our existence. And it is the atmosphere's oxygen that sparks our inner fire, permitting us to move, eat, and reproduce—indeed to live. Clean, fresh air gulped straight from the great aerial ocean is not just an old-fashioned tonic for human health, it is life itself, and thirty pounds of it are required by every adult, every day of their lives.[218]

Breath, the Yahwist recognized, has all to do with *nephesh*. The breath of life also bears its own cosmic breadth. Every breath, whether exhaled or inhaled, contains roughly 10^{22} gas molecules (a liter), comparable to the total number of stars in the universe. Each breath taken is also a breath shared. The molecules we exhale enter the turbulent atmosphere and become thoroughly mixed with all other gas molecules. The air that we breathe is shared by all air-breathing life on Earth, turning the ancient notion of *nephesh* into a collective source of life.

It is no coincidence, then, that the biblical narrative applies this breathlike *nephesh* to all the living creatures the groundling names in Genesis 2:19 and, later, to all the animals of the ark, all recipients of God's universal covenant (Gen 9:10, 12, 15, 16). So whatever it is, this "living *nephesh*" applies not just to humans but to all the other creatures formed from the ground in Genesis. If *nephesh* is a "living soul," then all animals have it! And if it is a "soul," it is not a metaphysical, spirit-as-opposed-to-matter kind of soul. Nothing in this text and elsewhere in Genesis indicates an immaterial, much less immortal, soul infused into a body.[219] The simple but remarkable consequence of God's creation of the human in 2:7 is the granting of life, specifically the imparting of breath. Like a newborn infant, the *'ādām* is given his first breath and, *viola*, becomes animated. The attainment of consciousness comes much later.

"Then the Eyes of Both of Them Opened"

The Yahwist's "dirty" picture of human genesis does not remain inert. With its evolving characters, the story builds to its climactic episode of disobedience. The couple's partaking from the tree of knowledge renders them in some sense self-transcendent, but not in any deified sense. Instead, they attain enough consciousness to be *self*-conscious. They recognize themselves in a mirror, as even chimpanzees and elephants do, but a mirror that reflects something quite different about themselves. They come to see themselves as vulnerable selves, exposed and deficient when measured against the divine. They realize they are "naked" and as such recognize their finitude and frailty. Their emerging consciousness does not come any more easily in the narrative than does consciousness from evolution's tortuous route toward complexity.

The dramatic turning point had been building all along within the narrative.[220] The groundling was placed in the garden to "serve it and preserve it" (2:15). The first human was given a function, a job, for which no moral assembly was required. At this point, human identity remained at a very "nascent stage."[221] The attainment of moral agency was yet to happen. The groundling's development, however, ensued when he was given an imperative (vv. 16–17), one that conjoined freedom and restriction with a defined consequence.

At the moment the prohibition is uttered by God, a new stage in human development is reached: choice becomes a meaningful factor for the first time in the narrative. Disobedience is now a possibility. Nevertheless, the choice to obey is not grounded "in a larger vision of the good."[222] The *ādām* remains, as it were, a child faced with a parental command accompanied by the threat of punishment. The next stage of the couple's development is occasioned by the cognitive dissonance introduced by the snake's countertestimony and the tree's desirability for acquiring wisdom. The resulting disobedience emerges out of a conflictive mix of desire and dilemma. By partaking from the tree, they gain a level of self-consciousness, an awareness of their vulnerability and of their newly acquired ability to make decisions on their own. In so doing, they have chosen adulthood, indeed, humanity with all its costs and complexities.

The narrative is not so much interested in the birth of consciousness per se as in the ripening of moral consciousness or conscience, the mature fruit of the tree.[223] But a central question lingers: Why would God forbid, but not prevent, the primordial couple from partaking the fruit, from gaining moral and cultural discernment? Perhaps, as some have proposed, the timing was all wrong: the fruit had not ripened or, more plausibly, the couple was not yet ready to receive it. But this is too easy a solution. The answer instead may lie in the nature of the consequences. In view of the evolving characters in the narrative, the fruit of consciousness and conscience *had* to be prohibited. There was no other choice. The fruit of knowledge was by necessity forbidden, for conscience, according to logic of the text, could not have emerged otherwise. The story's narrative logic bears a profoundly psychological insight. Growth in conscience begins with an act of disobedience followed by harsh consequences.[224] Conscience can only take root in the soil of guilt and regret. It is not artificially imposed upon the human psyche. Conscience, and thus human identity itself, is homegrown, a product of evolutionary and narrative development.

So who were "Adam and Eve" from an evolutionary perspective? The ancient narrative shows them to be a work in progress, covering a wide spectrum of human development. Contrary to one opinion, the characters represent not just the first hominid group to act religiously, say 50,000 years ago.[225] The Yahwist's view is not so narrow in scope. In light of human evolution, the primal couple in the garden represents the hominid developing through various challenging transitions: from specialized knowledge to cognitive fluidity, from gathering food to cultivating the land, from nakedness to clothing, from blind trust to moral consciousness. From child to adult. This short story is a wonder of conflation.

"You are Dust, and to Dust You Shall Return"

In addition to illustrating the painful transitions of human development in such dramatic simplicity, the Yahwist's story profoundly affirms humanity's natural source of origin, the ground. As geologists and paleoclimatologists are beginning to reconstruct Earth's early history, it is becoming increasingly clear just how important the "ground" was in the origin of life, beginning with the introduction of oxygen in the atmosphere. As continents formed, organic material was buried, allowing for a net buildup of oxygen. The ground, thus, made possible the *'ādām*'s first breath, the "breath of life" (Gen 2:7).

As with all life, humans are carbon-based creatures. The Yahwist's claim of humanity's "dusty" origins provides a necessary counterbalance to the near divine status claimed by the Priestly creationists behind Genesis 1. The second, but by no means secondary, creation account of the Bible coheres well with Darwin's own conclusion that despite humanity's "god-like intellect," humanity "still bears in his bodily frame the indelible stamp of his lowly origins."[226] However elevated in power over the rest of creation, however exalted in God's image (not to mention self-image), we "groundlings" are descended from the "ground of being,"[227] from stardust and "slime," the compost of life.

"YHWH Elohim Expelled Him . . . to Serve the Ground"

Humanity's evolutionary history is a story of changing environments, technological development, and repeated migrations. Transitions are key: from dense forests to expansive savannas, from small to complex societies, from Africa to the rest of the world (again and again). Hominid evolution is as much about remarkable anatomical and cognitive developments as it is about dead ends and extinctions, of adaptive successes and failures, as with every species. Some transitions are environmentally forced; others are, for whatever reason, taken out of choice, all the while the tree of human life remains firmly rooted and ever growing.

The garden story revels in messy transitions. God goes back to the drawing board when the animals do not satisfy the groundling's need for companionship, and thus human gender is born. The couple's expulsion into the harsh agricultural world is an inevitable transition, a necessary one even as it is a tragic one. They are tantamount to exiles, and their migration carries no hint of manifest destiny or triumphal procession. They leave and they change, for so has their environment. Agriculture becomes their adaptive strategy, and from agrarian subsistence emerges cultural and technological achievements as rich and varied as the ancient author can contain within a few verses: urban life (Gen 4:17), music (v. 21), and metal tool making (v. 22). As the narrative continues far beyond the garden, humans also learn to build an ark and, in Israel's case, construct a tabernacle, the movable apparatus of worship and the height of technological development from the Bible's perspective (Exod 25–40).

The redemptive message of the garden tale is that primal man and woman do not go forth alone. God's continued care accompanies the evicted couple, providing for their adaptive success within the larger, harsher world, beginning with clothing. Even the curse contains the seeds of something salutary. The presence of "thorns and thistles" (3:18) is a healthy sign of ecological succession, of the soil's long term re-fertilization. By casting roots, weeds and other woody shrubs aerate the soil and prevent erosion. They are signs of the land restoring itself, evidence of land that is ready for cultivation. When measured against true barrenness, the ground's curse is a blessing. Fallowing is next to flourishing.

"To Guard the Path to the Tree of Life"

The tree of life has the last word in this ancient tale, and so it is only appropriate in our rereading of the garden story to conclude with it. In addition to the two trees in the narrative, science provides another tree, one less forbidden. Though the "tree of life" within the garden remains inaccessible, the *arborvitae* science describes covers all of life. From its trunk to its budding branches, this universal tree reveals the familial relationships of all living things in the course of evolutionary development.

Although Darwin never presented a full phylogeny in any of his writings, the image of the tree of life did appear in his notebooks by 1837, a mere sapling,[228] and he waxed eloquently about its usefulness as a "simile" for evolution.

> The green and budding twigs may represent existing species; and those produced during former years may represent the long succession of extinct species. At each period of growth all the growing twigs have tried to branch out on all sides.[229]

But as with any tree, branches die and fall.

> From the first growth of the tree, many a limb and branch has decayed and dropped off; and these fallen branches of various sizes may represent those whole orders, families, and genera which have now no living representatives, and which are known to us only in a fossil state. . . . As buds give rise by growth to fresh buds, and these, if vigorous, branch out and overtop on all sides many a feebler branch, so by generation I believe it has been with the great Tree of Life, which fills with its dead and broken branches the crust of the earth, and covers the surface with its ever-branching and beautiful ramifications.[230]

The "great" tree illustrates evolution's fundamental principle, namely, vertical descent: if two species share a common feature, they inherited it from a common ancestor.

The first phylogenetic tree, resembling a towering oak, was published in 1870 by the German naturalist and embryologist Ernst Haeckel. It was based strictly on morphological features, such as anatomy and skeletal structure. With the discovery of DNA a century later, research shifted from morphology to biochemistry. With the recent automation of genome sequencing, phylogenetic reconstruction has become even more precise. As a result, the tree of life has grown "bushier." It has also become more circular in order to accommodate so many species (and no doubt many more to be discovered). At the trunk or center of this "highly resolved and robust tree"[231] is what biologists call LUCA, the "Last Universal Common Ancestor," a single-celled organism that perhaps thrived at extremely high temperatures.[232]

The tree of life remains the most suitable simile for describing the meta-narrative of life on Earth.[233] In the Genesis narrative, it sinks its roots outside the garden. In the couple's banishment, the garden's tree of life is exchanged for the human tree of life, as shown by the narrative's subsequent focus on humanity's genealogical advance toward the "table of nations" in Genesis 10:1–32.[234] In between the garden and the nations, the flood story of Genesis 6–9 tells not only of the near extinction of terrestrial life but also of life's resilient power to recover and repopulate the world, all in the span of two chapters. The primal couple's expulsion was also their transplantation. God remains the gardener, and God's good earth remains the ground of being, hosting everything from LUCA to Lucy to Henry Luce.

5

Behemoth and the *Beagle*

Creation According to Job 38–41

Tell me, what is it that you plan to do with your one wild and precious life?
—Mary Oliver[1]

We are saved in the end by the things that ignore us.
—Andrew Harvey[2]

What do Newton's bucket, Maxwell's demon, Einstein's elevator, and Schrödinger's cat all have in common? They are all "thought experiments," exercises of the imagination meant either to challenge or to encapsulate a theory about how the world actually works. As is well known, Albert Einstein facilitated the birth of modern physics by perfecting the use of *Gedanken* experiments. As a bored sixteen-year-old, Einstein imagined what it would be like to travel alongside a light beam.[3] The eventual result was his Theory of Special Relativity.

Of all the books of the Bible, Job comes closest to being a thought experiment. As a non-Israelite character, an outsider no less, Job is given the freedom to challenge traditional notions about God and the world. At the same time, Job's God is Israel's God, YHWH (1:20–21; 12:9), and Job himself is the paradigm of piety. Taking place in the land of Uz, a place nearly as elusive as Eden,[4] Job's story stretches the theological envelope in ways that no orthodox Israelite could imagine, for it reaches behind and beyond Israel's story. The narrative returns to the world of primordial beginnings even as it ventures far beyond human culture and control. The book of Job shares much in common with Genesis; indeed, it has been called "The Creation Story: Part Two."[5] Job the Gentile is the new Adam, and his story begins back, as it were, in the garden of Uz.

As a thought experiment, the book of Job is essentially a "What if?" story aimed at dismantling conventional views about human identity, God's character, and the moral construction of creation. What if God waged a bet on someone of unassailable character, even allowing for severe trials to test his moral mettle? What if this paragon of righteousness were to fall into unimaginable ruin and disgrace, all "for nothing"? How far could this model citizen push the standard notions of piety and yet retain God's approval? To what extent can the contours of human integrity be redrawn in the face of horrific suffering? And what kind of world would reflect such a reimagining of human and divine integrity? Enter Job's lumbering Behemoth (Job 40:15–24), assuming its rightful place alongside Schrödinger's quantum cat.[6]

Socio-Historical Background

Because Job lacks clear historical references, determining the book's background and dating is difficult.[7] Its literary roots are found in *The Babylonian Theodicy*,[8] dated around 1000 BCE, and *Ludlul bēl nēmeqi* ("I Will Praise the Lord of Wisdom"), at least a couple of centuries earlier.[9] These precursors, however, do not shed light on dating Job. The figure of Job achieved legendary status by the sixth century BCE (Ezek 14:14, 20), and the *book* of Job was likely codified soon thereafter, when Israel's exile remained in full, retrospective view.[10]

Job's misery on the ash heap possibly parallels Israel's misery on a devastated land. His desolation perhaps alludes to Israel's desolation. In any case, Job is an outcast exiled from his home and community and deported to only God knows where. On his ash heap, Job suffers the trauma of displacement, and in his trauma he is transported. He finds himself a stranger in a strange land, "a brother of jackals, and a companion of ostriches" (Job 30:29). Certain biblical traditions, in fact, describe the emptied, exiled land as repopulated by such animals.[11] But Job's landscape is much more than a scorched earth; it is creation thriving at the margins of human culture.

The Strange World of the Text

As the climax of this tale of trauma, God's answer to Job provides the most panoramic view of creation in all of the Hebrew Bible (Job 38:1–42:6). On its surface, the text serves to chasten Job and expose the limitations of his knowledge and ability (38:2; 40:2, 8–14). But never has a rebuke been so colorful and richly textured, even from God. God reproves Job by taking him on a scenic detour through creation's rugged, far-flung lands, a mind-bending tour of the vast domains of cosmology, meteorology, and zoology. God's answer features such a variety of particularities, from hail to hawks, that some scholars have compared these chapters to the ancient Near Eastern genre of a catalogue or list, as we shall see.

Translation

(38:1) Then YHWH addressed Job out of the whirlwind, saying:
(38:2) "Who is this that darkens (my) design[12]
 with words lacking knowledge?
(38:3) Gird up your loins like a man!
 I will question you, and you shall inform me.
(38:4) Where were you when I founded the earth?
 Say so, if you have understanding!
(38:5) Who determined its measurements? Surely you know!
 Or who extended a measuring line upon it?
(38:6) On what were its footings sunk?
 Or who laid its cornerstone,
(38:7) when the morning stars rejoiced together,
 and all the divinities[13] shouted for joy?

(38:8) And who shut in the sea with double doors,[14]
 when it burst out from the womb, gushing forth,
(38:9) when I made a cloud as its garment,
 and deep darkness as its swaddling band,
(38:10) and when I broke in upon it[15] with my boundary line,
 and set bars and double doors,
(38:11) saying, 'Only so far shall you come, and no farther;
 here your surging waves shall stop!'?[16]

(38:12) Have you at anytime[17] commanded the morning,
 and taught the dawn its place,
(38:13) to seize the earth by its skirts
 so that the wicked be shaken out?
(38:14) (The earth) takes shape like clay (under) a seal,
 and (its features) become dyed like a garment.[18]

(38:15) Withheld from the wicked is their light,
    and (their) uplifted arm is broken.
(38:16) Have you entered the sources of the sea,
    and roamed the recesses of the deep?
(38:17) Have the gates of death disclosed themselves to you?
    Have you seen the gates of deep darkness?
(38:18) Have you discerned the expanse of Earth?
    Say so, if you know all this!
(38:19) Where is the pathway to where light dwells?
    As for darkness, where is its place?
(38:20) Have you taken it to its domain,
    that you may discern the paths to its home?
(38:21) You must know, for you were born then,
    and great is the number of your days!

(38:22) Have you entered the storehouses of snow,
    or have you seen the storehouses of hail,
(38:23) that I have reserved for the time of adversity,
    for the day of battle and war?
(38:24) What is the way to where lightning is dispersed,
    or where the east wind is scattered upon the earth?
(38:25) Who has cut a channel for the downpours,
    and a way for the thunderbolts
(38:26) to bring rain upon a no-man's land,
    upon an uninhabitable desert,
(38:27) to satisfy the desolate wasteland,
    and to bring forth grass?

(38:28) Has the rain a father?
    Who has begotten the drops of dew?
(38:29) From whose womb has ice issued forth,
    and who has given birth to heaven's hoarfrost?
(38:30) The waters hide themselves like a stone,[19]
    the deep's surface becomes trapped (in cold).[20]

(38:31) Can you bind the chains of the Pleiades,[21]
    or loose the cords of Orion?[22]
(38:32) Can you lead out Mazzaroth[23] in its season,
    or guide the Bear with her cubs?[24]
(38:33) Do you know the heavenly ordinances?
    Can you establish their[25] rule[26] over the earth?

(38:34) Can you raise your voice to the clouds,
  so that an abundance of water inundates you?
(38:35) Can you dispatch lightning bolts,
  that they may come and say to you, 'Here we are!'?
(38:36) Who has given wisdom to the ibis?[27]
  Or who has given understanding to the rooster?[28]
(38:37) Who can count the clouds with accuracy?[29]
  Or who can pour[30] the water pitchers of heaven,
(38:38) for dust to flow into cast metal,[31]
  and clods to cling together?

(38:39) Can you hunt prey for the lion,
  or fill the appetite of the young lions,
(38:40) when they crouch in their dens,
  or lie in wait in a thicket?
(38:41) Who provides for the raven its food,
  when its brood cries out to God
   and wander about for lack of food?
(39:1) Do you know the time the mountain goats give birth?
  Do you watch the calving of the deer?
(39:2) Can you count the months they must fulfill?
  Do you know the time when they give birth,
(39:3) when they crouch down, bringing forth their young,
  and release their pain?[32]
(39:4) Their young thrive as they grow up in the open field;
  they go forth and do not return to them.

(39:5) Who has set the onager[33] free?
  Who has loosed the bonds of the wild ass,
(39:6) to which I have given the desert for its home,
  the salt land for its dwelling place?
(39:7) It laughs at the city's commotion;
  it does not hear the taskmaster's shouts.
(39:8) It roams the mountains for its pasture,
  searching after all manner of greenery.

(39:9) Is the auroch[34] willing to serve you?
  Will it spend the night at your feeding trough?
(39:10) Can you bind the auroch in a furrow with its rope?
  Will it harrow the valleys after you?

(39:11) Will you rely on it because its strength is so great,
      and leave your toil to it?
(39:12) Do you place your trust in it, that it will return
      and bring your seed-grain to your threshing floor?[35]

(39:13) The ostrich's wings flap wildly,[36]
      though its pinions lack[37] plumage.[38]
(39:14) Indeed, it leaves its egg on the ground,
      letting them get warm in the dirt,
(39:15) forgetting that a foot may squash it,
      or that a wild animal may trample it.
(39:16) It deals cruelly with its young, as not its own.
      In vain is its labor, done without fear,
(39:17) for God has made it forget wisdom,
      without allotting it understanding.
(39:18) When it rears up,[39]
      it laughs at the horse and its rider.

(39:19) Do you give might to the horse?
      Do you clothe its neck with thunder?[40]
(39:20) Can you make it leap like the locust?
      Dreadful is its majestic snorting.
(39:21) It paws (the ground) with force[41] and exults in its strength,
      charging forth toward battle.
(39:22) It laughs at fear and is not dismayed;
      it does not retreat from the sword.
(39:23) Upon it rattles the quiver,
      as well as the flashing spear and javelin.
(39:24) With trembling and rage it swallows the ground;
      it cannot keep still at the sound of the *shofar*.
(39:25) As often as the *shofar* blasts, it says, "Aha!"[42]
      From afar it smells the battle,
         the thunderous shouting of the captains.

(39:26) Is it by your understanding that the hawk flies,
      and spreads its wings southward?
(39:27) Is it by your mouth that the vulture mounts up,
      that it builds its nest on high?
(39:28) It dwells in the rock and makes its lodging;
      on the jutting crag is its stronghold.
(39:29) From there it searches for food,
      from afar its eyes perceive it.

(39:30) Its nestlings slurp up[43] blood,
>and where the slain are, there it is."[44]

(40:1) Then YHWH addressed Job, saying:
(40:2) "Will a faultfinder contend with *Shaddai*?[45]
>The one who reproves God must respond."

(40:3) Job answered YHWH, saying:
(40:4) "I am too small—How can I answer you?
>I place my hand over my mouth.
(40:5) I have spoken once, but I will not answer;
>even twice, but I will not add anything more."

(40:6) Then YHWH addressed Job out of the whirlwind, saying:
(40:7) "Gird up your loins like a man!
>I will question, and you shall inform me."
(40:8) Would you even annul my governance?[46]
>Would you condemn me so that you would be in the right?
(40:9) Do you have an arm like God's,
>and can you thunder with a voice like his?
(40:10) Adorn yourself with majesty and grandeur,
>and clothe yourself with honor and dignity.
(40:11) Unleash your overflowing rage;
>look at all who are proud and abase them.
(40:12) Look at all who are proud and subdue them;
>tread down the wicked where they stand.
(40:13) Hide them together in the dust;
>bind their faces in darkness.
(40:14) Then even I will give you praise,
>that your right hand can give you victory.

(40:15) Behold Behemoth, which I made with you.
>It eats grass like an ox.
(40:16) Behold its potency in its loins,
>and its power in the muscles of its belly.
(40:17) It lowers[47] its tail like a (falling) cedar;
>the sinews of its thighs are intertwined.
(40:18) Its bones are tubes of bronze;
>its limbs are like a rod of iron.
(40:19) It is the first of God's works;
>[Only] the one who made it can approach it with sword.[48]
(40:20) Indeed, the mountains yield food for it,
>where every wild animal plays.

(40:21) Under the lotus plants it lies down,
    in a covert of reeds, and in a marsh.

(40:22) The lotus plants cover it as its shade,
    the willows of the wadi surround it.

(40:23) Even if the river presses forth, it is not alarmed;
    it remains confident when the Jordan surges against
        its mouth.

(40:24) Can one catch it with his eyes?
    Can one pierce its nose with snares?

(41:1)[49] Can[50] you pull Leviathan out with a fishhook,
    or tie down its tongue with rope?

(41:2) Can you put a line through its nose,
    or pierce its jaw with a hook?

(41:3) Will it make many supplications to you,
    or speak meekly to you?

(41:4) Will it make a covenant with you,
    that you take it as a servant forever?

(41:5) Will you play with it as a bird,
    or bind it on a leash for your girls?

(41:6) Will traders bargain over it?
    Will they divide it up among merchants?

(41:7) Can you fill its hide with harpoons,
    or its head with fishing spears?

(41:8) Lay your hand upon it;
    you will not remember the battle!

(41:9) Indeed, any hope of it proves false;[51]
    just by the sight of it one is overwhelmed.[52]

(41:10) No one is fierce enough to stir it up.[53]
    Who then can stand up to it?[54]

(41:11) 'Whoever confronts me, I will repay.[55]
    Under the whole heaven, he is mine!'[56]

(41:12) I will[57] not silence its boastings,[58]
    [its] mighty discourse,[59] or its graceful[60] [. . .].[61]

(41:13) Who can strip off its outer garment?
    Who can penetrate the double coat of mail.[62]

(41:14) Who can pry open the doors of its face?
    All around its teeth is terror.

(41:15) Its back[63] (consists of) rows of shields,
    closed like a tight seal.

(41:16) One is so close to another
    that air cannot come in between them.
(41:17) Each cleaves to the other;
    they hold fast and cannot be separated.
(41:18) Its sneezes flash forth light,
    and its eyes are like the eyelids of the dawn.
(41:19) Out of its mouth shoot forth flaming torches;
    fiery sparks escape.
(41:20) Out of its nostrils pour forth smoke,
    as from a boiling, seething[64] pot.
(41:21) Its breath[65] kindles coals,
    and a flame issues from its mouth.
(41:22) Strength abides in its neck,
    and before it dances dismay.
(41:23) The folds of its flesh cling together,
    set firmly upon it and immovable.
(41:24) Its heart is cast like stone,
    cast like a lower millstone.
(41:25) The gods are terrorized at its rising;
    they are beside themselves[66] at (its) acts of devastation.[67]
(41:26) The sword that reaches it cannot prevail,
    nor can the spear, missile, or javelin.
(41:27) It counts iron as straw,
    bronze as rotten wood.
(41:28) The arrow cannot put it to flight;
    sling-stones are reduced to chaff before it.
(41:29) Clubs are counted as straw,
    and it laughs at the rattling of the lance.
(41:30) Its underparts are like sharp potshards;
    it spreads itself like a threshing sledge upon the muck.
(41:31) It makes the deep seethe like a cauldron,
    makes the sea like an ointment pot.
(41:32) It leaves the wake luminous behind it;
    one would count the deep as white-haired.
(41:33) On the earth[68] there is nothing like it,[69]
    a creature made without fear.
(41:34) It surveys all who are lofty;
    it is king over all the sons of pride."[70]

(42:1) Then Job addressed YHWH, saying:

(42:2) "I know that you can do everything,
>   and that no plan of yours can be thwarted.
(42:3) 'Who is this that hides counsel without knowledge?'
>   Yes, I did declare what I did not understand,
>   things too marvelous for me, which I did not know.
(42:4) 'Hear, and I will speak;
>   I will question you and you shall make known to me.'
(42:5) By the hearing of the ear I had heard of you,
>   but now my eye has seen you.
(42:6) Therefore, I waste away,
>   yet am comforted over dust and ashes.[71]

YHWH's answer to Job is couched as a challenge. References to "girding," which introduce both speeches, are meant to elicit from Job a robust response that could, in principle, teach the deity a thing or two. God taunts Job into taking on the position of teacher. But Job is unable to serve as God's superior pedagogue or combative partner. The teaching remains squarely on God's end, and the learning falls entirely on Job's end of the pedagogical spectrum, and no doubt by design, for Job barely gets a word in edgewise. (For God's 123 verses, Job has only nine!) God's answer discharges one question after another, like waves battering the shorelines of Job's worldview. Call it the pedagogy of the oppressor, for the first lesson learned is that Job's power is no match for God's. And for many interpreters that is the *only* lesson.

But there is much more to God's final answer. Its content is far too rich to be reduced to a single affirmation of divine omnipotence. God's response to Job exhibits an irreducible ambiguity.[72] The questions that God throws at Job are not just rhetorical; they are profoundly existential. God's questions have all to do with the nature of human identity and vocation.[73] Moreover, God delivers not only questions, but also information, and plenty of it. God's answer, as a whole, moves from question to information. The harsh questions pave the way for edifying content, which in turn mitigates their harshness. In God's rhetoric of rebuke, Job is granted nothing short of a revelation of creation, a harrowing vision of the Encyclopedia of Life.

Literary Background: The Catalogue

Scholars have long noted affinities between God's answer in the book of Job and ancient Near Eastern lists or catalogues.[74] Such catalogues provide "a list of terms, activities, names, etc., having a definite common denominator, usually noted in its introduction."[75] From Egypt to Mesopotamia, examples include lists of graves, kings, gods, types of people, tribes, nations, cities, sins, anatomy, cosmic elements, and animals.

In its pure form, the catalogue "conveys a large amount of information with great brevity."[76] But catalogues, particularly ancient ones, rarely convey information just for information's sake. As Michael Fox has argued, such lists may have served as writing exercises for developing scribes.[77] But not merely so: the Onomasticon of Amenope from Egypt (twelfth century BCE), for example, opens with the following introduction: "Beginning of the teaching for clearing the mind, for instruction for the ignorant and for learning all things that exist."[78] This taxonomic exercise, whether read or copied, offered a way to perceive the world with greater understanding, a way for "clearing the mind" to the world around. As will be seen, God's answer to Job does more than "clear" Job's mind; it transforms it.

God's Creation: A Whirlwind Tour

A mere catalogue Job is surely not. Rich in detail and peppered with questions, the climax of the book of Job is no dry taxonomy. Pathos drives God's recitation of creation, beginning with the farthest reaches of the cosmos and concluding with the tightly knit scales of the sea-dragon, from the farthest to the smallest scale of perception, from cosmos to chaos. According to Yair Hoffman, a primary characteristic of the catalogue is that the order of its items is "inconsequential."[79] Not so in Job's case. This taxonomy propels Job on a terrifying tour with a particular direction in mind.

It is no coincidence that the first question posed by God about creation lifts up the issue of place: "*Where* were you when I founded the earth?" (38:4), to be followed by similar questions: "*Where* is the pathway to where light dwells . . . *where* is the place" of darkness? (38:19). "What is the way to *where* light is dispersed or where the east wind is scattered?" (38:24). But Job knows not where. He has not explored the "gates of deep darkness" or cruised the earth's extremities (38:17b–18a), for all of it is a "no-man's-land" (38:26). But through the power of poetry, Job is transported precisely to these places. God's answer sets Job on a course of discovery and discernment, both cosmic and personal. God's answer is creation's revelation.

God reveals many items for Job's consideration, not to mention consternation, and they are all given specific placements. God's answer is replete with references to natural niches and environments. The rain falls on "desolate wastelands," the onager (wild ass) "ranges the mountains," the lions crouch "in their dens," the warhorse belongs to the bloody battlefield, the auroch has no

place on the cultivated field, the raptor has its nest on the "jutting crag," Behemoth lies lazily under the lotus trees, and the mountains are where the wild
things are. Every animal is given place, a habitat, even if it is left undeclared. As
everyone in Canaan knew, Leviathan dwells in the recesses of the deep, far
beyond the reach of fishnets.

But the language of place is simply one feature among many that provide
Job a close look at creation's extremities, a God's-eye view of regions untouched
or uncontrolled by human habit and habitation. Through rich detail, evocative
imagery, and the power of poetry, Job explores a new world of broadened horizons and alien creatures. Job is taken on a grand tour of creation's fringes, a
cosmic field trip of unimaginable proportions.[80] He is propelled to the very
margins of creation, driven from bounded security to sublime danger, to places
where beauty and horror converge.[81] He is flung far off to "the gates of death"
and dragged next to Leviathan's underbelly, zooming in close enough to view
the infinitesimal space that separates one scaly shield from another. Vast dis-
parities of scale are dramatically covered as Job is taken across incalculable
distances and depths. His cosmic tour is the ancient equivalent to *Powers of
Ten*.[82] Yes, Job's journey is an imaginary one, but it draws from, even as it
extends beyond, the *realia* of nature. This is language with a kick, or better, its
own propulsion unit.

Structure and Movement

God's answer consists of two speeches (38:1–40:2 and 40:6–41:34), each of
which is introduced with the challenge for Job to "gird" himself. The first challenge addresses God's cosmic "design" (*'ēṣāh*, 38:1); the second deals with
God's "justice" or governance (*mišpāṭ*, 40:8). The overall movement of God's
twofold answer is also telling: It begins with cosmic expanses and moves toward recounting various phenomena, meteorological and biological, and concludes with a detailed examination of one creature. As creation's purview
zooms from the cosmic to the particular, from the general to the detailed, God's
cosmic poetry runs counter to the narrative logic of the ancient *mythos* of creation as exemplified in *Enūma elish*, which begins with chaos. Tiamat's defeat
sets the stage for creation, which concludes with the gods' adulation of the
conquering creator god. God's speeches move Job in the opposite direction,
from the construction of a well-established earth, whose founding prompts celestial rejoicing, to the preeminent chaos monster, Leviathan. As Carol Newsom observes, "an element of 'uncreation' takes place . . . as Job is led to a
sustained and intimate encounter with the symbol of the chaotic."[83] And I
would add that it is here in Leviathan's abode where life itself began. Whether

measured by ancient Near Eastern myth (i.e., watery chaos) or evolutionary biology (e.g., photosynthetic cyanobacteria[84]), Job's final stop is, ironically, life's origin.

God arrives as a destructive whirlwind, a storm god. And yet, like Baʻal, such a terror-provoking manifestation of the divine also has its salutary side: a whirlwind in the wilderness delivers rain to the parched desert (see 38:25–27). God's terror is a *fertile* terror, and it sets the tone for all that follows. To Job, God's discourse edifies even as it terrorizes. And it begins with an admiring look at the earth's foundations (38:4–7): measurements are determined, lines are drawn, bases are sunk, and a capstone is laid. The celestial joy that erupts in response suggests that the earth is no ordinary edifice but a cosmic temple (cf. Ezra 3:10–13).[85] Lacking, however, is any "image of God," human or otherwise, to inhabit, much less rule, the earth.

Next are the sea's confinement (vv. 8–11) and the unleashing of light (vv. 12–15). The sea is a tempestuous newborn whose birth is facilitated by God, who takes on the role of midwife.[86] Here God swaddles the sea and sets firm boundaries against its "proud waves." Light, however, does what it is told: it shakes out the wicked from the earth's "skirts." At the break of dawn, light restores the terrestrial order after the chaos of night, and it does so day by day. The following section explores creation's depth and breadth. No details are given; Job's ignorance of vast domains remains. But more is disclosed about the meteorological realm. Snow and hail are stored for "the day of battle." Such precipitous violence has its salutary side: torrents are channeled "to satisfy the waste and desolate land." Creation at its most desolate bears life and beauty. And it is precisely in the desolation that Job discovers creation's dignity, for even there the grass is green and wild hearts beat, including his own.

But Job's tour is not over. Next is a veritable carnival of animals, specifically five pairs: lion and raven, mountain goat and deer, onager and auroch, ostrich and warhorse, hawk and vulture. With the exception of the raven and the warhorse, all the animals listed in God's answer were routinely and ceremoniously hunted by Egyptian and Mesopotamian kings. The royal hunts were not conducted for entertainment purposes, thrilling as they may have been. They were staging grounds for demonstrating the king's prowess on the battlefield, a symbolic exertion of royal power. By slaying these wild animals, the king was "fulfilling his coronation requirement to extend the kingdom beyond the city to include the wilderness."[87] In the lion hunt, specifically, the king identified himself as both the hunter and the lion; hence, the leonine carcass was never mutilated.[88]

In Job, these denizens of the wild are not trotted out for him to kill and thereby to prove himself. To the contrary, these animals are deemed untouchable. In a remarkably ironic turn, God begins with the lion (38:39), the quintessential predator of the wild and the most prized game of kings, and asks, "Can you hunt prey for the lion?" Can Job hunt *on behalf of* the lions? Job is not to gird up his loins against lions.

Nor are these animals to be named or defined in some way by Job, as in the *ādām's* case in the garden (Gen 2:19–20). Far from it, through the power of poetry Job is taken into the wild to behold their dens and nests, their mountain lairs and vast plateaus, their livelihoods in situ. Job is driven into the wilderness to encounter the beasts on their own turf, embarking on a cosmic *Wander* into the far reaches of creation. But Job has no *Wanderlust*, because for him the wild is the quintessential Other, an object of fear and loathing.[89] No naturalist is Job. Yet he discovers the wild to be full of alien life filled with inalienable value, denizens endowed with strength, dignity, and freedom. The mountain goat kids "go forth and do not return" (39:4); the onager freely roams beyond human reach (v. 5); the auroch resists domestication (vv. 9–12); the ostrich fearlessly flaps its wings (vv. 16–18); the warhorse exults in its thunderous strength (v. 22); and the raptors soar spying out their prey (vv. 26–30).

To borrow from Christopher Southgate's innovative use of language, all these animals "selve"; each expresses its own identity unhindered and most fully.[90] They live and move and have their being just as God intended. And for all these fierce and fully "selved" animals, God is their provider, hunting the lion's prey (38:39), responding to the raven's cry (v. 41), and directing the raptor's flight (39:26). God describes each one with such evocative detail that Job is afforded a point of view that lies utterly beyond himself, a perspective that is God's, but one that the animals also share. Job is invited to see the looming battle through the eyes of the warhorse, to spy out corpses through the eyes of the vulture, to roar for prey as the lion, to cry for food like the raven's brood, to roam free on the vast plains, to laugh at fear, and to play in the mountains. Job's Earth trek is no descent but an ascent to Nature.[91]

Job's cosmic tour has rendered him speechless, so he says (!). But God is not satisfied. And so the tour resumes. God's second speech profiles two magnificent animals that loom mythically large: Behemoth and Leviathan, perhaps drawn in part from the water buffalo (or hippopotamus) and the crocodile, respectively, both formidable creatures in their own right.[92] Whatever they are, these larger-than-life beasts are the quintessential embodiments of the wild, highly esteemed by God, much in contrast to their mythical equivalency with chaos and evil.[93] Nothing is said of God's intent to subjugate either Behemoth or Leviathan; freedom reigns for both these fearsome creatures. Behemoth is claimed as the "first (or chief) of God's great acts" (40:19), taking on the preeminent status of Wisdom in Proverbs 8 and cosmic light in Genesis 1! The tour concludes in Job 41, the only chapter in the Bible devoted entirely to an animal.[94] With Leviathan, Job takes the plunge into the depths of chaos and survives, much to his own astonishment.

Character of Creation

What kind of world does God paint for Job's edification? It is certainly no cozy cosmos. It is terrifyingly vast and alien. Far from being barren, creation teems with life characterized by fierce strength, inalienable freedom, and wild beauty.[95] Limits, to be sure, are set in place: the earth rests on stable founda- tions, the tempestuous sea is contained like a swaddled infant, and the dawn renews the earth with some semblance of order (38:4–15). Nevertheless, the world is not an object of divine micromanagement.[96] Land, sea, and sky are host to myriad life-forms, all alien to the human eye and untamable to the hu- man hand,[97] but all affirmed and sustained by God. God's world is filled with scavengers and predators, even monsters (cf. Gen 1:21), all coexisting, though never peacefully. The lions eat their prey; the vultures feast on the slain. This world is God's *wild* kingdom. It pulses with "pizzazz."[98]

Character of God

What kind of a God is behind this vibrant, brutal creation? For Job, God is king, but God's sovereignty is a different beast altogether. It is a kingship of care and freedom. Put in more contemporary terms, the God of Job exhibits *biophilia*: an "innate pleasure from living abundance and diversity."[99] Express- ing an unbounded and abiding astonishment for the lowly onager as much as for the regal Leviathan, God proves to be a bona fide, indiscriminate "biophile"! All the denizens of creation are found to be "unbearably beautiful" by God, notes Bill McKibben.[100] The God who harshly answers Job marvels lovingly over every entry in the catalogue of zoological *marginalia*. God chooses to ap- proach wildlife not with a sword, but with a word of admiration and an open hand.

In Job, the conquering spirit of the Babylonian Marduk is effectively replaced by the sustaining care of Israel's God. All the more, God exhibits a sense of "abiding astonishment" toward creation.[101] Like a loving mother, this God proudly displays creation as if every element and animal were her child, even Leviathan. Rare it is that we find in the Bible God rendering praise to creation. But so it is in Job. God is more than willing to be pulled in by nature's "gravitational attraction,"[102] to be enamored by nature's wonder and vitality, so much so that God takes on the creatures' point of view, their instincts and their joys, their adaptive strengths and their failures. God the sovereign king is God the parental provider.

Job's Place in Creation

What is Job's place within such a world? On the surface, Job seems bereft of place. He has been shown a no-man's-land. Moreover, no mention is made of humanity's creation, let alone dominion. This is no anthropocentric world that God so loves. The world is a hodgepodge of life in all its wondrous and repulsive variety, a world of "ultimate pluralism,"[103] with Job included in the mix: "Behold Behemoth, which I made with you (*'immāk*)" (40:15a). The clue is in the preposition: Behemoth is created *with* Job. This preposition is as key here as it is in Genesis 3. Call it the preposition of companionship. In the garden, the man was "with" the woman (3:6); the woman was given to the man "to be with" him (v. 12). But Job, the new Adam, is given a far stranger bedfellow. As the woman and the man share a common identity of flesh and blood, so Job shares an uncommon identity with this lumbering, fearless, playful creature of the wild. Job is no isolated creation and clearly not the apex of the created order. Rather, he is created with his monstrous twin, who receives the credit of being born "first" or "best" (*rē'šît*, 40:19). But unlike the biblical character of Jacob, Job has no recourse to steal his elder twin's birthright. Job finds himself to be a monster's companion, and by extension, a companion to all the monsters of the margin, the aliens of the wild.

This, then, is the crux of the Joban "thought experiment": Imagine, if you will, that you are kin to a monster. Not an evil twin, but a monstrosity nonetheless. Behold the monster whose loins *are* girded with strength and whose confidence does not shrink before the surge of chaos! Can Behemoth be, somehow, a model for Job? What does that say about Job and his relationship to the natural world, stock full of wild and alien creatures? Job once complained of being a "brother of jackals and a companion of ostriches" (30:29), ostracized by friends and family. But God has turned Job's lament into wonder by showing him that he is actually in good company! Through Behemoth, Job discovers himself to be created *of* the world, inextricably linked to all life, including the wild. This paradigmatic human exists alongside, beside, in tandem with all the creatures of the wild.

God's revelation of Job's place in creation transforms Job's identity. Job's worldview is infinitely expanded, but for what purpose? Having discerned the world anew and his place therein, Job is to take home something of the wild, a recognition of his connection to the wild. He is to find in himself something of Behemoth's strength and Leviathan's fearlessness when confronting human presumption, theological distortion, and rampant injustice.[104] But he is also to see himself among the brood of ravens that "cry out to God" for food (38:41). The frail and the fierce are both his kin. Job discovers himself to be a child of the wild.

Through God's answer Job has come to a greater knowledge of the world and, in turn, a deeper knowledge of himself in the world, as one who sits in "dust and ashes" yet also "surveys all that is lofty" (41:34).[105] More than partaking from a "tree of knowledge" in a garden, Job has been forced fed the vast world of knowledge in creation. The result is not a more powerful Job. Divinized he is not. To the contrary, Job has beheld the terrifying God of the whirlwind and God's astonishing, frightful world and, thus, welcomes death. Job finds genuine comfort in his languishing state: "Therefore, I waste away yet am comforted over dust and ashes" (42:6).[106] Job has finally gained the comfort that his friends had failed to offer, comfort gained from the fierce beasts of the wild, comfort from God. No longer the objects of contempt and fear, these allegedly underclass citizens of creation now form part of Job's new community. Job has found comfort with God and with God's wild and wooly world, the first step toward biophilia.

Biology and Biophilia

Human beings are animals that classify, and those who do it painstakingly well are the esteemed taxonomists among us. They are "the librarians of life; without them, nature's volumes are meaningless."[107] Distinguishing various species of plants and animals has been a necessity since the dawn of *Homo sapiens*, if not earlier, and it continues to be a matter of survival.[108] If the beginning of biblical wisdom is cultivating the "fear of the LORD" (Prov 1:7), the beginning of scientific wisdom is "getting things by their right names."[109] Job gets both.

Behind Job stands every reader, ancient and modern, who has been captivated by God's detailed "catalogue." Through this divine taxonomy, Job embarks on a journey of discovery about the world and himself, a journey whose endpoint sets the occasion for his restoration. But perhaps it is only at the end of God's address, with the world's frontiers laid bare, that Job's journey truly begins: his return home. Sometimes the most epic of journeys is the journey of self-discovery. How does Job lead his life at home for the next 140 years? God only knows, but let the reader imagine.

Tour of Taxonomy: Job and the HMS Beagle

I imagine Job's journey not unlike another taxonomic tour, one taken by someone who was much more willing though perhaps equally uncertain of what was
to be found. Unsure of his vocational aspirations, including the ministry,
Charles Darwin obtained passage on the HMS *Beagle* in 1831 at the age of twen-
ty-two and traveled around the world to regions that even today still seem remote, from St. Paul's Rocks and Mauritius to the Cocos Keeling Islands and,
most famously, the Galapagos Archipelago. His journey is recorded in his famous diary, *The Voyage of the Beagle*, published in 1845, which contains a detailed catalogue of biological diversity cast in gripping narrative form. In his
own words, Darwin discovered "temples filled with the varied productions of
the God of Nature."[110] Darwin's diary, like Job, is a field book of wonder.

In it one can read about the cuttlefish, near the Cape Verde Islands, whose
"chameleon-like power" provoked Darwin's amusement (35–36). There are the
musical frogs of Rio de Janeiro, whose choral harmonies prompted Darwin to
remark, "Nature, in these climes, chooses her vocalists from more humble per-
formers than in Europe" (53) At Maldonado, there is the giant *Hydrochaerus
capybara*, or water-hog, of the Rodentian order, "the largest gnawing animal in
the world," as well as the small mole-like tucutucos, whose spoken name imi-
tates the sound it utters from underground, confusing anyone who hears it for
the first time (68–69). Other notable entries include the noisy habits of the
jaguar near the Parana river (134–35), the flying spider at St. Fé, the graceful
habits of the cliff-dwelling condor in Patagonia (173–74), the fierce behavior of
the ostrich when confronted by a man on horseback at Bahia Blanca (100), the
murderous force of a hail storm at Sierra Tapalguen (120), the formidable
strength of horses "drenched with gore" in the great corrals at Buenos Ayres
(124), the "snowing butterflies" on the coast of Patagonia (153), the "brave" penguin on the Falkland islands (185), the "great monsters" of the Galapagos,
including the giant tortoises on whose backs Darwin was allowed to ride (334)
and perhaps most famously, the "hideous" yet "graceful" marine lizard (*Ambly-
rhynchus cristatus*) (335–37). What Darwin saw in those far-flung areas, at creation's "margins," makes Job's catalogue a modestly short list.

Reflecting back on his journey, Darwin concludes, "The map of the world
ceases to be a blank; it becomes a picture full of the most varied and animated
figures" (431). The same could be said of Job's new world. Darwin's five-year
journey, E. O. Wilson notes, served as the *Wanderjahre* in "the genesis of [Dar-
win's] scientific mind."[111] In Darwin's own words, "[N]othing can be more
improving to a young naturalist, than a journey in distant countries."[112] For the
career of a budding naturalist, such a journey was a revelation: "no one can

stand in these solitudes unmoved, and not feel that there is more in man than the mere breath of his body" (429). It was, to quote from Job 38:27, the "waste and desolate land" that gripped Darwin's imagination the most. Nearly ten years after his voyage, Darwin struggled for words to understand why.

> In calling up images of the past, I find that the plains of Patagonia frequently cross before my eyes; yet these plains are pronounced by all wretched and useless. They can be described only by negative characters; without habitations, without water, without trees, without mountains, they support merely a few dwarf plants. Why, then, and the case is not peculiar to myself, have these arid wastes taken so firm a hold on my memory? . . . I can scarcely analyze these feelings; but it must be partly owing to the free scope given to the imagination. The plains of Patagonia are boundless, for they are scarcely passable, and hence unknown: they bear the stamp of having lasted, as they are now, for ages, and there appears no limit to their duration through future time. . . . [W]ho would not look at these last boundaries to man's knowledge with deep but ill-defined sensations? (429–30)

Within the "free scope" of Darwin's informed imagination, the desolate takes precedence over the fertile, eliciting wonder and a stark sense of eternity. It was in the "foreign clime, where the civilized man had seldom or never trod" that provoked "extreme delight" for Darwin: to live "in the open air, with the sky for a roof and the ground for a table" (430). Darwin regarded his fascination with the "boundless" terrain as "a relic of an instinctive passion." Call it awe.

Multiversity of Biodiversity

From Job's perspective, God's answer is tantamount to a Copernican revolution. Job comes to realize that the world does not revolve around himself, not even around humanity. Creation is polycentric. It has its various centers or domains, each accommodating different forms of life yet all interlinked. Indeed, "each species is a small universe in itself," E. O. Wilson observes.[113] If so, then Earth itself is a multiverse! In Job, God reveals the parallel universes of life on Earth, species that had either remained invisible to Job or drawn his contempt.[114] But they all are connected to him. Job's Behemoth is very much alive, lounging next to him under the lotus trees. Job has come to see what was hitherto invisible to him, namely, his kin among the wild. He has discovered himself as truly "a brother of jackals and a companion of ostriches" (30:29) and, in so doing, grasped something of the biologist's dictum, "I link, therefore I am."[115]

Job has also discovered that God's creation does not run like clockwork. No, creation is characterized by wildness and spontaneity, stability and revolution, waste and resiliency, death and life. From rain forests to sandy deserts, life does more than go on, maintaining itself. Life goes *forth*, never to revert to the past, just as the mountain goat kids "go forth and do not return to [their parents]" (Job 39:4) and as species never return, genetically speaking, to their progenitors. God celebrates the potent resilience of life that is always going forth.

Biologists call this "going forth" speciation, the continual emergence of new species. There are at least two ways of defining "species." A species is a "population whose members are able to interbreed freely under natural conditions."[116] This is known as the "biological species concept." But there are problems with this definition: it cannot account for fertile hybrids, which are common among plants and fungi, and for species that reproduce asexually.[117] The alternative, the "phylogenetic species concept," defines a species as the smallest group of organisms that exhibits a "parental pattern of ancestry and descent."[118] But this definition may be too refined, for it results in an explosion of the current number of accepted species.[119] The lesson? The danger of the Procrustean fallacy, of trying to fit nature into a single concept of measurement: "nature is often too diverse to be described by a single" definition of species.[120]

Nature is also too dynamic. "Like stars in an expanding universe, species are always evolving away from all other species."[121] In a perfect world, every species would, given enough time, proliferate into multiple species and rise to the taxonomic rank of genus or higher. The longer a certain species evolves, the more it differs "genetically from the remainder of life."[122] Species "differ from one another in infinitely varying directions and distances."[123] Evolution, in short, favors diversity even as it presupposes the genetic linkage of all life. In evolution, *Vive la différence!* meets *Vive la connexion!* And it is here where Behemoth boards the *Beagle*. God's litany of animals affirms the diversity of life, each animal in its own niche, but each linked together by the common thread of life that wraps itself even around Job. Biologists call it DNA or RNA, the genetic foundation for all life on Earth.

As biologists know firsthand, diversity is what makes life resilient, even in the harshest environments.[124] Conflict and competition, coupled with natural selection, have helped give rise to greater complexity of form. The rise of multicellularity, for example, was in response to predation.[125] Having withstood the test of time, including at least five major mass extinctions spread over the last half-billion years,[126] biodiversity remains "the key to the maintenance of the world as we know it."[127] But the triumph of life over death always takes time. For each major extinction, millions of years were required before a "complete" recovery was achieved.[128] As for a looming sixth disaster, unprecedented for its cause (*Homo sapiens*), such a lengthy recovery period is little comfort for Earth's "wise" citizens.

Job and Evolution

The great evolutionary biologist J. B. S. Haldane (1892–1964) was asked by a cleric what biology could say about God. He allegedly replied, "I'm really not sure, except that the Creator, if he exists, must have an inordinate fondness of beetles."[129] Indeed, beetles, with their 400,000 species, make up close to 25 percent of all known animal species.[130] Although beetles are not listed in Job's catalogue of the wild, they easily could have been. Through the eyes of the biologist, God's answer to Job constitutes a powerful testimony to biodiversity, concluding with a phylogenetic surprise: Job is related to Behemoth and, by extension, to all the animals of the wild. From Job's perspective, the figure of Behemoth is humanity's "missing link" to nature's diversity. Job and Behemoth share DNA, the genetic code for life on Earth. Anatomically, the fact that the developing human embryo exhibits early in its gestation a tail and gill-like features is additional testimony to our evolutionary linkage to species of very different environmental niches.[131] The genetic relatedness of all organisms has, like Job's Behemoth, revealed numerous phylogenetic surprises. The closest relatives of whales, for example, are ungulates such as pigs, antelope, and deer.[132] Dolphins are more closely related to humans than to sharks. The chicken is the closest living relative to the fiercest predator ever to roam the land, the *Tyrannosaurus rex*.[133]

For biologists, the common genetic code that translates information from DNA to amino acid indicates the common origin of all living things. The abiding astonishment that characterizes God's wild kingdom and Job's connection to it is, I submit, made all the more astonishing in light of evolution's tree of life. The phylogenetic tree of life remains the most effective and perhaps marvelous way of conveying the historical interconnectivity of all life as we know it.

Life in Extremis

Evolutionary biologists have identified three major branches or domains from which the "tree of life" on Earth developed: Archaea, Eubacteria, and Eukarya. First delineated by Carl R. Woese in 1977 (with help from DNA sequence coding for ribosomal RNA), these domains reduce the traditional five kingdoms of life[134] to three, one of which is a branch that had been previously thought to be bacterial in form.[135] Eukarya designates the latest and most complex domain of life. It includes every form of life that is familiar to the nonbiologist, from plants to mammals. Nevertheless, the tree of life is dominated by unicellular microorganisms such as bacteria. Microorganisms dominated roughly 3 billion years of evolutionary history, whereas animals, plants, and fungi have existed for less than a billion.[136] Archaea, given its name, designates the oldest branch, which includes (but is not limited to) single-celled organisms that can thrive in extreme environments. They are extremophiles, "organisms that love to live, and live to love in extreme conditions," such as in boiling water, high acidity, extreme radiation, and extreme pressure.[137] Astrobiologist Chris Impey affectionately refers to them as microbial "superheroes":

> There's Thermophile, who emerges from the inferno unscathed.
> There's Psychrophile, who shrugs off extreme cold. There's Endolith,
> who does his best work encased in rock. There's Acidophile, who
> energizes by bathing in battery acid. And there's Barophile, who
> withstands pressure that would bring a lesser superhero to its knees.[138]

Modern genetics suggests that life itself began in extremis. It began with heat-loving, self-replicating protocells when the earth was ruled by flowing magma, crashing comets, fierce lightning, bursting steam, and surging sulfur.[139] So much for Darwin's "warm little pond."

Unnoticed by casual observers, extremophiles continue to thrive on our "familiar" planet. In 1977 geologist Jack Corliss and his two crewmembers guided the submersible *Alvin* over a mile and a half beneath the Pacific Ocean near an undersea ridge off, in fact, the Galapagos Islands.[140] There they discovered the first deep sea vents, which generate high pressure and superheated water beyond the boiling point. In and around them teemed countless life forms oblivious to the sun but sustained by geothermal energy unleashed from below the earth's surface and from heat continuously produced by the radioactive decay of unstable isotopes, some of which last for billions of years. Corliss discovered near these vents tube worms of six to ten feet in length flourishing amid large colonies of bacteria, all dependent not on sunlight and photosynthesis but on "chemosynthesis," energy produced by chemical reactions. Science stranger than science fiction!

The label "extremophile" is, of course, a relative term; it redefines what is normal. Of course, most of the animals listed in God's answer to Job, except for those now extinct such as the auroch, are not so alien to us, hunters and zookeepers that we are. For Job, however, they were extremophilic, these wilderness-loving creatures. Job's transformation in part involved an overcoming of his extremophobia, his revulsion of fierce landscapes, those deemed godfor-saken and, thus, "human-forsaken."[141] But Job was wrong about the "godfor-saken" part. With its resolute focus on creation's extremities, God's answer depicts the "outside" without any familiar image to impose. The world is nei-ther a cosmic temple nor a lush garden nor a playhouse for child Wisdom. No, the world is a wilderness, uncultivated and untrammeled, and it is valued as such. To borrow from the Wilderness Act, Job is merely a "visitor who does not remain."[142] For the twentieth-century preservationists, the wilderness was "big and fierce, and people were weak and small."[143] In the wild, Job finds himself also of small account. Humility is in order. But the wilderness, both Job and Darwin also realize, is a place of wondrous discovery, even self-discovery.

Leviathan Effect

Job is shown glimpses of nature at its wildest. And glimpses they remain. Job's venture into the wild is not permanent; there is no hankering to return to the wild whence humanity came. Job is no Enkidu of Gilgamesh fame.[144] As the poet knows—as God knows—Job will die if he remains in the wilder-ness. Job's call to the wild is mercifully brief. Job was not meant to exist cozily with Leviathans, or with wolverines for that matter.[145] They are simply allowed to be what they are in the wild. E. O. Wilson refers to the "Grizzly Bear Effect" known by environmental ethicists and evoked, I would claim, by the book of Job in its own way:

> We may never personally glimpse certain rare animals—wolves, ivory-billed woodpeckers, pandas, gorillas, giant squid[s], great white sharks, and grizzlies come to mind—but we need them as symbols. They proclaim the mystery of the world. They are jewels in the crown of the Creation. Just to know they are out there alive and well is important to the spirit, to the wholeness of our lives. If they live, then Nature lives.[146]

To regard Leviathan and Behemoth also as symbols of human well-being is an arresting thought. But that is precisely where Job takes us, away into the wild to behold them and back home to ponder them. These crown jewels of creation should only be glimpsed at most, for any greater length of exposure to the wild would constitute an invasion. Visitors at most we remain for their sake and for ours. And so we must stand back in awe, keeping our distance, knowing that our flourishing depends on theirs. Call it the "Leviathan Effect."

Viewed from the perspective of the wild, Job's world remains truly alien, of "foreign clime," to quote Darwin, and by extension so also the world closer to home. Yes, one can agree with Terence Fretheim that "with God there are no alien creatures, no outsiders."[147] Granted, no creatures stand outside the orbit of God's providential care. But alien they remain, utterly strange and fully wild. God's answer to Job conveys something similar to what Michael Cunningham says about a good novel. "Any good novel helps us understand difference. It's about trying to break through our sense of foreignness and isolation. I can walk down the street in New York, and the people I see might as well be lizard women from other planets. We're so strange to each other."[148]

It takes not only flights of novelistic fancy to recognize our strangeness. Astrochemistry also helps. The very stuff of life, such as carbon, originates from the nuclear furnaces of stars and is carried forth by comets, the "dirty snowballs" of interstellar space.[149] Life is derived from recycled star dust. As Harold Kroto, Nobel Prize winner and professor of chemistry at Florida State University, enthusiastically notes, "So if you want to see an alien, just look at yourself in the mirror."[150] As Job beholds Behemoth, he also sees something of his own alien self. And it is precisely in his alienness that he finds kinship. "How to Be an Alien without the Alienation"—now that's a book that Job himself could write! But you would not find it in the self-help section.

Job's "New" Animals

If Job were alive today, what would God show him? What would God show us, we who have explored the land's frontiers and have either held Earth's more exotic denizens in zoos and aquariums or brought them to extinction? No need to show us E.T. Perhaps God would beckon us, like Job, to take the plunge deep into the watery depths, to Leviathan's abode, where life of a completely alien sort thrives.

"Job," God might say, "fasten your seatbelt[151] and let us travel, you and I, into the dark, cold depths of another world, free from the propellers and harpoons of the surface, free from the 'toil under the sun.'[152]

"Behold, Job, the deep, dark sea where among my creatures great and small flourishes the carefree Sea Cucumber. Its simple body and low metabolism are well suited for my watery abyss. After scavenging the sediment for food, it blissfully swims away, undulating gracefully toward other organic pastures.[153]

"Behold the enigmatic *Grimpoteuthis*. Humans call it the Dumbo Octopus, though they are quite confounded about what it does in the deep. It simply rests on the bottom, wrapped in its mantle.[154] Job, do you know what it does there sitting so still and quietly in the dark? Answer me, Job, for surely you know! No? All right, then, I'll let you in on a secret: It's meditating on the Torah!

"And this little squid of mine, I'm going to let it shine. My Jewel Squid has the most sophisticated lighting system of any creature of the deep, complete with filters, reflectors, and eyelids, by which it controls the duration and intensity of its luminosity. It can vanish into the darkness in an instant.[155] But all you have, Job, are fragile receptors, with blind spots.

"Behold, Job, the Pacific Viperfish with its long and pointy fangs. Its teeth are so prominent that they pose a danger to its eyes. If the Viperfish misjudges the size of its prey and impales an animal that is too big, it may find itself unable to release it, destined to die along with its last supper.[156]

"But my favorite creature of the deep is the one that humans disparagingly call *Vampyroteuthis infernalis*, 'the vampire squid from hell,' so named because it so repulsed its first discoverers. But it is my mascot of the deep: half-squid and half-octopus, dating back to 200 million years ago.[157] Oh, yes, you were born before then, weren't you, Job? This creature can do something no other complex creature can: it can dwell quite happily in the oxygen-depleted layer of the ocean because of its special respiratory blood pigment. Being the slowest cephalopod of the sea doesn't hurt either.

"But, Job, let us go back in time to another creature named *Tiktaalik* by humans.[158] But I will simply call it by its nickname, 'Snout.' Its story has been told by one of my favorite writers (next to Moses, of course), the anthropologist Loren Eisley: Snout's story begins 'as such things always begin—in the ooze of unnoticed swamps, in the darkness of eclipsed moons. It began with a strangled gasping for air.'[159]

> On the oily surface of the pond, from time to time a snout thrust upward, took in air with a queer grunting inspiration, and swirled back to the bottom. The pond was doomed, the water was foul, and the oxygen almost gone, but the creature would not die. It could breathe air direct through a little accessory lung, and it could walk. In all that weird and lifeless landscape, it was the only thing that could. It walked rarely and under protest, but that was not surprising. The creature was a fish.
>
> In the passage of days the pond became a puddle, but the Snout survived. There was dew one dark night and a coolness in the empty stream bed. When the sun rose next morning the pond was an empty place of cracked mud, but the Snout did not lie there. He had gone. Down stream there were other ponds. He breathed air for a few hours and hobbled slowly along on the stumps of heavy fins.
>
> It was an uncanny business if there had been anyone there to see. It was a journey best not observed in daylight, it was something that needed swamps and shadows and the touch of the night dew. It

was a monstrous penetration of a forbidden element, and the Snout kept his face from the light.[160]

"Job, behold the Snout:

> Even when the pond dries up,
>> it is not alarmed.
> It remains confident,
>> even above the face of the waters.
> Do not mock the Snout, Job,
>> for in three hundred and seventy-five million years,
> its face would be your own."[161]

The world according Job is filled with strange beings, all fiercely independent yet all integrally related. God's answer to Job takes up what is wildly alien, considered peripheral and extreme by any human measure, and places it front and center. The bond forged in creation between Job and Behemoth, the bond forged in evolution between Job and Snout, requires Job to affirm his own life in extremis, to embrace his identity as *Homo alienus* and his connection with all aliens, to revel in his freedom, wild thing that he is, and to step lightly on God's beloved, vibrant Earth. God's challenge to Job to "gird up his loins" and be catapulted into the margins of life, to take the plunge into the depths of chaos and come up for air, to muster the courage to start a new family, is now fulfilled. Such was Job's baptism, his terrifying and edifying immersion into Nature.

6

The Passion of the Creator

The Manifold Nature of Nature in Psalm 104

The treetops are another world, smelling of bark,
a stratum of freer air and larger views.

—Wendell Berry[1]

I looked into His eyes and then knew what to say to any angel who might
serve as a sentry to God: No creature should be turned away.

—St. Thomas Aquinas[2]

The most extensive creation psalm in the Bible, Psalm 104 is a
fitting match for Job 38–41. It is tantalizing to imagine Job himself
having composed the psalm in response to God's revelation.[3] But
regardless of authorship, whether implied or actual, Psalm 104
and Job 38–41 form a remarkable pair. Kindred texts they are. The
psalm focuses just as resolutely as does Job on the manifold nature
of creation, Leviathan included. One is, in effect, the mirror image
of the other. As God's answer turns Job's home-based world into
something vast and alien, the psalmist turns the terrifying wilderness into a habitat for all God's creatures. The psalm, moreover, carries a different yet complementary viewpoint: creation is
seen not from the creator's perspective but from the creature's,
specifically from the standpoint of *Homo laudans*, "the praising
human."[4]

Translation and Text

(1) Bless YHWH, O my soul!
  O YHWH, my God, you are exceedingly great;
    clothed are you with honor and majesty!
(2) Who enwraps himself with light as with a garment;
    who unfurls the heavens as a curtain;
(3) Who sets his upper chambers upon the waters;
    who makes the clouds his chariot;
      who moves about on the wings of the wind;
(4) Who makes the winds his messengers.
    Fire and flame[5] are his ministers.

(5) Who establishes[6] the earth upon its foundations;
    it shall never, ever totter.
(6) As for the deep, its covering[7] was like clothing;
    [its] waters stood over the mountains.
(7) At your blast they took to flight;
    at the sound of your thunder they fled with fright
(8) —as the mountains rose and valleys sank—
    to the place that you established for them.
(9) The boundary you set they shall not pass,
    so that they may never again cover the earth.
(10) Who sends forth springs into the wadis;
    between the mountains they flow,
(11) giving drink to every wild animal,
    breaking the onagers of their thirst.
(12) Beside them[8] the birds of the sky make their home,
    raising (their) voice among the foliage.
(13) Who waters the mountains from his lofty abodes;
    from the fruit of your hands the earth is well satisfied.
(14) Who makes the grass grow for cattle,
    and plants for human cultivation
      to bring forth food from the earth:
(15) wine, which cheers the human heart,
    oil,[9] which makes the face shine,
      and bread, which sustains the human heart.

(16) The trees of YHWH are well watered;
    the cedars of Lebanon, which he planted,

(17) where the birds make their nest.
    The stork has its home in the junipers.
(18) The high mountains belong to the mountain goats;
    the crags are refuge for the coneys.
(19) He made[10] the moon for its seasons,
    the sun to know[11] its time for setting.
(20) You bring on the darkness, and it is night;
    in it creeps every animal of the forest.
(21) The young lions roar for their prey,
    seeking their food from God.
(22) When the sun rises, they withdraw,
    and to their dens they retire.
(23) Humans go forth to their work,
    to their labor until evening.

(24) How manifold are your works, O YHWH!
    You have made them all in wisdom!
        The earth is stock full of your creatures![12]
(25) There is the sea, both vast and wide.
    There are the creeping things beyond count,
        living things small and great.
(26) There go the ships,
    and Leviathan, with which you fashioned to play![13]
(27) All of them wait for you,
    to provide their food in due time.
(28) You give to them, and they gather it up;
    you open your hand, and they are well satisfied.
(29) You hide your face,
    and they are terrified.
    You take away their breath,[14]
        and they expire and return to their dust.
(30) You send back your breath, and they are (re-)created.
    You renew the face of the earth.

(31) May YHWH's glory endure forever!
    May YHWH rejoice in his works!
(32) Who looks at the earth, and it trembles;
    he touches the mountains, and they erupt in smoke.

(33) I will sing to YHWH as long as I have life;
    I will sing praise to my God while I still live.

(34) May my meditation[15] be pleasing to him;
    I shall rejoice in YHWH.

(35) May sinners cease from the earth,
    and the wicked be no more.

Bless YHWH, O my soul.
Hallelujah!

Psalm 104 explodes with poetic energy. There is nothing serenely "medita-tive" about it (cf. 34a). The poem bursts at the seams with joy as it celebrates creation's manifold nature. And yet for all its poetic revelry, the psalm exhibits a well-defined movement. First, the psalm delineates the broad structures or domains of creation before proceeding to detail various life forms and their habitations. Evident is a "downward" movement from the divine realm (vv. 1–4) to the earthly domain, including the watery depths (vv. 6–9, 25–26) and the land (vv. 10–18), with emphasis on God's provision for all life (vv. 27–30; cf. Gen 1:29–30). The hymn concludes with expressions of praise to God for all crea-tion, except for one little "glitch" (vv. 31–35).

Social Context

As with many psalms, Psalm 104 does not readily divulge its historical context. It is pure poetry, setting its focus on the world of nature, not on Israel's history, and in a strikingly novel way. It offers an unabashedly positive view of the natu-ral world that includes the wilderness, traditionally considered dangerous and chaotic. Instead of "Lions and tigers and bears, O my!" we have "Lions and tigers and bears, Amen!" (along with the coneys, onagers, and mountain goats). The psalmist celebrates the world of the wild and the God who sustains it all.

But for all the wide-eyed wonder over nature's goodness and God's grace, a dark cloud looms on the horizon. The psalm acknowledges the ever-present possibility of famine (v. 29), as well as earthquake and volcanic activity (v. 32). Perhaps the psalmist's community experienced firsthand a depletion of natural resources or a natural disaster. Regardless, the psalm ends on an admittedly minor chord. After celebrating the sheer diversity of life, the psalmist exhorts God to vanquish the wicked (v. 35a). While the psalm includes even the mon-strous Leviathan within the orbit of God's providential care, it has no room for the wicked. For many readers, this imprecation is a "damned spot" on an oth-erwise perfect poem. But for the ancient listener, calling God to exterminate the wicked made sense in a less than perfect world. By cursing the wicked, the psalmist transfers the evil chaos traditionally assigned to mythically monstrous figures such as Leviathan and places it squarely on human shoulders. Conflict in creation, the psalmist acknowledges, is most savage among the distinctly human beasts.

We do not know whom the psalmist had in mind regarding the wicked. But whoever they were, the wicked apparently had a nasty habit of violating creation's integrity in something of the same way that the imprecation in v. 35 blemishes the psalm's picturesque poetry. Put positively, this grim petition rescues the psalm from seeing the world through rose-colored spectacles. The psalmist acknowledges, moreover, both predator and prey. Here is an authentic assessment of creation as it stands, not as it once was in some pristine state or as it will be in a future fulfillment. It is a world in which the purveyors of chaos are not mythically theriomorphic—monsters made in the image of animals— but monstrously human.[16] In comparison to other psalms that target the wicked, it is quite remarkable that Psalm 104 says so little about the wicked.[17] Its primary focus is not on the wicked but on the creator and creation.

Character of Creation

As a whole, Psalm 104 presents a panoramic sweep of creation from the theological and the cosmological to the ecological and the biological, all bracketed by the doxological. Because giving praise is foremost in the psalmist's mind, the psalm aims not to provide information about how the world works but to motivate the reader to praise God and, as we shall see, to sustain God's joy in creation (v. 31b). The psalm begins with God (vv. 1–2). Metaphorically cast as a cloak, light is not the first act of creation but an attribute of God (v. 2a). In *Enūma elish*, the gods don "mantles of radiance" or effulgent "auras" (e.g., Tablet I 68, 138). So also YHWH: every dawn could be construed as an act of self-clothing.

A nearly identical metaphor is employed to describe the creation of the heavens in v. 2b. Unlike the solid image of the firmament featured in Genesis 1:6–8, heaven in Psalm 104 is likened to unfurled fabric. The celestial sphere is much more pliable than what is depicted in Genesis. Similarly so in Psalm 102:25–27: Heaven and Earth are associated with clothing, but clothing that God cannot afford simply to shed without donning something new. No naked God is the creator. According to the psalmist, creation is not God's body,[18] but neither is it disposable rags.

In describing the earth's creation, Psalm 104 approaches the mythically dramatic. As in Genesis 1, the waters cover the earth's surface. But while the earth remains firmly fixed, the waters "flee" at the sound of God's thunderous rebuke. No resistance is registered by the waters. Nevertheless, the waters do pose a problem: like the temperamental aquatic newborn in Job 38:8–11, they require restraint (vv. 7–9). Their containment makes possible the provision of flowing streams for quenching thirst, providing habitation, and ensuring the earth's fertility. The combination of stream and soil results in the provision of life and enjoyment. By providing grain for bread, grape for wine, and olive for oil, plants sustain life and impart joy (v. 15).

As in Job, the psalm's primary focus is on animal life: mountain goats, storks, coneys, lions, and finally Leviathan all populate Earth's various domains. All are lovingly detailed in a tone of rapturous praise to the creator. The psalmist's catalogue of the natural order is matched only by God's extensive answer to Job. But for all that is comparable between these two taxonomies of the biosphere, there is one major difference. Whereas YHWH's response to Job makes scarce reference to botanical life,[19] trees have standing in Psalm 104:

> The trees of YHWH are well watered;
>> the cedars of Lebanon that he planted.
> There the birds build their nests;
>> the stork has its home in the fir trees. (vv. 16–17 [cf. v. 12])

The psalmist lingers admiringly over the mighty cedars of Lebanon, which in his day were the prized commodity of imperial regimes. Armies from Mesopotamia would march westward, conquering cities and territories in their path, to get to these cedar forests, cut them down, and use the timber for various building projects. These trees once grew in dense forests on the slopes of Lebanon's mountains. Few remain today.[20]

The psalmist, however, prizes these trees not for their lumber but for their majestic stature and their hospitality: the cedars are literally for the birds! This seemingly minor detail is representative of how the psalmist views creation as a whole. Most commentators have marveled over the central theme of provision in the psalm, and appropriately so.[21] God provides drink to the wild animals (v. 11), "waters the mountains" and "the trees" (vv. 13, 16), causes "grass to grow for the cattle" (v. 14), provides bread, wine, and oil for human beings (v. 15), and supplies "prey" for the lions (v. 21) as well as food for all creatures "in due time" (v. 27). God's "open hand" and "renewing breath" are evocative images of such provision (v. 28).

However, in addition to the theme of divine provision in the psalm is another central feature, for which the cedar trees offer but just one example. The clue lies in God's first act of creation (vv. 2b–3a). In the beginning God created a home, a habitat for divinity and, in turn, established domiciles for every living creature: streams and trees for the birds (vv. 12, 17), mountains for the wild goats (v. 18a), and rocks for the coneys (v. 18b). Even the waters have their "appointed" place (vv. 8–9). The lions have their dens, just as humans have their homes (assumed in vv. 22–23), and Leviathan has the sea (v. 26). The earth is not just "habitat for humanity" but habitat for diversity. "The earth is stock full of your creatures!" (v. 24), so the poet marvels. Indeed, human beings are scarcely mentioned until v. 23 (cf. vv. 14–15), and only then in consort with the lions, whose only noted difference from humans is that they take the nightshift to pursue their living (v. 22). "Even lions, the psalmist acknowledges, have to make a living."[22]

The psalm, thus, views creation in thoroughly ecocentric terms; the earth is created to accommodate myriad creatures great and small, people included. The earth is host and home to all living kind, and as such it is a source of joy. The earth provides wine "to *cheer* the human heart, oil to make the face shine" (v. 15a). The sea, home to innumerable marine creatures, is a playfield for both God and Leviathan (vv. 25, 26b). In short, creation is cast in the *imago habitationis*, and joyfully so.

Whereas Job 38–41 extends creation outward toward the fringes, Psalm 104 brings creation home from the margins. The animals of the earth are given homes, their native habitats, whether they be mountain crags, majestic trees, or four-pillared houses. Their fierce freedom, limned so vividly in Job, is now coupled with the domesticity of dependence. The world so conceived by the psalmist is not so much a free range as a spacious home, and its inhabitants all share the earth as their common habitat. Psalm 104, in short, is a fanfare for the common creature.

God the Playful Provider[23]

The God of Psalm 104 is the provider par excellence. Providential care is God's passion. Toward the end of the psalm, the speaker exhorts God to "rejoice" in creation, to find joy in creation's abundance, diversity, and its capacity for accommodation.

> May YHWH's glory endure forever;
> may YHWH rejoice in his works! (v. 31)

The language of unabashed joy is rarely attributed to God in the Hebrew Bible.[24] With the possible exception of Isaiah 9:17, the Hebrew verb *śmḥ* ("be joyful") never takes God as its subject in the biblical corpus.[25] Typical subjects of "rejoicing" in the psalms range from the individual speaker (104:34) and the worshipping community (126:3) to the ends of the earth (97:1). While creatures are to rejoice in the creator (9:2; 32:11; 96:11–13; 104:34), nowhere in the psalms is God said, much less exhorted, to rejoice.[26] Except here. In Psalm 104, the creator is to rejoice in the creatures.

The Joyful God

Novel to this biblical psalm is the claim that creation is sustained not by God's covenantal commitment but by God's unabashed joy. According to Genesis 9, God solemnly pledged to never again destroy the created order by floodwaters, as sealed by the sign of the rainbow (vv. 8–17). Yet, conceivably, God in sovereign freedom *could* opt out of such a covenant and renege on the promise. In the pointed words of Jon Levenson, "Humanity's only hope is that God will spurn that option, fail to exercise his freedom, and consider himself bound by his word to Noah. Creation has become a corollary of covenant."[27]

Perhaps in recognition of the problem, the psalmist prefers the language of divine delight over that of covenantal fidelity. The solemn formulation of a self-restraining order is replaced with the psalm's joy-filled poetry of praise. The psalmist is, after all, a poet, not a legal expert. But the poet does so at a cost: psalmic exhortation allows for the frightful possibility that unilateral covenant-making attempts to preclude. If, God-forbid, the creator were to *stop* enjoying creation, the world would collapse or wither away. The psalmist's commendation of divine joy in v. 31b smacks of urgency. By ceasing to rejoice, God could at any moment turn creation back into a quivering mass of chaos. The possibility of cosmic collapse in the psalm is attributed not to divine wrath set against an obstinate creation, not to threatening chaos poised to overtake creation, but to something much less dramatic but much more possible, namely, God's lack of joy in creation.

In Genesis, God's anger, provoked by creation's wickedness, unleashed the floodwaters to cleanse the earth of violence (Gen 6:11–13). In the troubling aftermath, God came to accept the intractable "inclination (*yēṣer*) of the human heart" to do "evil," and it was out of such resignation that God promised to "never again curse the ground because of humankind" (8:21; cf. 9:11, 15), an assurance that rests entirely upon divine restraint. For the psalmist, by contrast, restraint is not what ensures the world's perdurability. God has a much more active, personal involvement in creation. Enter Leviathan. While Job devotes more than one whole chapter to this formidable beast,[28] the psalmist ascribes only half a verse. Nevertheless, Leviathan is not cheated in this psalm.

The Playful God

Leviathan elicits more than any other creature God's rapturous joy, and like all creatures featured in Psalm 104, Leviathan too has its home in the created order. The vast sea accommodates a multitude of living beings, including the greatest of them all, Leviathan, the monster of the deep. Ship and Leviathan serve as a contrasting pair in v. 26. Both are bona fide creations, one "man-made" and the other divinely wrought. One skims the surface; the other dwells underneath. Both share the sea, but only one actually inhabits it, and does so at God's good pleasure.

As discussed in chapter 2, Leviathan's reputation as a formidable creature of the abyss precedes its appearance in Psalm 104 (cf. Ps 74:12–15; Isa 27:1). Leviathan is a Hydra, the Semitic version of the multiheaded sea monster, a creature not for play but for mortal combat: its defeat is considered a necessity for the world's sake and for God's. Not so in Job; not so in Psalm 104. The psalm renders an entirely positive profile of this fearsome creature in two ways. First, reference to this sea monster is literarily distanced from the one hint of creation's fleeting resistance to God in the psalm, namely, the waters in vv. 6–9. God's "rebuke" or "blast" sends the waters fleeing to a designated spot. No hint of combat is present. Second, the relationship between God and Leviathan succinctly described in v. 26b lacks any hint of animosity. Instead of God's mortal enemy, Leviathan is God's playmate!

Reference to God "playing" (śḥq) is rare in the Bible.[29] The verb's field of context is wide: Hagar and Isaac are caught *playing* by a scornful mother (Gen 21:9); Isaac and Rebekah are caught in *foreplay* (or worse) by a peeping king (Gen 26:8). In these examples, "play" designates a mutual form of engagement. Most suggestive is Wisdom's play in Proverbs 8:

> I was daily [YHWH's] delight,
>> playing (śḥq) before him always,
> playing (śḥq) in his inhabited world,
>> and delighting in the human race. (Prov 8:30b–31)

Although God is not the subject of the passage, Wisdom's "playing" is understood as reciprocated by God, since God's "delight" is found in Wisdom.[30] Not coincidentally, Psalm 104 refers to creation fashioned in wisdom (v. 24). Creation's differentiated form and abundant diversity of creatures are signs of God's wisdom and the objects of delight, both God's and the psalmist's.

The closest parallel to Leviathan in the psalmic text is found in Job 41:5 [Hebrew 40:29], part of God's answer to Job.

> Will you play (*śḥq*) with it as a bird,
> or bind it on a leash for your girls?

Because the question constitutes God's challenge to Job, who is powerless, it is understood that playing with Leviathan is something only God can do. In Psalm 104, delight, rather than defeat, is God's design for Leviathan, and such delight, exercised in the form of play, binds God to the world to ensure its continued existence.

It is Leviathan that brings out God's playful side. But godly play is no isolated moment in God's engagement with the world. To the contrary, it supports all creation. Play makes for creativity.[31] Were this monster of the deep to reassume its traditional role as primordial adversary, then God's delight would cease and the ancient script of chaos battle (*Chaoskampf*) would be replayed. The joy of play would be replaced with violent struggle, like children turning an innocent game of cops and robbers into something far too serious.[32] From the psalmist's perspective, it is the "wicked" who refuse to play and choose instead to struggle against God and the created order. They are the purveyors of chaos, not Leviathan. They are, to quote a colloquial phrase, the "fun-suckers."

Humanity's Place

The psalmist discerns a common denominator among all creatures, both alien and familiar, wild and domestic, human and nonhuman. As the tenants of God's cosmic mansion, all creatures are recipients of God's providential care. Humanity constitutes only one class of occupants, and there is no "first class"! There are, however, boundaries to be respected, such as that of the waters within their "appointed" place (vv. 8–9). Human labor is consigned to the day, not the night, which is reserved for the animals of the forest, including the lions, to exercise their rightful living. By contrast, the wicked, the object of the psalmist's scorn (v. 35a), ply their trade at night and, in so doing, become predators of their own species (see Job 24:14–17).

While respect for boundaries, both temporal and spatial, is stressed in the psalm, a generous degree of freedom is granted to humanity: the ships operate on the sea, the habitat of Leviathan, no doubt at their own risk. Absent, however, is any hint of human dominion. Vegetation is shared by both cattle and human beings. The trees are for the birds, not for lumber (vv. 12, 16–17). Dependence upon God proscribes the hoarding of resources and the encroachment of domains. By recognizing that God provides for all, the psalm generates a sense of gratitude and awe, and it is out of such awe that the psalmist crafts this "meditation" (v. 34). Psalm 104 is an exercise in joy, one that the psalmist encourages even God to share in (v. 31b). As plants are shared, wine is distributed, and all "are filled with good things" (v. 28), so joy is passed from creator to creature and from creature to creator. As the choirmaster of praise, the psalmist calls readers to take on humanity's true nature not simply as *Homo sapiens* but as *Homo laudans*, the praising human, in the hope that God remains the *Deus ludens*, the God who plays to sustain creation.

The Science of Biodiversity

Comparable to the psalmist's wonderment is Darwin's astonishment that concludes *On the Origin of Species*: "[F]rom so simple a beginning, endless forms most beautiful and wonderful are being evolved."[33] Both the psalmist and the biologist are awestruck by the sheer diversity of life, consisting of at least 1.8 million species and counting.[34] Armed with a microscope, the psalmist would have also marveled over the more minute forms of biological life, down to the unicellular, all vibrant and thriving, all bounded and interdependent. Manifold are the forms of life; manifold, too, are the scales of life.

Big Bang versus Big Bore

Place and provision, according to Psalm 104, are the fundamental features of creation that ensure the continuance of life. At base, Earth is a hot zone of habitation. Physicists and biologists alike marvel at the salutary conditions for life in "our" biosphere, from Earth's location in relation to the sun to the atmosphere's protective blanket to the resilient power of evolution. They marvel because in the grand scheme of things the conditions for life are wildly improbable. Indeed, Earth's "bio-friendly" qualities can be traced back to the very beginnings of a universe that is "uncannily fit for life."[35]

The variables that set the conditions for life could easily have been otherwise. Our universe exists within a tiny latitude of cosmic "parameters that permit matter to organize itself into galaxies, stars, and planets, and for those objects to last for billions of years."[36] Had the rate of cosmic expansion soon after the Big Bang been incrementally smaller, then the universe would have recollapsed. If the expansion rate had been greater by a miniscule figure, then the universe would have inflated far too quickly for galaxies and carbon-based life to form. Decreasing strong nuclear force at the Big Bang by a mere 1 percent would have resulted in the formation of only hydrogen, the simplest of all chemical materials. The Big Bang would have rendered a Big Bore. As it is, the cosmos, with its remarkable diversity and density of galaxies, stars, nebulae, black holes, and planets, "seems to be balanced on a knife edge."[37] Such a cosmos would enthrall any observer, human or divine.

Such a "finely tuned" process of cosmic evolution, however, did not involve a perfectly balanced interaction of forces. Cosmologists have determined that as the universe began to cool soon after the Big Bang, it exhibited a slight but distinct preference for matter over antimatter: for every one billion antibaryons present in the early universe, there were a billion and one baryons (e.g., protons and neutrons). As particles and antiparticles annihilated each other in puffs of radiation, the result was a remnant baryon. With odds far worse than Pickett's Charge, the lone particle survivor for every billion "battles" provided the means for the material formation of the universe as we know it. Our material universe began with remnant matter. When it comes to the wonders of cosmogony, perfect balance is highly overrated. Our cosmos evolved from leftovers.

But one need not go back 13.7 billion years to find creation at "knife's edge." Closer to home (and billions of years later): if Earth were at all closer or farther from the sun, its life-sustaining water would have vaporized or remained frozen, thereby eliminating the prospect of advanced life from the start. Of all the planets in our solar system, only Earth has a distance "just right" (93 million miles from the sun) for water to remain liquid, thereby making its surface a "haven for life."[38] All of these "accidents of physics and astronomy," according to physicist Freeman Dyson, have yielded an "unexpectedly hospitable place for living creatures to make their home in."[39] It is precisely this kind of world that the psalmist marvels over, a world made exquisitely inhabitable for many and varied forms of life. But little did he know how delicate an operation it was for such a world to be created, astrophysically, geologically, and biologically. Theologically, the psalmist attributed the hospitable, accommodating character of creation to the "knife's edge" of divine pathos.

Anthropic versus Biotic Principle

In light of the many precisely "determined" variables that set the cosmos on its life-producing, self-sustaining evolutionary course, it seems that our universe is a rare breed indeed. This observation serves what is frequently called the "anthropic principle,"[40] although "anthropic *approach*" would be more accurate.[41] Such evidence, whether it is the delicate balance of fundamental forces, the delicate imbalance of antiparticles and particles immediately after the Big Bang, or the specialized position of the earth in relation to the sun, is cited to advance the claim that the universe was purposively "finely tuned" to set the conditions for intelligent life on Earth. This "principle," in its strong form, is developed by working backwards from the current state of affairs—human flourishing and perception (hence the name "anthropic")—to the Big Bang for the purpose of inserting a full measure of teleology or purpose into the cosmic equation and thereby pointing to a cosmic designer responsible for such an arrangement. In the famous words of the astronomer Fred Hoyle, it all seems like a "put-up job."[42] One problem, however, is that the argument as stated presumes life as we know it as the only form of life obtainable. The possibility of life elsewhere in the universe remains open, if not probable.[43] It takes a leap of faith, boosted by a species-centric elitism, to argue that the only "goal" of creation is to produce rational carbon-based life.

To the anthropic argument the psalmist delivers a decisive correction, or better expansion. Humankind is nowhere featured in the psalm as the dominant species on the animal planet, let alone the culmination of creation. To be sure, Psalm 104 acknowledges that the earth is an eminently habitable zone, but for whom? Certainly not just for human beings. If there is a "principle" at work in creation according to the psalm, it is inclusively "biotic" rather than anthropic. Each species of life has its home and sustenance in creation. The earth is a residential neighborhood of creatures whose lifestyles and traits are quite different from each other. Yet they all coexist, even amid predation. If there is a perfection or ideal presumed in the psalmist's world, it is the perfection of diversity, the diversity of life and habitat. As life's species are varied and numerous, so also are their niches, from towering trees and flowing wadis to mountainous crags and the deep, dark sea. In God's cosmic house there are many dwelling places, each fit for each species. Humanity's place in the variegated order of creation is as legitimate as that of any other species, along with the coney and the onager. There is living room for all, according to the psalmist, even if the lions have control of the TV remote at night.

And so it is on the animal planet, a world that remains largely undiscovered. "Earth is biologically still a mostly unexplored planet."[44] While about 1.8 million species have been officially catalogued, most experts believe that around 10 million species actually exist.[45] Indeed, considering all the microbes and other tiny organisms awaiting discovery, some suggest 100 million or more![46] And one must not overlook the fact that the fossil record, sparse as it is, reveals many more species having thrived on Earth in the past.[47] Even mammals may not be fully counted: a new species of monkey was recently discovered in Tanzania (*Lophocebus kipunjji* in 2005). At this point, Earth's biodiversity remains almost immeasurable, but perhaps not for too long as the number of extinctions eventually overtakes our expanding count of new species.[48]

Plants and Provision

In the catalogue of life, the inclusion of plants distinguishes Psalm 104 from Job's taxonomy. Whereas both the psalmist and the Joban poet populate the earth with the animals of the wild, only the psalmist acknowledges the essential value of plants. Plants are a primary means of God's provision for life: they "bring forth food from the earth," providing wine, oil, and bread (vv. 14–15). They also warrant admiration, such as the majestic cedars of Lebanon (v. 16), and provide shelter (v. 17). The psalmist gives plants their due.

So also does science, all the more so. The theme statement of the Twelfth International Congress on Photosynthesis in Brisbane, 2001, was "Photosynthesis: the plant miracle that daily gives us bread and wine, the oxygen we breathe, and simply sustains all life as we know it."[49] The miracle behind the botanical gift of life begins with photosynthesis, the harnessing of sunlight to convert carbon dioxide into biomass. The process is remarkably efficient and bewilderingly complex. It takes only a matter of seconds for a plant to capture the energy of a photon, process it, and store it in a chemical bond,[50] whose byproduct is oxygen. Thus, we can add to the psalmist's list of botanical benefits the most basic, essential source of sustenance for most animals, oxygen. And the list continues to grow: "not only do plants provide us with fuel, food, shelter, and medicines that sustain the human way of life, but they also uplift and inspire us," observes paleoclimatologist David Beerling.[51] So did Lebanon's cedars for the psalmist. Plants help regulate the sustaining cycles of life, such as the cycling of carbon dioxide and water, the rate of rock erosion, the composition of the atmosphere, and how the landscape absorbs or reflects sunlight. Without plants, carbon dioxide would be fifteen times greater, global temperatures would rise 10°C, and oxygen "would be pinned to an asphyxiating 10%."[52] The psalmist wasn't kidding about plants.

Fit for a Niche

The psalmist marvels not only at the plethora of flora and fauna but also at the various habitats in which they flourish. The psalm assigns a home fit for every animal. Fitness for survival and reproductive success, biologists have observed, is not simply the result of adaptation driven by natural selection, not just the consequence of random genetic variations selected by the environment. Organisms not only respond to their respective environments; they also, in many cases, transform them. They develop niches. Israeli biologists Eytan Avital and Eva Jablonka have identified "social learning" as one way in which animals adapt to *and* alter their environments, thereby becoming agents of their own evolution.[53]

The beginning of social learning is the acquisition of information. Through their study of birds and mammals, Avital and Jablonka note that animals can acquire information from individual trial and error as well as from the behavior of others within their species. The songs of birds provide a particularly rich example. The male blackbird, for example, learns and modifies its song from the feedback of other blackbirds as well as from various acoustic sources.[54] There is clearly a genetic basis for the distinctive song of a particular species. Nevertheless, "plasticity is the hallmark" for many species of songbird.[55]

In addition to learning, higher-order animals can transmit information to the next generation.[56] Both acquisition and transmission make such learning "social." Avital and Jablonka conclude that social learning is, in addition to natural selection, "an important agent of evolutionary change."[57] It is an "additional tier" of evolutionary inheritance that is not specifically gene based.[58] Whereas natural selection is "basically gene selection," social learning is a matter of cultural evolution.[59] And yet it remains thoroughly Darwinian: such variations in patterns of behavior among animals are heritable from one generation to the next. Indeed, social learning is as much an "inheritance system" as DNA. "Learning defines the direction and the general nature of change, while [natural] selection fine-tunes it."[60] Through social learning, new habits are formed and new habitats become niches.

To be fit for a niche is, thus, not simply a matter of passive genetic adaptation. An animal's environment is not just an external given to which a species adapts itself. It is also a matter of formation through the species' own agency. An animal's specific environment is its ecological niche, "the habitat as it is experienced and constructed by the animal itself."[61] Environment, hence, is a "species-specific concept," even an "individual-specific concept."[62] Introduced into the discourse of biology by Charles Elton in 1926, an animal's "niche" is its interactive "place in the biotic environment"; it describes "the status of an animal in its community, to indicate what it is doing and not merely what it looks like."[63] A niche, in short, "includes address *and* occupation."[64]

Not only do species evolve; they also create their environments. Organisms can take the initiative by selecting a new environment or transforming a familiar one, with genetic and anatomical changes emerging as a consequence of their actions. Such was the case with the "pioneering" fish that ventured forth onto land, blazing the evolutionary trail for land-dwelling animals, as well as with hoofed land mammals returning to the water to become the ancestors of whales.[65]

Whereas genetically based variation that leads to adaptation is "blind," the evolutionary agency of social learning is "guided by goals."[66] As they achieve fitness within their respective environments, organisms are not passive agents.

They can "actively determine and construct their environment."[67] The result is what F. J. Odling-Smee calls "niche construction."[68] A classic example of cultural niche construction, identified by anthropologist William Durham, is the domestication of cattle, which established a new environment resulting in certain groups of humans gaining the gene-based ability to absorb lactose, as one finds among northern Europeans and the wandering pastoralists in the Congo basin.[69] "The evolution of lactose absorption is an example of harmonious and simple co-evolution between genes and culture" (15). Such co-evolution is by no means limited to humans, however. There are many ways in which animals change their environment, thereby affecting their own evolutionary development. Termites, for example, create a new environment by constructing mounds for shelter and climate control. Mounds, burrows, and dams are all examples of "niche construction." In many ways animals "adjust the environment to themselves" (113), thereby "fundamentally affect[ing] the rate and the direction of genetic changes" (29).

A more mundane example, perhaps, but equally illustrative is the one given by the psalmist himself: birds building nests (v. 17). Alfred Lord Wallace, who ranks with Charles Darwin as one of the greatest biologists of the nineteenth century, was fascinated by the complexity and variety of bird nests. Although he explained them by natural selection, he did not think that nest-building was an instinct (305). Rather, he considered experience to play an important role in determining varieties of nest building. Recent studies have in fact found "many examples of birds adjusting their nest material, their building method and their nest location to local circumstances" (306). A pair of house martins in Britain, for example, used wet cement instead of the usual mud to construct possibly the strongest nest ever built (ibid.). Although learning by observation is not necessary for the normal weaving behavior of male birds, the speed of learning is higher for males who live with their parents and their peers than for those who are isolated (308–9). Nest-building, in short, is an interaction of both "innate and learnt" behaviors (308). The animals in Psalm 104 are all active agents in turning their environments into habitats, their places into niches, and with God's blessing.

The Human Niche

Ever since Darwin, evolutionary biologists have stressed the continuity between *Homo sapiens* and nonhuman animals. "[T]he difference in mind between man and the higher animals, great as it is, is certainly one of degree and not of kind," so claimed Charles Darwin in 1871.[70] Credit is due the ancient psalmist for acknowledging a certain parity of existence between human being and wild animal, as in the example of humans and lions (104:20–22). Both earn their living, and rightfully so. Darwin's statement is far from claiming that "we are psychologically and cognitively simpler than we believe we are—that we are psychologically more like 'lower' animals."[71] Quite the contrary, Darwin regarded animals as more complex, and thus more similar to human beings than previously thought. Biology does not claim a demotion of human capability but rather an elevation of animal capability, including ours. Nonhuman animals are capable of social learning and, even, of culture.[72]

Speaking of learning: the human species still has much to do. Today the earth's biodiversity is rapidly diminishing. According to the National Audubon Society and the American Bird Conservancy, habitat destruction and global warming have put more than a quarter of birds in the continental United States in jeopardy. As of the end of 2007, 59 species of birds in the continental United States are on the "red list" of greatest concern, and 119 more are seriously declining or considered rare.[73] More broadly, the rate of species extinction currently underway has reached an all time high since the demise of the dinosaurs, between 1,000 and 10,000 times faster than anything within the past 60 million years.[74] Many plant species are also on the endangered list. The recent report from Botanic Gardens Conservation International in London (BGCI) states that climate change could kill off half of the earth's botanical species.[75] Especially vulnerable are plants that grow on islands and mountainsides because they have "nowhere to go" as the climate shifts.[76]

The Sixth Great Extinction is looming. But no volcanic or meteorological activity is to be blamed for it. The psalm, in fact, yields a clue even here. Because species are matched by their habitats or niches, then the destruction of habitats invariably threatens inhabitants, whose adaptations to their environment have been naturally selected or, one might say, "finely tuned." But far from finely tuned or naturally selected, environmental degradation caused by human activity is not habitat-specific. With the accumulation of greenhouse emissions and the buildup of toxic waste in rivers, lakes, and oceans, not to mention deforestation and overexploitation, the biosphere as a whole is threatened. Humanity, Earth's consummate niche constructor, is going down in evolutionary history as Earth's preeminent niche destroyer.

Reading Psalm 104 Naturally

In the face of anxiety, the psalmist's recourse is to render praise to God. But what kind of God does the psalmist praise for creation's sake? Taken to its extreme, the anthropic principle, as Richard Dawkins gleefully points out, turns

God into a "Divine Knob-Twiddler,"[77] a deity that tweaks and twiddles the cosmic parameters to bring forth life. Such an image is not much different from the little man behind the curtain projecting the image of the "great and terrible" Wizard of Oz. Not so with the psalmist: the God behind heaven's "curtain" is the great and gracious creator of all. Anthropic principles, whether strong or weak, only go so far in the poetry of doxology, no more than they do for science.[78]

The God to whom the psalmist renders praise is the God whose pathos for life sustains all of creation. God is the providential provider in Psalm 104. Creation in the psalm is not so much a beginning point as it is an ongoing process conducted moment by moment, what theologians call *creatio continua*. The psalmist lodges this "doctrine" in the very heart of God, filling it with divine delight. The world is sustained by God's love and joy for creation. However, much in contrast to the psalmist's theological revelry, science identifies purely natural mechanisms by which life on the earth is sustained. From the Pauli exclusion principle to the perpetual carbon cycle, life continues apace by means of myriad lawful forces, cycles, and feedback loops, all internal to the world. No deity is needed to keep things running.

The only "external" source of energy critical for most life on the earth is solar. The sun provides a continual flow of energy, which is then converted, for example, by photosynthesis. The psalmist, too, recognized the life-sustaining quality of the sun by identifying light as part of God's very nature, from which all else flows. The sun, of course, is not God for either the psalmist or the scientist, but the scientist does not require God to explain the sun, which has about 7.6 billion years left before its demise incinerates the earth. For the psalmist, the sustaining power of the sun illustrates God's gracious orientation toward creation. In describing God's provision for all life, the psalmist does not point to anything that is peculiarly miraculous, anything that interrupts, disrupts, or contravenes the rhythmic regularities of nature. God's passion for creation is *naturally* realized. Through natural, "ordinary" means, through nature's wondrous workings, God sustains the panoply of life, enabling each animal and plant, each species and ecosystem, to develop and function according to its capacities and vitalities.

Although the scientist and the psalmist will forever differ regarding the *ultimate* source of creation's continuing vitality, both share one viewpoint that calls for immediate action. Life's flourishing has become tenuous. We can celebrate the anthropic principle for all we want, but the principle of life on Earth is now under siege. The disruption of nature's cycles that regulate the temperature of the earth's surface, both land and ocean, is bearing catastrophic results. As science is monitoring the impact of human industrial activity upon the biosphere, the psalmist acknowledges creation's vitality existing "on edge," namely, on God's personal investment in creation, which can be withdrawn anytime.

We are destroying precisely that which the psalmist celebrates and commends to God's enjoyment: habitats and their diverse inhabitants.[79] By eliminating habitat and inhabitant, we are diminishing creation's rich diversity, reducing creation to one big godforsaken bore and, in so doing, turning God's "Joy to the World" into God's grief for the world.

Through poetry the psalmist hopes to sustain God's biophilia or love for all things biological, a love as passionate as it is steadfast. But the ancient poet also acknowledges a tipping point within God's heart: God's passion for creation cannot be taken for granted. In response, the psalmist does whatever it takes to hold God's interest in creation by lifting up creation's vitality and diversity, all for the sake of God's delight, a delight exercised for the world's sake.

Read ecologically, the psalm claims God's biophilia as a model for humanity's role and presence in the world. Delighting in creation has nothing to do with exploiting the world for the common greed. Rather, it has all to do with receiving the world's abundance for the common good, a sufficiency to be shared, not hoarded. God does not value the world for its uniformity, as one totalizing habitat for humanity. The destruction of natural habitats across land and sea takes us, in the eyes of the psalmist, one step closer to diminishing God's joy. Preserving natural habitats for all of God's creatures is crucial. The grave irony is that totalizing humanity's habitat on Earth is tantamount to destroying humanity's habitat on Earth.

Equally important is finding ways to celebrate and enjoy creation's goodness. Ecology, the psalmist would remind us, is an exercise of joy. Delight celebrates the abundance provided in the world without turning the world into a commodity. The creatures to whom God extends life-giving provision do not respond by savagely hoarding what is received. Rather, they live and they die, fully dependent upon the God in whom they had their genesis, fully dependent upon creation through which they are sustained.

Psalm 104 places God's joy squarely upon the shoulders of human responsibility. God's delight in creation requires reciprocal engagement on the part of the creature. Call it God's "covenant" of play, of mutual engagement, whereby divine joy invites human response. God's active delight in creation only heightens human agency in behalf of creation, for it all comes down to this: to feed the flame of biophilia, both God's and ours, we must preserve and sustain creation's biodiversity. If Leviathan falls, then so do we all. If creation's wondrous variety is diminished, then the psalmist's worst fear is realized: creation left to wither away. It is incumbent upon God's most powerful creatures to ensure that divine delight is sustained so that the world be sustained. As long as the psalmist rejoices in God and God rejoices in creation, the delight shared between creator and creature continues to sustain the world. May it continue to be so.

7

Wisdom's World

Cosmos as Playhouse in Proverbs 8:22–31

Somehow the universe has engineered, not just its own awareness, but also its own comprehension.

—Paul Davies[1]

Wisdom is the never-ending journey of beginning.

—Ruth Byron Jones[2]

One of the most exquisitely crafted poems in all of Scripture is found in Proverbs, a singularly evocative passage that has captivated readers for centuries, from ancient sages and church fathers (and heretics) to feminists and ecologists. The text has been fought over in the Christological disputes of the past and the theological controversies of the present. Through no fault of its own, the text bears a bruised legacy, and we would do well to drop some of the interpretive baggage, weighty as it is, that the text has gathered over the centuries before entering its vivid world with eyes wide open, come what may. But before exploring the passage itself, we must first delve into the larger world of Proverbs in which the passage has its home.

Socio-Literary Context

Proverbs is a collection of collections of didactic sayings, from extended lectures to pithy apothegms, all brought together for the purpose of cultivating wisdom. The quest for wisdom, according to

Proverbs, is open-ended, ongoing, and never finished. Even the aged have much to learn (1:5).[3] The book, moreover, adopts a self-critical posture, indicated no less by the diverse and often contradictory sayings contained therein (e.g., 26:4–5). To take up one of its own metaphors, wisdom is a "pathway" (2:8–9; 3:17; 8:20), and as a path is formed by the passage of many feet, so wisdom is cultivated communally by those who have gone before as much as by those who currently make the trek. Wisdom's path "is like the light of dawn, shining brighter and brighter until full day" (4:18). But the "full day" that ushers in all knowledge and insight never arrives within any given lifetime.

Most scholars agree that the *book* of Proverbs was finalized in the late Persian period of Israel's history, that is, during a time of relative growth and stability. The framing units of chapters 1–9 and 31:10–31, the latest compositions of the book, reflect a prosperous, urban setting, filled with bustling street corners, marketplaces, city gates, and at least one rather spacious home.[4] These chapters provide a snapshot of urban affluence. Wisdom, not coincidentally, has her home in the city (9:1–6). She is not ensconced in a secret garden or situated beyond human reach (cf. Job 28). To the contrary, she is found in the hustle and bustle of city life (1:20–21; 8:1–3). If wisdom is a "tree of life," as claimed in 3:18, she is a distinctly urban arbor.

In addition to "her" rich metaphorical heritage, wisdom covers a range of skills, competencies, values, and virtues, all catalogued in the opening verses of Proverbs—a dense introduction that invites the reader to forge ahead and enter into Wisdom's world.[5]

> (1:2) To know wisdom and instruction,
>   to understand words of insight,
> (1:3) to gain effective instruction,
>   as well as righteousness, justice, and equity,
> (1:4) to teach prudence to the inexperienced,
>   knowledge and discretion to the young.
> (1:5) Let the wise also hear and gain erudition,
>   and the discerning acquire skill,
> (1:6) to understand a proverb and a figure,
>   the words of the wise and their enigmas.
> (1:7) The fear of YHWH is the beginning of knowledge;
>   fools despise wisdom and instruction.[6]

For ancient (as well as modern) sages, wisdom covers the ethical ideals that promote the communal good and the personal ideals that promote individual standing within the community. Wisdom incorporates both instrumental and moral values: prudence and problem solving, justice and peacemaking, self-enhancement and self-relinquishment, technical, literary, and social skills, are all wrapped in a single package.

In addition to the myriad skills and values referenced in Proverbs, there is at least one more aspect that the sages found foundational for wisdom: reverence of God. Wisdom's beginning point is "the fear of YHWH" (1:7, 29; 2:5). "Fear" does not involve terror, the kind of fear that either paralyzes or provokes conflict. Rather, wise "fear" leads to healing and wholeness.[7] Godly "fear" does not debilitate but empowers.

With reverence of the creator as its starting point, the search for wisdom is oriented toward the created order:

> YHWH founded the earth by wisdom;
>> by understanding he established the heavens;
> by his knowledge the deeps broke open,
>> and the clouds drop down dew. (3:19–20)

Here, wisdom is instrumental in the creation of the cosmos; it is reflected in creation's integrity and intelligibility. The sages discerned order, beauty, and wonder within the natural world. For them, the wisdom by which God established creation, the wisdom reflected in nature, is the same wisdom found in the bustling marketplace, city gates, and street corners. In Proverbs, cosmic Wisdom makes her home in the day-to-day world of human intercourse.

On a more down-to-earth, practical level, the book of Proverbs addresses the urgent need to educate youth in the civic duties and privileges that attend the passage into adulthood. Proverbs is all about growing up, about maturing. The first nine chapters are filled with voices of instruction and temptation. The voice of the parent, particularly that of the father, resounds throughout,[8] as well as Wisdom's voice, each addressing the reader as child (literally "son").[9] Also heard are the voices of violent peers and of the seductive "strange woman" (or "foolish woman"), who beckon the reader to follow their tempting words.[10] Amid all these competing voices, the reader is faced with a choice: whether to take the path of wisdom onward toward life or to follow the crooked, seductive path of foolishness toward disgrace and death. Wisdom's own counsel is aimed at directing the reader on the first path, the path of life that has no end but whose milestone can be found in the great creation poem in Proverbs 8.

Creation According to Wisdom

Unlike most other creation accounts in the Bible, Proverbs 8 is given not by an anonymous narrator or poet, such as in the stories of Genesis, but by one speaking unabashedly in the first person. Creation's "I"-witness is Wisdom. Proverbs 8 is her hymn of self-praise,[11] but praise with a purpose. Wisdom lifts her voice above the fray of conflicting voices, all vying for attention, to persuade the reader of her inestimable worth. By claiming her intimate association with both the creator and the creature, Wisdom hopes to capture the reader's allegiance and keep him or her on the right path.

The creation poem marks the pinnacle of Wisdom's discourse. In the first half of chapter 8, she speaks of her inviolable integrity (vv. 6–11), as well as her benefits (vv. 12–16) and preeminent authority (vv. 17–21). Personified as feminine,[12] Wisdom seeks, no less, to woo her audience. The poem in vv. 22–31 is the culmination of her self-presentation. There she bears testimony to her own genesis within the cosmic sweep of creation's genesis.

Translation

> (22) YHWH had[13] me (as) the beginning of his way,
>   the earliest of his works of yore.
> (23) Of old I was woven,[14] from the very beginning,
>   even before the earth itself.
> (24) When the deeps were not existent, I was birthed.
>   When the wellsprings were not yet laden with water,
> (25) when the mountains were not yet anchored,
>   before the hills themselves, I was brought forth.[15]
>
> (26) Before [YHWH] made the earth abroad,
>   and the first clods of soil,
> (27) when he established the heavens, I was there.
>   When he circumscribed the surface of the deep,
> (28) when he secured the skies,
>   and stabilized the springs of the deep,
> (29) when he assigned the sea its limit
>   (lest the waters transgress his decree),
>   when he inscribed the foundations of the earth,
>
> (30) I was beside him growing up.[16]
>   I was his delight day by day,
>   playing before him every moment,
> (31) playing with his inhabited[17] world,
>   delighting in the offspring of 'ādām.

Wisdom's own account of creation lacks the symmetrical character of Genesis 1 and the linear drama of Genesis 2–3. But the account more than compensates for such "deficiencies" with its poetic flair. Its fluid, artistic quality breaks all sense of linear progression. The heavens are not really created *after* the mountains (vv. 25, 27), as the presented "order" would suggest. Wisdom's revelry, moreover, is not bound to chronological constraints. She plays "every moment" (v. 30c), and in her poem, the arrow of time seems to point in all directions.

Nevertheless, timing is crucial at least in one respect. Repeated throughout the poem is Wisdom's primordial "preexistence": Wisdom was birthed[18] prior to earthly creation, and thus she is preeminently first and foremost of all God's creation. If there is anything that assumes chronological ordering in this timeless recitation of creation, it is Wisdom's genesis. She is conceived in v. 22, gestated in v. 23, birthed in vv. 24 and 25, present before creation in v. 27, and actively "growing up" and "playing" in vv. 30–31. The world's genesis is told strictly from the standpoint of Wisdom's genetic primacy. And yet creation matters for Wisdom. As Terence Fretheim aptly notes, *"wisdom needs a world to be truly wisdom."*[19] Although her origin is sharply distinguished from the world's creation, she does share an intimate bond with the "inhabited world." But I would add to Fretheim's insightful remark: in order to be truly wise, Wisdom first needs a world to be truly playful, to be truly a child.

So what kind of world does Wisdom need? One that is fully and multidimensional, replete with heights and depths, wellsprings and mountains, soil and sky, a world that is circumscribed, secured, and stable, a world whose foundations are firmly established. When the world was not yet, there was only begotten Wisdom, daughter of God. But with the world fully created, Wisdom can now play.

God the Architect and Parent

As for creation, the agency behind it is entirely God, the father of Wisdom. As an architect, God works by carving, anchoring, stabilizing, establishing, circumscribing, securing, and setting boundaries. The mountains serve as weight-bearing pillars that hold aloft the heavens, preventing cosmic collapse. God sets the cosmic infrastructures and boundaries firmly in place, all to maintain the world's stability. The universe is a cosmic construction zone in which God builds an inviolably secure place, a world firm enough to withstand the onslaught of chaos, whose boundary is established by decree. It is a world devoid of structural flaws, a world carefully designed for habitation. But whose?

Strikingly absent is any specific reference to creaturely life, except for a glancing reference in the final verse. The cosmos seems to be all bricks and mortar, with life only a by-product or endnote. But not quite. From beginning to end, the hymn pulses with the vibrancy of Wisdom, her passion, and her play. And in Wisdom's eyes God is not just a cosmic contractor. The deity of design is also a doting, playful (not to mention single) parent. As far as Wisdom is concerned, God is a parent first and a builder second.

Wisdom the Playful Child

Every step and facet of creation is graced by Wisdom's joyful presence. She is ever-present physically "beside" God before, during, and after creation. She is preeminently alive as much as she is uniquely engendered. Wisdom is life *in principium*. The poem nowhere suggests that Wisdom collaborates with God in the task of cosmic construction; she is no child laborer. But neither is she a passive observer. Far from being a spectator, Wisdom remains a player throughout, and her play serves double duty. Wisdom's activity engages *both* God and the world in the mutuality of play, holding creator and creation together through the common bond of delight. She is no child left inside. Rather, she is let loose in creation to explore and play. Wisdom is "delight" of the world.

At God's side before and during creation, Wisdom is in the know. She is the one and only "I"-witness to God's creation; she has "grown up" with creation and can thus sing about it with a deep sense of intimacy and joy. Wisdom's hymn is itself a tangible testimony to her continued delight in creation and in God. She is God's full partner in play, and creation is hers to enjoy. Wisdom is no mere instrument of God's creative abilities; she is more than an attribute, divine or otherwise (cf. 3:19). Wisdom is fully alive, interdependent and interactive with God and the world. All the world was made for her, and her delight affirms it all.

Yet Wisdom offers more than childlike delight. After her poetic revelry, we find Wisdom directly addressing her listeners:

> (8:32) And now, my children, listen to me:
> "Happy are those who keep my ways.
> (8:33) Hear instruction and be wise,
> and do not neglect it.
> (8:34) How happy is the one who listens to me,
> watching daily at my gates, waiting beside my doors.
> (8:35) For whoever finds me finds life
> and obtains favor from the YHWH;
> (8:36) but those who miss me injure themselves;
> all who hate me love death."

In her admonition for life, Wisdom addresses her audience as "children" and in so doing casts herself now as mother. By figuring herself as a child at the dawn of creation, Wisdom establishes a level of mutual identity with the child who is addressed throughout the first nine chapters, the child who serves as the stand-in for the reader.[20] But now that bond has turned maternal in 8:32, and so has her discourse. Wisdom is as much a nurturer as she is a playmate, beckoning her children to come and listen.

Wisdom also takes on the role of architect and host. In the very next chapter, Wisdom builds a house of seven pillars (9:1–2). Could it be that by witnessing God at work in creation Wisdom herself has learned to create? The juxtaposition of chapters 8 and 9 in Proverbs would suggest so. Wisdom is not only God's playmate but also God's apprentice, and by the time she is grown, Wisdom is ready to ply her trade as edifying teacher. She builds her house and invites her students to come and partake of her nourishing fare, her food for thought (9:1–6). In so doing, Wisdom instructs them on the foundational pillars of community, "righteousness, justice, and equity" (see 1:3), but with the excitement of one in love with creation.

Wisdom and the Character of Community

As the world was created for Wisdom's delight, so the world is to be conducive for Wisdom's flourishing. At the same time, Wisdom's delight, like God's joy in Psalm 104, makes possible the world's flourishing. Such delight takes on a peculiar form and function in Proverbs: it is distinctly edifying. As an object of human appropriation,[21] Wisdom, with her delight, informs humanity's role and place in the world. Her play-filled development, in fact, mirrors human development, for she is created in the *imago nati*, in the image of the growing child, and the whole world is created in the *imago domus*, as Wisdom's home. But her "edifice complex" exists not just for herself but also for creation's inhabitants, her playmates. By making her home in the city, Wisdom also makes her home in the human heart (see Prov 2:10). Her position in the world sets the context and catalyst for those who desire to grow in wisdom. Growth in wisdom does not diminish childlike wonder. Far from it: as a child Wisdom shares her wonder, and as a mother she nurtures it.

And so all the world's a stage for Wisdom's play. As the book of Proverbs reflects a certain cultural liveliness, so creation according to Wisdom offers an engaging model of acculturation. Wisdom's witness to creation testifies to the security and rich complexity of a world that sustains her growth and inspires her delight. Just as children develop fully within secure and enriching surroundings, growing into adults who can go forth in confidence to establish their own place in the community, so Wisdom matures and ventures forth to build her house and to host the community (9:1). At the same time, her own growth suggests that even the aged can rekindle the childlike delight needed to continue on the path that Wisdom sets forth. Wisdom's world coheres well with a prophet's urban vision:

> Thus says YHWH of hosts: "Old men and old women shall again sit in the streets of Jerusalem, each with staff in hand because of their great age. And the streets of the city shall be full of boys and girls playing (√śḥq) in its streets." (Zech 8:4–5)

Playing in the streets, playing in the cosmos: such is Wisdom's vocation. Born of wonder, Wisdom's play shapes and sustains the just community, her beloved community. Wisdom's homage to God and creation highlights the inhabitable and, hence, political (from *polis*, "city") nature of the cosmos, a world full of fully living agents, all thriving and playing together. Wisdom's domicile is a home that accommodates a city. It is a "cosmopolis."[22]

Scientia and *Sapientia*

Far from being a scientific treatise, Wisdom's ode to cosmic joy delights in poetic metaphor. Wisdom speaks of her "growth" in relation to divinely established creation and of creation as her divinely constructed "playhouse." Such language seems worlds apart from that of science, although perhaps not so far from science fiction. The riveting, final image of the cosmic child in Arthur C. Clarke's and Stanley Kubrick's sci-fi movie classic *2001: A Space Odyssey* (1968) could easily have been Wisdom!

According to the ancient text, the cosmos does not exist apart from Wisdom. References to the circumscribed sea and anchored mountains, to the earth's foundations and secured skies, all stress creation's stability and complexity, all for Wisdom's delight. The cosmos does not hang by a thread; it is not on the verge of collapse. That Wisdom is present at creation (8:22–29), that God "founded the earth by wisdom" (3:19), poetically claims the cosmos as wondrously intelligible. The universe exhibits a rational order, despite its innumerable variables and complexities. As physicists would point out, spacetime remains absolute, and its speed limit is universally observed (i.e., the speed of light). Creation exhibits lawful regularities—some bewilderingly complex, some bafflingly simple—that can be mathematically expressed.

Cases in point: Newton's inverse-square law, which measures the attractive force of gravity between two bodies, and Einstein's iconic equation $E = mc^2$, which describes the conversional equivalency of matter and energy. As a model of greatest simplicity, there is Hubble's Law, which posits that galaxies twice as far away from the observer are moving away twice as quickly, galaxies three times away move three times as fast, and so on. The cosmos is characterized by regularity and uniformity, frequently expressed by nothing more than mathematical constants. The laws of physics work in the same way everywhere in the universe. Physicists do not have to go back to the drawing board for every different location in space they are studying.[23] The "laws" of physics are omnipresent throughout our four-dimensional spacetime world. The cosmos suffused with wisdom is a cosmos that is comprehensible, and that itself is a wonder!

Cosmic stability and security ensure Wisdom's flourishing, analogous perhaps to Earth's stable support system for carbon-based life. But there is no "anthropic principle," at least in any strong sense, at work in Proverbs.[24] Humanity is barely mentioned. The poem's primary focus is on Wisdom. If one insists on imposing a teleological principle for creation's construction in Proverbs 8, then "sophic"[25] would be more apt. In her testimony, Wisdom pushes the "sophic principle" of creation: the cosmos is finely and firmly constructed for Wisdom's sake, for her growth, delight, and play, far beyond humanity's. The cosmos exists in delightful intelligibility.

Wisdom's Quantum Connection

Another connection with science lies not in creation itself but in Wisdom's relationship to it. Her playful spirit fills the "bricks-and-mortar" structure of creation, as reflected, virtually so, in creation's own dynamic interdependence. Such is explicitly noted in the ancient Greek translation known as the Septuagint (LXX) of v. 30:

> I was beside him *binding together.*[26]

Early in the interpretive history of this most enigmatic verse, the Greek translators assumed a distinctly cosmological outlook in defining Wisdom's relationship to creation. Something like gravity, Wisdom was considered the binding, dynamic force behind creation, ensuring its interconnectedness.

But whether one accepts the Greek translation or not, Wisdom's all-encompassing play, the Hebrew text suggests, interconnects all creation, dynamically so. At the quantum level, interconnectedness transcends even space itself. Beginning with Einstein, and later confirmed experimentally, researchers have confirmed "an instantaneous bond between what happens at widely separated locations."[27] Two objects can be far apart in space but behave as a single entity. In our everyday world, space implies independence among separated objects. But on the quantum level, "intervening space, *regardless of how much there is,* does not ensure that two objects are separate."[28] Welcome to quantum "entanglement."

Einstein was one of the first to recognize this counter-intuitive phenomenon, but he was also highly suspicious of its veracity. But through the experimental reasoning of the Irish physicist John Bell in 1964, researchers by the early 1980s gave experimental confirmation of quantum entanglement by measuring the spins of subatomic particles (clockwise or counterclockwise) once conjoined but then separated by as much as thirteen meters only to find them perfectly aligned.[29] These particles are "like a pair of magical dice, one thrown in Atlantic City and the other in Las Vegas, each of which *randomly* comes up one number or another, yet the two of which somehow manage always to agree."[30] But in the quantum realm they are not magically bound but hopelessly "entangled." No amount of space can overcome their connectedness. What Einstein called "spooky," the ancients would have perhaps called "playful."

More fundamental, Wisdom's "play" resonates with the quirkiness of the subatomic level of reality, where uncertainty is the name of the game. Wisdom's subatomic dance is more improvisational than choreographed. Whereas the universe on the grandest of scales, according to the theory of general relativity, exhibits uniformity, on the quantum level it is unpredictable. As our imaginary microscope zooms down to 10^{-33} centimeters (i.e., Planck length), space and time seem to "dissolve into a boiling mess" or what John Wheeler of Princeton refers to as "space-time foam."[31] Space consists of a "foam-like topology of bubbles connected by tunnels . . . that are continually forming and closing in Planck time."[32] At this level, Wisdom's dancers are many: more than two hundred subatomic particles have been discovered, all moving in a blur.[33] Their energetic dance exhibits an irreducible duality: sometimes they behave as particles, sometimes as waves. At this level, energy suffuses even a vacuum. Empty space can "give birth to short-lived particles that can spring in and out of existence on a time scale controlled by the Uncertainty Principle."[34]

In his or her own way, the ancient poet captures, virtually so, this paradox: the orderliness of the universe contrasted with its liveliness. To the physicist, the world exists on two very different (and so far irreconcilable) planes: the level of macroscopic order, in which the laws of both Newton and Einstein have their predictive values, and the quirky world of quantum physics. Reconciling both realms, whether experimentally or mathematically, remains elusive. Nevertheless, accounting for the explosive birth of the universe from its smallest beginnings demands a resolution. The ancient poet finds both Wisdom's liveliness and creation's orderliness as more than just coexistent; they are mutually supportive. Wisdom's passionate play is manifest within the setting of cosmic order, and cosmic order is set to serve Wisdom's lively involvement. Yet amid these two contrastive levels, of play and stability, a certain "historical" primacy is evident. In the beginning was playful Wisdom, just as one could say about the birth of the cosmos. In the beginning were lively quantum fluctuations. Both modern physics and ancient poetry discern a primordial bonding between these two disparate realms.

Wisdom's Cosmic Dance

On the macroscopic level as well, Wisdom's play summons, as it were, all to join in her cosmic dance. Her play throughout the universe introduces cosmic movement within the steadily accelerating expansion of space. Astronomers frequently refer to stars and galaxies "dancing" in relation to each other, such as the Perseus Galaxy cluster, wherein "whole groups of billions of stars orbit each other in a grand celestial waltz."[35] One thinks, again, of Kubrick's *2001: A Space Odyssey*, in which a docking procedure between a rocket and a space station is accompanied by the "Blue Danube Waltz" of Johann Strauss. So also the celestial populace of the cosmos: binary stars twirl around each other at dizzying speeds, galaxies gyrate like slow-motion hurricanes, star clusters slam dance into each other. Such complex cosmic movements are all choreographed by gravity, which binds the universe together while rippling through it, warping space and time.

That is all to say, nothing stands still in the universe. "Stationarity" does not exist. "Everything in the Universe pulls on everything else,"[36] including our solar system, which takes a mere 240 million years to orbit the Milky Way's center, even as our galaxy is eventually destined to partner up with the Virgo cluster. The sun, too, dances to its own rhythm as influenced by the gravitational pull of its varied planets. "The Sun's net motion . . . amounts to a superposition of orbital dances, each with a different cycle period of repetition."[37] Jupiter, for example, changes the sun's velocity by about forty feet per second as it performs its twelve-year orbit, thereby prompting the sun to perform its own little orbit. While the planets twirl dizzyingly around the sun at varying speeds, the sun prefers to slow dance. So it is with most stars. By measuring the movement of a particular star's "dance" (through Doppler shifts), astronomers can determine the relative size of its planets and their orbits. DeGrasse Tyson and Goldsmith call this the "star dance method."[38] In the cosmic scheme of things, gravity is not a "downer"; it is none other than the choreographer of the cosmos.

From quarks to quasars, the lively interactions that occur on every order of magnitude the ancient sage could very well attribute to Wisdom's cosmic play. A "still" night, for example, is an illusion. A long gaze into the night sky is merely a strobe's flash capturing one brief instance of a protracted cosmic dance whose steps we are just beginning to figure out. Through gravity, the universe is filled with movement and interaction, collisions and collapses, stellar deaths and births. Through nuclear and electromagnetic forces, particles or waves collide and separate, interfere or amplify, exchange energy and matter, wink in and out. It is enough to keep any child of Wisdom, regardless of age, hopelessly enthralled.

Wisdom's Growth in the Cosmopolis

It may be no coincidence that gravitational theorist Lee Smolin likens the universe to a city, a place of "endless negotiation, an endless construction of the new out of the old, . . . where novelty may emerge without violence."[39] The metaphor is not new. The great Jewish philosopher Philo of Alexandria (ca. 20 BCE–ca. 50 CE) draws from Genesis 1 to show how creation is like a great cosmic city, a *megalopolis*.[40] Older still, the sages of Proverbs celebrate Wisdom's revelry in the inhabited cosmopolis. Similarly, Smolin inscribes the universe as a thriving city, freely mixing the cultural and, if one reads closely, the biological.[41] Of course, it's all metaphor, but not *just* metaphor. While lacking explanatory power, a metaphor still points to levels of truth that elude the grasp of scientific theory and mathematical formulas.

Case in point: "Wisdom," so neuropsychologist Warren Brown states, "is, by nature, not conscious of itself."[42] But according to Proverbs, she most certainly is conscious, albeit metaphorically. The poetic conceit of self-conscious, playful, growing Wisdom makes the rather obvious point that the exercise of wisdom presupposes consciousness and conscience. In addition, Wisdom's personified nature is a clear case of metaphor cast with a didactic purpose: Wisdom's growth in God's cosmopolis parallels human growth in community.[43] Her ontogeny recapitulates human ontogeny.

How do the sciences describe human growth in wisdom? From a neurobiological perspective, human growth involves heightened cognitive development, which for Warren Brown includes an integrated use of both hemispheres of the brain, right and left. For example, a "deficit in interhemispheric transfer" is well documented in the condition of Agenesis of the Corpus Callosum (ACC), a case of "split brain" syndrome.[44] The corpus callosum is the major "pathway between the cerebral hemisphere[s]."[45] Damage to it, Brown has found, leads to impaired social functioning, poor social judgment, and diminished problem solving abilities.[46] Comparable to ACC is frontal lobe damage to the brain, manifested behaviorally by the disassociation between thought and affect, as well as between words and actions. Individuals who have suffered frontal lobe damage, as in the famous case of Phineas Gage, exhibit a lack of proper decision-making ability.[47] Brown and others continue to identify the neurobiological properties necessary for cognitive development without reducing growth in wisdom to strictly physical roots. Their studies confirm that the exercise of wisdom involves an integration of cognition, emotion, and social functioning.

From an evolutionary perspective, growth in wisdom is interpreted within the context of adaptation. If wisdom is a matter of existing optimally within the order of creation, then even the notion of biological "adaptation" should find some resonance with growth in wisdom. At the most basic level, organisms develop traits through natural selection in order to better "fit" with their environment and thereby achieve "achieve functional efficacy,"[48] particularly for reproductive success. But this is not wisdom at work, which requires active, self-conscious discernment. Descent with modification strictly confined to random genetic mutations and natural selection involves no initiative on the part of the organism.

Learning, however, does. The Harvard evolutionary biologist Henry Plotkin refers to the kind of learning that is associated with intelligence as a secondary biological "heuristic," a means of discerning new ways of fitting, of generating new adaptations to the environment in the face of change.[49] Learning cannot be reduced to the primary biological heuristic of genetic adaptation.[50] Learning, moreover, exhibits a much shorter response time to new environmental stimuli than does genetic adaptation, which takes generations of reproductive succession.[51] "Once intelligence has evolved in a species, then thereafter brains have a causal force equal to that of genes."[52]

There is also a "tertiary heuristic" that is distinctive of the human intellect, according to Plotkin,[53] and its basis is entirely cultural. Sometimes referred to as "meta-cognition," it represents the cultural function of the intellect, the most complex form of cognition, for "culture is the most complex thing on earth."[54] It is a heuristic that rises above the ability to solve problems (i.e., learning) and the biological need for reproductive success (i.e., genetic adaptation).[55] Meta-cognition is not just another domain-specific intelligence, or another facet of the mind's "modularity." It is, rather, a distinctly integrative operation; meta-cognition "integrate[s] multiple representations of the world," resolves social conflict, and fosters cultural innovation.[56] Cognitively, meta-cognition brings together various modes and modules of mental intelligence,[57] not unlike Steve Mithen's notion of "cognitive fluidity."[58]

The biblical concept of wisdom clearly shares much in common with "meta-cognition," for it too draws from many different dimensions of human judgment. Indeed, wisdom casts the net even wider to include not only cognitive skill and social insight but also religious and moral valuation, as indicated in the opening verses of Proverbs. Proverbs raises the heuristic function of wisdom to a level beyond self-enhancement and cultural adaptation (Prov 2:4; 3:14; 8:10, 19). Wisdom seeks both the common good and the common God; it fosters reverence of the creator of all and cultivates "justice, righteousness, and equity" (1:3). The myriad proverbs in the book present various, even conflicting, views about life and the world, all to develop the reader's critical, meta-cognitive intelligence, that is, to cultivate wisdom.

In addition to the strictly cognitive elements involved in the growth of intelligence, developmental psychology has shown that emotion plays a critical role in the acquisition of knowledge. Maturity is not only measured cognitively; it can also be evaluated in terms of emotional sensibility. Emotions shape how we perceive the world. Fear finds the world filled with threat; depression overlooks the good things in life; bliss neglects the storm clouds on the horizon.[59] "Simply put, emotions are there in order to tell us what to think about; our hearts not only try to rule our heads, but should perhaps be allowed to do so."[60] Drawing from Aristotle and contemporary moral theory, Nancy Sherman identifies various interrelated roles that emotions play in the development of the self: they serve as "moral antennae," "modes of broadcast," and motivations for action.[61] "A world without humor, laughter, playfulness, as well as aggression and fear, would simply be impoverished, let alone unrecognizable as human."[62] Emotions are "intentional states"; they form part of the individual's character as much as do belief and reason.[63] Emotion, thus, takes its rightful place alongside meta-cognition as a critical component of wisdom. Biblical wisdom is no bloodless abstraction. Wisdom is as fully emotive as she is cognitive. It is by her that kings rule and children play (Prov 8:15–16, 30–31).

Wisdom's Maternal Play

In just a few short verses, Wisdom's childlike delight shifts to motherly love as she addresses her "children" with the lessons of life (8:31–36). The transition, though abrupt, is deliberate. Her maternal side is no accident, even if taken metaphorically, for Wisdom fills a critically important maternal role in human development. When it comes to the evolution of learning, the mother takes the lead.

The maternal role of teaching is practically universal among mammals, from the simple fact that the young must be nursed by the mother for its survival. Up until the time of weaning, mother and infant form an intimate bond that, at least among primates, involves a period of "voluntary isolation."[64] While such isolation affords protection from "cannibalism, predation, and disturbance," it also "enhances the mother's impact as a 'teacher' at a time when the capacity of her offspring to learn is at its peak."[65] In primate evolution, the relative size of the neocortex increases as higher cognitive functions are reached. Biologists have noted that the size of the neocortex is proportionally related to group size: as group size increases, so does that of the neocortex.[66] But there is a clear gender distinction to be made in this dramatic evolutionary development: it is the size of the *female* group, not the male group, that makes the difference. Because females are the primary caregivers and because they constitute the "stable nucleus" of the group among primates, it is the "size of the female group that exerts the greatest selection on the evolution of intelligence."[67] Given this "cultural" fact, biologists Eytan Avital and Eva Jablonka conclude that the "maternal transmission of behaviors may have driven the evolution of intelligence in primates."[68] And it continues to do so.

Identifying another evolutionary feature of the maternal bond, Ellen Dissanayake locates the origins of art in the special "rhythms and modes" that characterize the social interactions between mother and infant.[69] The evolutionary development of artistic creativity, she claims, began with the inborn capacity and need for "mutuality" between mother and infant, whose bonding sets the stage for intimacy. Because human infants are helpless for a much longer time after birth than infants of other species, they require longer attention and care.[70] This evolutionary distinction occasions greater mutual interaction between human mother and child, thereby providing the foundation for various capacities, including "belongingness," "finding and making meaning," "acquiring a sense of competence through handling and making," and "elaborating these meanings and competencies."[71]

Such acquired capacities are the "legacies of mutuality,"[72] and it is the last capacity, "elaborating," that according to Dissanayake gives rise to the arts. "Elaborating" is "making special," a characteristic of human nature that extends well beyond the aim of survival.[73] The arts of "chant, song, poetry, dance, and dramatic performance . . . [are] multi-media *elaborations* of rhythmic-modal capacities" developed in the mutual bonding between mother and child.[74] In short, "baby talk" between mothers and infants is more than meets the eye and ear. The communication that transpires uniquely between mother and child is rendered both bodily and verbally. It is "exquisitely" aesthetic.[75] Its rhythmic, "patterned sequences" are mutually sustained by the facial, verbal dance between infant and mother. In other words, play has its pedagogy.

Dissanayake's profound thesis lends added significance to child Wisdom's relation to the God who gave her birth and to mother Wisdom's relationship with her "children." The intimate, playful bond between God and Wisdom, like that of mother and child, enables Wisdom's growth. As an artist herself, indeed a poet (8:4–36) and architect (9:1), Wisdom demonstrates the results of her growth and development, her learning outcomes, as it were. And specifically as a mother, Wisdom nurtures the growth and development of her children. Not only does Wisdom's mythic preexistence before creation place her as teacher above all teachers; her maternal side confirms her central pedagogical position. In primate development, it is the mother who remains the primary teacher.

Read in this light, biblical Wisdom's figuration as both child and mother takes on added developmental, evolutionary, and cultural depth.[76]

Revisiting Wisdom's World

Taking my cue from the scientific quest to identify the physical and cultural conditions that are conducive for cultivating wisdom, I return to Proverbs with the question: What in Wisdom's world fosters her growth and, in turn, a person's growth in wisdom?

By her own testimony, Wisdom revels in a world that is both secure and interactive, a world of discovery and delight. There is no chaos lurking around the corner, or under the bed. It is a world in which fear is banished and joy reigns. As a child actively acquires wisdom by interacting with her environment, so Wisdom actively engages creation in her exercise of delight. It is easy to imagine Wisdom's joy stemming from her discovery of the world's wondrous complexities and her interactions with its marvelous inhabitants. Wisdom's delight derives from the dancing cosmos and fluctuating quanta, and everything in between. It is also our delight.

The joy of discovery is not the only source of Wisdom's delight. Were that the case, Wisdom's growth would simply be an exercise in cognitive development, a mere mental ascent to the meta-cognitive level. Her development is also suffused with the pleasure of play at every dance step along the way, "at every moment." The world is not just an object of Wisdom's delight; it is a living, active subject in its own right. Sapiential "play" is a *shared* enterprise. As in the case of God and Leviathan in Psalm 104, play requires partnership, and Wisdom has two partners: God and creation. Her world is more relational than referential. Who else, in addition to the "offspring of *'ādām*," occupies creation for the sake of Wisdom's delight? Frolicking coneys, roaring lions, breaching whales, and flapping ostriches? They, too, inhabit creation, and thus have a right to play. And then there is God, with whom Wisdom shares a particularly intimate relationship. As God's partner in play, she is "beside" the creator of all as she is beside herself in joy.

To live in Wisdom's world is to experience the joy of discovery, the delight of discernment, and the thrill of edifying play. To live in Wisdom's world, the sages claim, is to walk the path she forges as a child. Wisdom's path is the journey of discernment in which what is discovered and what is revealed come together. As Wisdom's growth begins in joy, may the wide-eyed delight of children never be lost on the wise. For in Wisdom's eyes there really are no grownups. The quest for wisdom is ever ongoing, and progress on the path will always be marked with baby steps.

8

The Dying Cosmos

Qoheleth's Misanthropic Principle

[T]he whole temple of man's achievement must inevitably be buried beneath the debris of a universe in ruins.

—Bertrand Russell[1]

For everything moves, my friend.

—Galileo (via Bertolt Brecht)[2]

Ecclesiastes contains the most unconventional perspective on creation in the Bible. In fact, the entire book, except for its ending, is so unorthodox that it has been called the "strangest book in the Bible."[3] It is also the most enigmatic.[4] Little is known about the author. Properly speaking, he bears no name. His allegedly royal pedigree ("son of David") vanishes after the first two chapters.[5] The only clue to the author's identity comes from his self-designated title *qōhelet*, which means something like "assembler." Here, too, mystery reigns. In context, the verbal root can mean one of two things, or perhaps both: "convene" an assembly consisting of, perhaps, students,[6] or "collect" things such as wisdom sayings (see Eccl 7:27). Both roles, in fact, suit Qoheleth well. Like an auditor, Qoheleth takes an inventory of life by collecting and codifying the "data" of experience, both individual (his own) and collective (tradition). And like a teacher, Qoheleth candidly shares the results of his work to an expectant audience.

In either role, Qoheleth presents himself as the consummate examiner of life, the Bible's most rigorous empiricist. Employing his intellect to examine reality,[7] the sage frames his insights largely as observations.[8] Like Socrates, he is driven by the conviction that "the unexamined life is not worth living." And who most qualified to examine life itself but Qoheleth! With the best of credentials and the greatest of expectations, the sage begins his heroic quest for the meaning of life. He returns, however, empty-handed, much to his grave disappointment.[9] Like Saul Bellow, this sage attaches an addendum to the Socratic motto: "But the examined life makes you wish you were dead."[10] In the end, Qoheleth finds more to dissemble than to assemble.

Socio-Historical Background

The book of Ecclesiastes wears its socio-historical context on its sleeve. The presence of foreign loan words in the text indicates the Persian period (ca. 539–333 BCE) as the book's earliest possible dating. This was a time of social reconstruction and economic growth but also of great uncertainty, all spilling into the Hellenistic age inaugurated by Alexander's conquest of the region in 333 BCE. Qoheleth's observations reflect the social anxieties of his day. In contrast to the largely subsistence, agrarian-based economy of preexilic Israel, the economy of the Persian period became increasingly commercialized beginning in the fifth century. A standardized monetary currency was introduced for the first time in order to facilitate commerce between Egypt and Persia. An aggressive system of taxation was enforced under Persian hegemony. Consequently, a new market-driven economy of global proportions emerged, complete with myriad entrepreneurial opportunities. Yet such rapid growth did not benefit everyone. To be sure, those who had extensive capital outlays possessed unprecedented opportunity for cultivating greater assets. But those of lesser means were at a distinct disadvantage. The prospect of financial gain in the Persian and Hellenistic periods was both alluring and elusive; the rewards were great, but so were the risks.[11]

Creation According to Qoheleth

It is no surprise, then, that Qoheleth opens his cosmic reflections with a leading question about toil and gain (1:3). Although the answer is deferred for more than a chapter (2:11), the sage builds up to it by offering his observations on creation.[12] Not a creation account per se, 1:2–11 nevertheless provides a probing "snapshot" of the cosmos. Ecclesiastes, moreover, concludes with further cosmological reflections in 12:2–7. Creation, thus, frames Qoheleth's reflections on the purpose and meaning of life. Despite his dispassionate tone,[13] the sage is no disinterested observer of the world. His initial question about material gain strikes at the heart of human purpose (1:3). Toil is the effort exerted for gainful living, and questioning its value places all creation in question. As creation has a cosmic stake in the pursuit of gain, so Qoheleth has a personal stake in creation's purpose.

Creation without Pause and Effect: Ecclesiastes 1:2–11

With that said by way of introduction, we now catch our first glimpse of Qoheleth's cosmic purview in Ecclesiastes 1:2–11.

> (1:2) "Vanity of vanities," says Qoheleth,
>> "vanity of vanities! All is vanity!"
> (1:3) What gain[14] does one get from all the toil
>> at which one toils under the sun?
>
> (1:4) A generation goes, and a generation comes,
>> but the earth remains ever the same.
> (1:5) The sun rises, and the sun sets,
>> panting[15] to the place where it rises.
> (1:6) Blowing[16] to the south and rounding to the north,
>> round and round goes the wind, and on its rounds
>>> the wind returns.
> (1:7) All streams run into the sea,
>> but the sea is never filled.
>> To the place where the streams flow,
>>> there they flow again.
>
> (1:8) All words are wearisome,
>> more than one can express.
>> The eye is not satisfied with seeing;
>>> the ear is not filled with hearing.
>
> (1:9) What has been is what will be,
>> and what has been done is what will be done.
>>> There is nothing new under the sun.
> (1:10) If there is a thing of which it is said, "See this; it is new!"
>> it has already been, in the ages before us.
> (1:11) There is no remembrance of those before (us),
>> nor of those who will come after.
>> There will be no remembrance of them
>>> by those who will come afterwards.

Generations come and go, the sun rises and sets, the wind blows hither and yon, and the streams flow perpetually, all the while Earth and sea remain unchanged. The sage observes the weary "revolutions" of the sun, whose "panting" to the place of its rising perhaps plays a part in the wind's unceasing circumambulations. In any case, sun, wind, and streams are all set in constant motion, all returning to where they began and, without pause, continuing on. There is no breather for the sun. The perpetual cycles exhibit neither beginning nor ending, much less a "new" beginning.

For all the constant motion that characterizes the cosmos, one would think that some measure of progress is attained. But no. Even as the millennia pass, any semblance of progress is a mirage. Activity abounds, but nothing is achieved. Like a hamster in a wheel, no destination is reached. Such frenetic movement amounts to only running in place. All this cosmic kinesis is for naught. Ever in motion but never changing, the cosmos is uniformly indifferent to human living, from birth to death. This is a world without pause and effect, a world without history and, as we shall see, without a future.

The same can also be said of human agency, according to Qoheleth. As the sea is never filled, so human yearning ("eye" and "ear") is never satisfied (v. 8b). For all their efforts, the people of past generations will be forgotten by those of subsequent generations (v. 11). Indeed, the same fate applies to every generation. Nothing of significance is left for posterity. Establishing one's legacy is a futile venture. As the world turns, as the cosmic wheels spin, life is, to quote the poet Edna St. Vincent Millay, not "one damn thing after another—it's one damn thing over and over."[17] Both natural and "man-made" history are doomed to repetition, much like the sun and the wind. The past is the future; "there is nothing new under the sun" (1:9). Any "new" thing is simply a replay of the past, at most a slight variation. Change lies only in the tunnel vision of the beholder. Whether true or not, Qoheleth's claim of life's sameness indicates just how far the sage is able to step back and paint a picture of totality, a still life on a peeling canvas.

For Qoheleth, the cosmos moves on its own frenetic inertia, with human history mirroring its lifeless movements. Any sense of wonder passes as creation presses on in wearisome repetition. The melodious music of the spheres is, in Qoheleth's ears, mere cosmic cacophony. No progress is forthcoming, no upcoming crescendo. Accompanying the lack of newness in cosmic history is the lack of memory in human history. Our amnesia of the past is incurable (1:11). With the passing of each generation, memory is by and large wiped clean (cf. 9:5). Qoheleth is not so much claiming that human beings are entirely oblivious to the past as he is undercutting their deepest and vainglorious aspirations to secure a permanent place in history. A life oriented toward ensuring its legacy only pursues the wind, for the past cannot be remembered anymore than the future can be controlled. Likewise, the sage concludes that creation itself has gone bust. For all the energy expended in creation, nothing is gained and everything is to lose.[18] The same goes for the human pursuit of gain. Like cosmos, like humanity.

Creation according to Qoheleth is emptied of *telos* and filled with toil, a cosmos without direction and deprived of its own genesis. Elsewhere in the Bible, genesis and purpose are inseparably wedded. But there is nothing, properly speaking, *creative* about Qoheleth's cosmos. Indeed, God appears not to be involved. Whereas the great creation traditions of Genesis, Psalms, Isaiah, and Job boldly claim the world as *created*, wrought by a beneficent deity, Qoheleth's cosmology, for all intents and purposes, excludes cosmogony. As there is no beginning, there also seems to be no point. Qoheleth's world is a creation void of creation, and *hebel* is its name (1:2; 12:8).

Hebel or "vanity" is the most used word in Ecclesiastes, and for good reason. It is the book's single-word thesis. Throughout his reflections, Qoheleth presents one example after another of life's *hebel*, from the cosmic to the personal. The word itself conjures the image of "vapor," something entirely insubstantial, perhaps even noxious.[19] "Vapor of vapors; all is flatulence" could be the sage's literal cry of despair in 1:2. And yet *hebel* bears a rich and varied function in Qoheleth's discourse. There are many forms of *hebel* in Ecclesiastes, for the term can be translated in a number of related ways: futility, transience, absurdity, farce, and shit all have been proposed. But regardless of its specific nuance within a given context, it is indubitable for Qoheleth that *hebel* happens, inexorably. The incessant cycling of the elements is for him the stellar example of "vanity," a cosmic exercise of futility that eventually proves ephemeral.

Death of Creation: Ecclesiastes 12:1–7

If the world in chapter 1 is a cosmos running in pointless perpetuity, then the world according to chapter 12 is a cosmos running on empty. As the generations come and go (1:5), so creation as a whole will eventually go, never to return. For Qoheleth, creation may not have had a beginning, but it surely has an ending.

> (12.1) Remember your creator in the days of your youth,
>   before the days of trouble come,
>   and the years draw near,
>   when you will say, "I have no pleasure in them";

(12.2) before the sun darkens, even the light,[20] as well as the moon
      and the stars,
         while the clouds return with rain;

(12.3) in the day when the guards of the house tremble,
        while the strong men cower,
      and the women who grind stop because they are few,
        while those who look through the windows see dimly;
(12.4) when the double-doors on the street are shut,
        while the sound of the mill is low,
      and one rises at the sound of a bird,
        while all the daughters of song are brought low;
(12.5) when one is afraid of heights,
        and terrors are in the road;
      the almond tree blossoms,
        the locust drags itself along and desire fails.
      Yes, humans go to their eternal home,
      and the mourners will go about the streets;
(12.6) before the silver cord is snapped,
      and the golden bowl is broken,
     and the jar is broken at the fountain,
      and the vessel[21] broken at the cistern,
(12.7) and the dust returns to the earth as it was,
      and the life-breath[22] returns to God who gave it.

Creation's demise is marked by cosmic darkening (12:2) and the cessation of life (v. 7). The end of natural history marks a return to creation's preexistent state before the *'ādām* was created from "dust" and infused with God's breath (see Gen 2:7). Between darkness and dust, the sage employs other images ranging from the domestic and the commercial to the natural. Some interpreters, from rabbinic times to the present, have understood these images as allegorical references to the aging body.[23] Reference to grinding women (Eccl 12:3) suggests tooth loss. The cowed "strong men" (v. 3) allegedly represent the bent back. The blossoming almond tree points to gray hair. Failing eyesight, insomnia, deafness, physical imbalance, and impotence are all considered to be allegorically featured in this coded narrative.

A strictly allegorical reading, however, tends to overlook the sheer variety of images employed, most of which can stand very well on their own. Qoheleth has chosen such a diverse array in order to demonstrate how death affects *all* areas of life, from the cosmic and the commercial to the domestic and the individual. The "panting" sun of 1:5 suffers burn out in 12:2. The toiling self of 4:8 is dead and buried in 12:5–7. The world's end is no apocalyptic over-throw.[24] It happens with gradual darkening and diminution. The perpetual cycles and swings of creation's regularity unwind remorselessly. The world passes away in cosmic dissolution and bodily deterioration. The kinetic leads to the kenotic. Energy expended is energy dissipated. Call it existential entropy. *Hebel*.

In between these cosmic bookends (1:3–11; 12:2–7) are various reflections on creation. Qoheleth, for example, finds no ultimate distinction between animals and humans.

> I said in my heart, "As for human beings, surely God has tested them[25] to show[26] that they are themselves[27] only animals. For the fate of humans and the fate of animals are the same. As one dies, so dies the other. They all have one breath, and humans have no advantage over the animals, for everything is vanity (*hebel*). All go to one place; all are from the dust, and all will return to dust. Who knows whether the life-breath (*rûaḥ*) of humans ascends on high and the life-breath (*rûaḥ*) of animals descends to the netherworld? (3:18–21)

Qoheleth has no doctrine of the *imago Dei*. Humans *are* animals, as evidenced in their ignominious death. All animals share common breath or "spirit" (*rûaḥ*) in life, and their fate is equally sealed in death. As for any distinction emerging after death, Qoheleth remains utterly agnostic. Regardless, humans "have no advantage over the animals."

Life without such "advantage" is life bereft of purpose or, conversely, life beset by "time and accident."

> Again I saw that under the sun the race does not belong to the swift, nor the battle to the strong, nor bread to the wise, nor wealth to the intelligent, nor favor to the skillful, for time and accident[28] befall them all. For, indeed, one cannot predict his (or her) time. Like fish taken in a cruel net, and like birds caught in a snare, so humans are snared at a time of calamity, when it falls upon them suddenly. (9:11–12)

Life cannot be planned, let alone controlled. Instead, life afflicts the living in ways that are cruelly unpredictable.

From Qoheleth's perspective, life's unpredictability remains constant in a world that seems to recycle itself *ad infinitum*, if not *ad absurdum*. Qoheleth's world is replete with paradox. On the one hand, the cosmos in its incessant cycles is wearingly predictable. On the other hand, life within that world is woefully unpredictable. Regularity and chance, repetition and surprise, stability and frailty all have their place in Qoheleth's ambivalent world. This sage would be hard pressed to declare creation "extremely good" (Gen 1:31).

As the Pendulum Swings: Ecclesiastes 3:1–9

By way of summary or synthesis we come to the most well-known passage in Ecclesiastes, the one that describes life's "seasons":[29]

> (3:1) For everything there is a season,
>   and a time for every matter under the heavens:
>
> (3:2) a time to bear[30] and a time to die;
>   a time to plant and a time to uproot what is planted;
> (3:3) a time to kill and a time to heal;
>   a time to break and a time to build;
> (3:4) a time to weep and a time to laugh;
>   a time to mourn and a time to dance;
> (3:5) a time to throw stones and a time to gather stones;
>   a time to embrace and a time to refrain from embracing;
> (3:6) a time to seek and a time to lose;
>   a time to keep and a time to throw away;
> (3:7) a time to tear and a time to sew;
>   a time to be silent and a time to speak;
> (3:8) a time to love and a time to hate;
>   a time of war and a time of peace.
>
> (3:9) What gain does the worker have from toiling?

If the life of the cosmos in chapter 1 runs like a spinning wheel going nowhere, human life in chapter 3 resembles a swinging pendulum. Life and death, love and hate, war and peace are the poles within which all of life oscillates. In an age of peace, one can count on the advent of conflict, and vice versa. Like the sun's "revolutions" and the wind's circumambulations, life swings incessantly back and forth, never stationary but never advancing. The modulated "swings" of human activity match the perpetual "cycles" of cosmic conduct. And so life meanders indifferently between gain and loss, prosperity and adversity. The lesson for Qoheleth? "In the day of prosperity be joyful, and in the day of adversity consider. God has made the one as well as the other, so that mortals may not find out anything that will come after them" (7:14). Qoheleth concludes this

"seasonal" poem by raising the same question posed at the beginning of his
cosmic poem (3:9; 1:3): What gain is there in all the toiling, in all the oscilla-
tions? Apparently none.

Qoheleth's self-contained poem (vv. 1–8) is given theological commentary
in the subsequent verses. Two verses in particular mark the sage's attempt to
arrive at a comprehensively cosmic perspective.

> I have seen the business that God has given to everyone to be busy
> with. He has made everything apt for its time. Moreover, he has put
> (a sense of) timelessness[31] into their minds, such that they cannot
> determine what God has done from beginning to end. (3:10–11)

From the clockwork vagaries of human life, coupled with the relentless cycles
of the cosmos, Qoheleth discerns an inscrutable elegance to the total picture.
An entirely positive term, the Hebrew word for "apt" (yāpeh) can also mean
"beautiful" in other contexts.[32] Here is an elegance that extends beyond human
understanding. A generation comes, a generation goes, and it is all suitable,
elegant, apt. But to whom? Only God.

This picture of constant undulation embraces both life and death, giving
equal value to each in the grand scheme of things. God has ordained them
both. The swings and cycles are suitable only in God's mind, which remains
closed to human observation, let alone adjudication. The beginning and the
end lie beyond human ken. Hence, the sense of "timelessness" implanted by
God comes with a confounding limitation: it clouds the human mind, preventing any discernment of God's purpose. The wearying cycles of the cosmos
described in the chapter 1 and the modulated swings of life given in chapter 3
together create a sense of timelessness that obscures rather than illumines,
concealing what God is truly up to.

Theologically, Qoheleth holds a high view of the deity, so high that in the
sage's eyes God ordains everything that happens, yet remains vastly inscrutable
(see 3:14–15). All attempts to grasp God, like grasping gain, result in failure. The
elusive God is the norm. One may recall the primal couple's attempt to grasp
divine status. The result was expulsion. For Qoheleth, the result of striving for the
unattainable is tired resignation. Neither bone-chilling terror nor benign indiffer-
ence, only wearied bewilderment is Qoheleth's response to the inscrutable apt-
ness of God's ways. God "shows the world a steely countenance," notes Michael
Fox in his study of Ecclesiastes.[33] Try as he might, the sage cannot detect a glimpse
of divine expression, at least from the vantage point of creation's big picture.

As for creation's big picture, the sage has sapped all sense of wonder from
it. Gone is the rapturous amazement with which his biblical peers viewed the

world. Gone is the sense of mystery that elicits astonishment and praise. To be sure, Qoheleth acknowledges a certain mystery about the creation, but it is the kind that prompts vexation rather than wonderment. It is the kind of impenetrable mystery that the sage, perhaps like any good scientist, would rather eliminate, but cannot. For him, mystery has darkened rather than illuminated the world. Qoheleth's awe of God and the world oscillates between exasperation and resignation.

The Recycled World of Science

Despite, or perhaps because of, its rather unorthodox take on creation, Ecclesiastes offers much that resonates with science, while science, in turn, underlines and broadens the sage's view of the world. The world according to Qoheleth is one that is cycled and recycled. Such a perspective leads to a radically broad view of the past, something that in the history of modern science geology was the first to claim. Nature's perpetual recycling prompted the Scottish geologist James Hutton (1726–1797) to remark: "we find no vestige of a beginning—no prospect of an end."[34] The cyclical present can be projected indefinitely into the past, and the result is in an overwhelming sense of the world's agedness, even "timelessness." To quote another geologist of the past, Archibald Geike (1835–1924), "the present is the key to the past."[35] Specifically, the cyclical processes that define the physical present also define the past. The primacy of the present leads to the precedence of the past and, as we shall see, the dissolution of the future. Qoheleth finds nothing significantly different between past and present, hence no beginning. Qoheleth's world, in other words, is closed, locked without a key.

Precedence of the Past

Geology, of course, is not the only science that has discovered deep connections between past and present. The recycling of the past into the present is also fundamental to biological evolution, a process of change based entirely on what "has already been," to quote the sage (Eccl 1:10). Sounding a bit like Qoheleth himself, paleontologist Neil Shubin points out that anything that is "innovative or apparently unique in the history of life is really just old stuff that has been recycled, recombined, repurposed, or otherwise modified for new uses."[36] Our bodies, for example, are the legacy of our past, and many of the quirks and glitches of human anatomy can be traced back evolutionarily to something old. The strange, circuitous routes taken by the nerves in our body are testimony to our convoluted past.[37] Knee problems plague human beings because of our watery origins: fish were not designed to walk.[38] Hiccups can be traced back to gill-breathing among amphibians, specifically tadpoles.[39] Hernias in males owe their origin to the descent of the gonads, which created a weak spot in the body wall.[40] Shubin likens the human body to a Volkswagen Beetle that has been "jerry-rigged" to run at 150 miles per hour. Similarly, our bodies can be tweaked and modified only so far before significant problems arise. Evolutionary innovation is mere modification and combination of "just old stuff." Nothing is truly new.

Cycles of Life

In addition to the primacy of the past, Qoheleth identifies various cycles that seemingly operate in perpetuity. Science, in fact, discerns numerous cycles at work. Although the sun does not revolve around the earth (contra Eccl 1:5), cyclical regularity is the norm of life, whether measured scientifically with precision instruments or gauged visually with the naked eye. Generations of life come and go, and as they return to the earth ("dust"), they help make possible new generations to thrive, ensuring the delicate balance of life. Everything about life on the earth is sustained by the perpetual recycling of carbon atoms. The carbon cycle reaches the heights of the earth's atmosphere even as it plumbs the depths of the hydrosphere and geosphere, all by means of a network of interlinking pathways that include plants, seawater, animals, sediments, and underground fossil fuel. Photosynthesis ($6CO_2 + 6H_2O \rightarrow C_6H_{12}O_6 + 6O_2$) and aerobic respiration ($C_6H_{12}O_6 + 6O_2 \rightarrow 6CO_2 + 6H_2O$) make possible the mutual exchange of carbon dioxide and oxygen, enabling animals and plants to flourish together. The carbon cycle spans life from breath to death, all in an intricate balance.

Think, too, of the water cycle. By the sun's heat, H_2O evaporates from the surface of the ocean, its vapor rising into the atmosphere where it forms clouds. Through precipitation, water returns to the earth's surface and begins its long journey back into the sea or a lake. Or water can flow on the surface in the form of rivers or underground as subterranean streams. In the end, though, water always finds a way, eventually returning to the sea, where the sun's heat lifts it back into the atmosphere in the form of vapor. And for all the recycling that water undergoes, the total amount on Earth remains roughly constant.[41] The ancients were unaware of water evaporation.[42] Qoheleth had no clue that the basic image behind his leitmotif, *hebel*, is what makes possible the water cycle. Nevertheless, the sage notes that the sea never fills. He would not have guessed, however, that it is actually rising.

Or take the interstellar generation of stars: as old stars die, cataclysmically, new stars are born to take their place. Within Orion's constellation, for example, is a stellar nursery. The fuzzy blob of light in the center of the sword is a nebula, "a cosmic cloud of gas studded with bright young stars."[43] Such gas is composed of remnants of former stars dispersed by supernovas. A time for stellar death and a time for stellar birth.

But the perpetual recycling of cosmic gas and dust cannot last forever. Likewise for life. So many processes that generate and sustain life do so by taking life, with new generations built on the death and decay of previous ones. It is unlikely that we would be here without the mass extinction of the dinosaurs some 65 million years ago. The alternations between life and death are written into the very heart of evolution. A time for extinction and a time for recovery. Regardless of whether much of life on Earth as we know it will survive the crushing weight of our carbon imprint, possibly leading to a sixth great extinction, in the far grander scheme of things, the recycling of life will ultimately be interrupted by cosmic forces that lie beyond anyone's control. There are many ways that the "cosmos wants to kill us," notes astrophysicist Neil deGrasse Tyson, from killer asteroids to the sun's explosive death to galactic collisions.[44] And one could also include a nearby supernova, black hole formation, or the collapse of a massive star.[45] It's only a matter of time, the ancient sage would remind us.

Time and Motion

Time is a central theme in Qoheleth's more abstract musings. In view of creation's relentless cycles and life's perpetual swings, time seems to lack direction. It bears no obvious "arrow." Time is simply there, past, present, and future, all bound together in one encircled whole. Without the apparent progression of time, Qoheleth questions the notion of change itself. The implication is that time, too, like the cosmic spheres locked in their incessant revolutions, remains static, a notion so alien to ordinary experience yet so true from the perspective of modern physics.

Time is a paradox. The laws of physics do not recognize the progress of time; the past and the future share an equal footing.[46] "The laws of physics that have been articulated from Newton . . . up until today, show a *complete symmetry between past and future*."[47] In principle, an object can trace its trajectory in reverse under the right physical conditions. In this respect, time resembles space, which also has no arrow indicating up or down, right or left. Together, space and time constitute a packaged whole, and the world of spacetime wholeness is a world devoid of history.

To describe this totality, Brian Greene employs the image of a loaf of bread, which can be sliced in any number of ways depending on the observer's perspective (as determined by her relative velocity), but as a whole remains impervious to change, that is, "eternal and immutable."[48] Such a view of time (with space) is, however, counterintuitive to everyday experience. We perceive time as flowing or flying. But it does neither according to physics. Time simply is. Although each slice represents a particular "now" for a given observer, the loaf as a whole remains the same. To press the analogy further, each slice is equally edible. For Einstein, the distinction between the past, present, and future is "only an illusion, however persistent."[49] We live in "an egalitarian universe in which every moment is as real as any other."[50]

To be sure, events are experienced differently in time between two observers in relative motion to each other. Their "nows" are different. And greater separation between them yields greater deviation between their respective "nows." The greater the distance between an object and observer, the longer it takes for the observer to receive the light of the object. We see stars as they were in the distant past. The Andromeda Galaxy that we see "now" is the galaxy as it was 2.4 million years ago.[51] The recent gamma ray burst in the constellation Bootes (classified officially as GRB 080319B) was just as real to us on the morning of March 19, 2008, as it was 7.5 billion years ago when it "actually" occurred, over half the lifespan of our universe. Light, which is in constant motion, serves as "a cosmic time capsule."[52]

In the cosmic loaf, the past does not fade away and the future is not waiting to happen. Past, present, and future all coexist and equally so, with each moment no more valid than any other. The physicist's depiction of singular spacetime deepens the vantage point of the "timelessness" claimed by Qoheleth (3:11). As cosmic space exhibits the "stunning power of symmetry,"[53] so also time, which displays its own uniformity.[54] Like Qoheleth's Earth firmly set amid the swirling cycles, time remains forever fixed.

Within the fixed loaf of spacetime, as in Qoheleth's worldview, change makes no sense. Change is not a quality that one would attribute to any slice within the loaf. "The concept of change has no meaning with respect to a single moment in time."[55] Moments just are; they do not change. "A particular moment can no more change in time than a particular location can move in space."[56] Every moment is, as it were, "forever frozen in place."[57] Like the physicist of today, the sage of yesterday recognized in his own way the static nature of time, "the frozen river."[58] From the physicist's standpoint, time remains the ultimate mystery. There is nothing in physics that explains why we remember the past but cannot read the future. Qoheleth claims, with a logical consistency pressed to the point of experiential absurdity, that we cannot even remember the past. According to the ancient sage, our lives are bounded by the stunning symmetry of amnesia and ignorance, amnesia of the past and ignorance of the future.

Nevertheless, time does not stand still; life is not frozen in place. Perhaps it is no small consequence that Qoheleth illustrates time as a series of perpetual swings and cycles. In Einstein's theory of general relativity, spacetime is also externally dynamic, even as it remains internally static.[59] In either case, time and space are interdependent. Time can dilate as much as length can contract and density can increase, all in relation to an object's relative velocity. The decay rate of radioactive particles, for example, is much slower when they are set at high velocities approaching the speed of light. At such speed, their "internal clock runs slowly."[60] The faster the speed, the slower the time. Time, thus, is very much tied to motion.

Time and Chance

Accident or "chance" also plays a crucial role in Qoheleth's closed world as it also does in quantum physics, the realm of greatest uncertainty. At the smallest measurable level of physical reality, the orderly precision of both Newtonian and Einsteinian physics breaks down, and the messiness of probability takes center stage, where reality seemingly "participates in a game of chance."[61] Known as the Uncertainty Principle developed by the German physicist Werner Heisenberg, subatomic particles lack sharply defined quantifiable values. An electron, for example, cannot have a definite position and velocity at the same time. An observer can measure one or the other, but not both simultaneously. Electrons, thus, exist as diffuse "clouds" of probability surrounding a nucleus rather than as orbiting particles. The principle also applies to energy: an electron cannot have a definite value of energy at a definite moment in time. Whereas the law of energy conservation is fully operative on the "ordinary" level, it can be suspended, albeit momentarily, on the quantum level. Energy can change, spontaneously and unpredictably, from one moment to the next: the shorter the interval, the greater the fluctuation.[62] Photons can suddenly pop into existence out of nowhere, only to vanish thereafter.[63] Any particle, in principle, can borrow "energy from nowhere, as long as the loan is paid back promptly."[64] Call it chaos with a credit card.

Nature, thus, seems to have a built-in limitation on precision. Quantum physics can only predict the probability or odds of a particular outcome in the subatomic realm, but not the outcome itself. Hence, chance remains forever fixed. Particle physics, eighteenth-century Continental philosophy, and an ancient sage's musings actually share something in common, a sense of nature's indeterminism. The Scottish philosopher David Hume (1711–1776) disclaimed any notion of a necessary connection between cause and effect. No one can predict with absolute certainty the consequence of any given cause. There is always the slight chance that the sun will not "rise" tomorrow morning ("panting" as it goes!), or that I will be able to walk through a brick wall. In quantum reality, anything is possible, improbable as it is.

Chance is also a part of time. If time does not flow, then does it have an arrow?[65] If it does, it is inferred from "the second law of thermodynamics": all closed systems evolve toward higher states of entropy or disorder and, consequently, toward greater energy dissipation.[66] Heat flows from hot to cold, not the reverse. A noxious odor spreads across a room; "farts do not magically reconverge on the offending emitter."[67] In light of thermodynamics, then, time is irreversible. Since the Big Bang, the universe has been inexorably dispersing energy and moving toward greater disorder. But it is also from entropy (acted upon by gravity) that planets and galaxies form. It is from "chance" that the evolution of life continues apace.[68] Yet it is also from entropy that both planet and platypus will meet their demise.

Time, thus, not only has an arrow, it also bears a "claw."[69] The unidirectional dissipation of energy, yielding death and decay, is evidence of time's unidirectional arrow, and "entropy never fails."[70] On the cosmic scale, gravity is "the most efficient generator of entropy in the known universe," particularly evidenced in the formation of black holes, "the greatest reservoirs of mayhem the universe has ever known."[71] Astronomers tend to describe the behavior of a black hole in thoroughly mythic terms, such as a cosmic "monster" with an insatiable appetite so great that not even light can escape its "mouth" or "event horizon." And yet black holes, as recently discovered, also have a hand in the growth of galaxies, even seeding their formation.[72] Disorder generating order. But in the long run disorder wins. In the distant future, measured in billions of years, the universe will come to know these monsters of cosmic mayhem all too well. And darkness shall reign (see Eccl 12:1–2). The "progress" of time is far from teleological; in fact, it is distinctly "dysteleological."[73] The world according to Qoheleth bears no purpose.

To add to the prospect of such eschatological misery is the frightening uncertainty of cosmic collapse that can strike at any given moment and without warning. Not comets or asteroid collisions, but something more universal: "true vacuum formation."[74] This scenario entertains the distinct possibility that the present state of the universe does not correspond to a true vacuum state, that is, a state of lowest possible energy. Drawing from the work of Sidney Coleman and Frank De Luccia, physicist Paul Davies points out the chilling possibility that the universe as we know it may have developed within a false vacuum that could at any moment decay, pitching the cosmos into an even lower energy state.

The harrowing result is sudden cosmic decay at some random location in space that begins as a tiny bubble of true vacuum, destabilizing the surrounding false vacuum that has upheld the universe for so long. Its expansion would approach the speed of light, engulfing a larger and larger region of the false vacuum, a sort of anti-cosmic inflation. The growing bubble wall dividing the energy difference between the two vacuum states would "sweep across the universe spelling destruction to everything in its path,"[75] in short, "an instant crunch." When it comes to destruction, anything is possible, as Qoheleth himself recognized: "Like fish taken in a cruel net, and like birds caught in a snare, so mortals are snared at a time of calamity, when it suddenly falls upon them" (9:12). No anthropic principle is celebrated here. Indeed, quite the opposite: Qoheleth laments a "misanthropic principle" that is continually at work.[76] The possibility of a cosmic vacuum suddenly engulfing the universe takes Qoheleth's "vapor" (hebel) to a new cosmic level, the cosmos collapsing in a puff of radiation.

Entropy and End

If not by sudden demise, the cosmos will most certainly suffer the more gradu-ated form of cosmic decay known as "heat death." In 1856, the German physi-cist Hermann von Helmholtz made what is probably the most depressing pre-diction in the history of science: the universe is dying.[77] But it was not unprecedented. More than two thousand years earlier, the sage Qoheleth made just as startling a claim. When cosmologists extrapolate into the future, they read, as did Qoheleth, the universe as a story of ultimate futility. Like our indi-vidual bodies, the universe is doomed to decay. Entropy may not win any given day, but it will billions of years from now.

Perhaps the greatest paradox within our solar system is this: the sun that has given life to the earth, and continues to do so unfailingly, will eventually deliver death to our planet. The sun has been in existence close to five billion years and will last, according to latest estimates, for another 7.6 billion years.[78] Once the sun exhausts its core source of hydrogen, it will begin to burn the hydrogen in its outer shell. The core will then contract and rise in temperature as the hydrogen-burning shell continues to expand. The resulting red giant will be 256 times larger and 2,730 times more luminous than the present sun. As the sun balloons outward, its gravitational grip will be weakened, and Earth will retreat to where Mars is now. But eventually our planet will be overcome by tidal forces brought about from its own gravity, causing the sun to bulge toward it and eventually drag Earth back toward its engorged state.[79] But by then it will not matter, for in a mere billion years from now the sun's increased brightness will have boiled away the earth's oceans, turning our planet into a lifeless rock. Eventually, the sun's core will collapse to become a very dense, degenerate remnant—a white dwarf.

On a far grander scale, the universe itself, cosmologists conclude, will either collapse under its own weight in a fiery "big crunch" or expand forever, dissipating itself to oblivion. It all depends on which cosmic force ultimately wins: the attractive pull of gravity or the repulsive force of dark energy (Einstein's "cosmological constant"). The continuing shape of the universe is marked by a celestial tension between contraction and expansion, a cosmic tug-of-war. In point of fact, both will win out, but at different levels. Gravity, the weakest of nature's forces at the atomic level, becomes most dominant on the galactic scale. Because gravity is cumulative, like compound interest, its overall effect on planets and stars is irresistible. Because of gravity, a star is eventually crushed, and for many of the larger stars, a black hole is the result.

To make matters worse, many galaxies have black holes located at their centers, as indicated in the rapid movement of stars around these galactic cores.[80] Our galaxy, the Milky Way, with its center located in the constellation of Sagittarius, seems to be no exception ("Sagittarius A*").[81] Like a moth to a candle, stars are drawn toward these black holes and eventually swallowed up. The fate of any object that passes beyond the point of no return suffers the painful process of "spaghettification," a "stretching and squeezing process" that results in the complete obliteration of all matter.[82] Moreover, "the violence of the infall process should in some cases be great enough to disturb the entire structure of the galaxy."[83] Such a cosmic "feeding frenzy" is, in time, the destiny of most galaxies.[84] And, in due time, all black holes will evaporate, disappearing in a puff of radiation and leaving behind (according to the "Hawking effect") a trace of photons, neutrinos, electrons, and positrons.[85] Such is the force of gravity, which initiates an "orgy of cannibalism."[86] Wisdom's cosmic play in Proverbs 8 is Qoheleth's dance of death.

But not all matter is destined to fall into black holes. There will be the lonely neutron stars, black dwarfs, and rogue planets that will have escaped the gravitational force of black holes, destined to wander across vast intergalactic stretches. Will these fateful remnants, along with the universe as a whole, come together in a fiery "Big Crunch"? Astronomers have for decades tried to measure the deceleration of the universe, but to no avail. It stood to reason that once the inflationary burst was over (after 10^{-34} seconds), gravity would have begun to put the brakes on the expansion of space. But observations in 1998 of the luminosity and redshift of Type Ia supernovae suggest that on the largest of astronomical scales, the universe continues to expand, and with a heavy foot on the cosmic pedal. Recent calculations indicate that around the midpoint of its life (between five and seven billion years old), the cosmos shifted gears: deceleration from gravitational attraction was overcome and the universe began to speed up its expansion and continues to do so.[87] We live in a "runaway universe."[88]

Given this cosmic "new" direction, the eventual outcome is a vast, empty, lonely universe. As the universe expands, there is more space, and with more space there is more force pushing galaxies outward with increased velocity from all points of reference. As galaxies approach the speed of light, they approach a horizon of darkness and vanish from view as if "falling into a black hole," with "the most distant galaxies disappear[ing] first as the horizon slowly shrinks around us like a noose."[89] What would be visibly left in a 100 billion years or so will be six galaxies, including the Milky Way, according to Lawrence Krauss and Robert J. Scherrer.[90] The stage would be set for cosmic ignorance: "Observers in our 'island universe' will be fundamentally incapable of determining the true nature of the universe."[91] By then dark energy would be entirely unobservable. And ultimately "our" provincial neighborhood of six galaxies will collide and collapse.

The finality of cosmic evolution is, to put it mildly, gloomy. Paul Davies prefers "eternal death" over "eternal life" to describe the "bleak sterility of a universe that has run its course."[92] Such dismal imagery matches Qoheleth's conclusions in the final chapter. His reflections, as they span bookend to bookend, are marked by the sustained tension between the seemingly perpetual cycles of life (1:4–11; 3:1–8) and eventual cosmic dissolution (12:1–7). In Qoheleth, entropy overcomes life and the static runs down the cyclical. The exercise of life—its growth in complexity and wisdom, its self-generating, evolving, autopoietic playfulness[93]—is merely a diversion, at most a postponement of the inevitable: the march toward death and dissolution for a tightly closed universe. *Hebel* always wins.

Humanity Gone to *Hebel*? Qoheleth Meets Steven Weinberg

Qoheleth's skepticism regarding creation's purpose is pressed all the more by the bigger picture science offers of cosmic beginning and extinction. To quote Steven Weinberg's oft-quoted conclusion in his book *The First Three Minutes*, "The more the universe seems comprehensible, the more it also seems pointless."[94] For Weinberg, the only solace that "lifts human life a little above the level of farce" is to understand the universe through scientific research.[95]

The quest to understand was also Qoheleth's mission, one that left him empty-handed before an indifferent God and a dying universe. In this regard, Weinberg and Qoheleth are kindred spirits.

But even as soul mates, the physicist and the sage must part ways. What lifts human life a little above the level of *hebel* is not just the unending quest to understand, according to Qoheleth. Despite our expanding grasp of the cosmic picture, we still run headlong into the wall of "pointlessness" when it comes to discerning anything resembling an overarching purpose. Both the ancient sage and the modern physicist acknowledge that. Qoheleth's cosmology is so thoroughly anti-creational from a biblical perspective yet so elegant from a broadly scientific and existential perspective that one might expect a nihilistic reduction of human purpose.

Far from it! Qoheleth's negative cosmology provides the foil rather than the source for human purpose. The cosmos, in his eyes, mirrors the human tendency to strive relentlessly for the unattainable. Creation is fashioned in the *imago futilitatis*, as human beings, in turn, are caught up in their own relentless cycles of toil driven by envy and deprived of joy (4:7; 5:13–17). Here, humans cast themselves in the *imago operarii*, in the image of the toiler. Like Sisyphus and his rock, they are locked in an unending cycle of work with no prospect of gain or lasting accomplishment.

But there is another way. Instead of endless toil for unattainable gain, Qoheleth commends a life of rest and enjoyment in a world without pause and effect. If meaning cannot be gained from the broadest scope of human inquiry, it can still be had within narrower fields of focus, all the way down to the single moments of joy. The seemingly mundane pleasures of eating, drinking, and enjoying one's work are for Qoheleth the *highest* goods in life:

> So I commend enjoyment, for there is nothing better for people
> under the sun than to eat, and drink, and enjoy themselves, for this
> will go with them in their toil through the days of life that God gives
> them under the sun. (5:15)

Qoheleth commends such pleasure seven times throughout his discourse.[96] The sage, moreover, finds that he has the highest backing for commending them. These simple joys are to be received and enjoyed for what they are, namely, God's gifts (5:19; 8:15). And so at the mundane, rather than cosmic, level one can discern an occasional smile breaking God's "steely" countenance. According to the sage, it is in those moments of enjoyment, few and fleeting though they are, that the hand of providence is most clearly evident. In those moments, *hebel*'s shroud is momentarily lifted, God is revealed in generosity, and humans are shown their true identity, made in the *imago acceptoris*, in the image of the recipient. For Qoheleth, those moments are precious singularities.

The static cycles and ultimate demise of the cosmos expose the human striving for gain as a pointless venture, like the pursuit of wind. Continual grasping leads only to continued dissipation, a miniature version of cosmic entropy, so claims Qoheleth (with a little help from astrophysics). In contrast, receiving life's simple gifts with joy and gratitude is of greatest significance, because the gifts are apportioned by God no less.[97]

Receptivity, however, is no passive matter for this "assembler" of wisdom. It is a conscious choice, an exercise of the will that cuts against the grain of the universe. While the sun and the streams continue their perpetual "cycles," getting nowhere, while much of the human race is caught up in a relentless, self-relinquishing quest for gain, Qoheleth offers sage advice that is nothing short of revolutionary. Not just countercultural, such advice is counter cosmic! Against a toiling, fading cosmos, Qoheleth's commendations burn with something close to moral urgency.

Qoheleth's commends, in short, a "nonprofit" existence. Like the smile that breaks out on Sisyphus's face, according to Camus, as his boulder tumbles back into the valley,[98] Qoheleth finds joy in receiving rather than toiling for God's gifts. Qoheleth's message is both a protest against creation's pointless, runaway pace and an affirmation of creation's simple, life-sustaining provisions: food, drink, and fellowship. And one could also add the sun's gentle warmth, the streams' purifying flow, and the wind's refreshing breezes. Once their life-sustaining benefits are realized, such "wearying" cycles are no cause for lament. Little did Qoheleth know that the cycles that he found so wearisome are the very cycles that support and regulate life. Even in a runaway universe, human beings can make a good run of it with the few moments that they have.

9

The Fabric of the Cosmos

The Emergence of New Creation
in "Second Isaiah"

[H]ere is a secret that never makes the headlines: We have taken apart the
universe and have no idea how to put it back together.

—Albert-László Barabási[1]

All creation has an instinct for renewal.

—Tertullian[2]

There is good reason why Isaiah and Ecclesiastes were not set side
by side in the Bible. Had these two books bumped up next to each
other, a fight would have no doubt ensued, at least in the mind of
the reader, a clear case of canonical dissonance. As it is, they are
safely separated, but not entirely.[3] They can still shout at each other
across the canonical divide. In the one corner is the cynical sage
with his rigorous denial of anything new. In the other is the exuberant prophet of the exile with his vigorous affirmation of everything new. The sage's cynical question, "Is there a thing of which it
is declared, 'See, this is new'?" (Eccl 1:10a) is met head on with the
prophet's ringing testimony: "New things I now declare!" (Isa
42:9). Through Qoheleth's eyes, the unwavering cycles of the cosmos preclude anything novel. For the prophet of the exile, the old
is finished and gone; behold, new things spring forth! In Isaiah,
God *continues* to create, and novelty is the norm.

The language of creation pervades what many readers consider
the greatest body of prophetic poetry in the Bible: "Second Isaiah,"

chapters 40–55.[4] The author is as much a poet carefully crafting stanzas as he is
a prophet boldly announcing a new reality. Unlike other creation traditions, there
is no self-contained account that effectively summarizes the author's perspective.
Not confined to a single passage, creation language is interwoven into the prophet's historically bound pronouncements. Creation does not confine itself to the
primordial past but extends into, invades even, the present. Included in God's
creation is Israel's formation, and Israel's redemption is folded into God's continuing work in creation. In "Second Isaiah," creation and history are inseparably
wedded.

> Thus says God, YHWH,
>> who has created the heavens and stretched them out (*nṭh*),
>> who has hammered out (*rqʻ*) the earth and what emerges from it,
>> who has given breath to the people upon it,
>>> and spirit to those who walk in it.
> I am YHWH,
>> I have called you in righteousness;
>> I have taken you by the hand and kept you;
>> I have given you as a covenant to the people,
>>> a light to the nations. (Isa 42:5–6)

The God who created the universe is the same God who sustains and commissions a people. Like fabric the heavens are unfurled (cf. Ps 104:2); like sheet metal
the earth is hammered out. For all its hymnic grandeur, the creation language of
v. 5 serves to introduce the bold announcement in v. 6, which apart from v. 5 has
ostensibly nothing to do with creation and all to do with Israel's history: Israel is
called to serve the nations. But joined to v. 5, Israel's commission becomes a
genuine extension of God's creative activity. Israel, after all, is commissioned to
be a "light" (v. 6), and such a calling is indeed something "new" (v. 9).[5]

Socio-Historical Background

Chapters 40–55 of Isaiah, like the first chapter of the book, presume a situation
of devastation. The rhetorical landscape on which the poet stands is a smolder-
ing, barren wasteland, a land of judgment and fire (42:24–25; 49:19; 51:3). It is
the landscape of exile. Zion is left in ruins, and those deported must eke out
their existence on foreign soil, the land of their oppressors, Babylonia. Historically, the poet recalls the horrific events of 587 BCE: the conflagration of Jerusalem, including its temple, and the deportation of a significant portion of the

population conducted by the Babylonian captain of the guard (2 Kgs 25:8–9; Isa 64:11).

> Nebuzaradan carried into exile the rest of the people who were left in the city and the deserters who had defected to the king of Babylon— all the rest of the population. But the captain of the guard left some of the poorest people of the land to be vinedressers and tillers of the soil (2 Kgs 25:11–12; cf. Jer 52:15–16).

This was, in fact, one of three deportations conducted within a span of fifteen years. The book of Jeremiah gives deportation numbers that cover the first deportation in 597 BCE, the second in 587 BCE, and a third in 582 BCE (unattested in 2 Kgs), totaling 4,600 (Jer 52:28–30). In 2 Kings, however, we find a substantially higher number of deportees, but only two deportations (2 Kgs 24:5–16).[6] Such numbers, however, pale in comparison to the Chronicler's claim that the land lay "empty" of inhabitants (2 Chr 36:21). The lack of extrabiblical data and the presence of conflicting numbers within the biblical tradition have prompted one historian to extrapolate figures from the Assyrian deportations of nearly a century and a half earlier and conclude that "some 25 percent" of Judah was deported under Babylonian hegemony.[7]

Regardless of the extent of population loss in the land, the social and religious loss was from the exiles' point of view nothing short of devastating. They had no king and no land. Wracked by doubt over YHWH's power, many exiles concluded that the Babylonian deity Marduk had defeated YHWH or that, as Ezekiel reports the *vox populi* in Judah, "YHWH has forsaken the land, and YHWH does not see" (Ezek 9:9). "Second Isaiah" employs a special term to describe the result of such desolation: *tōhû* or "waste,"[8] comparable to what is described in Genesis 1:2 (*tōhû wābōhû*), except without the waters. *Tōhû*, moreover, overlaps with the picture Qoheleth paints at the conclusion of his book, an empty shell of darkness and death into which all life will enter. Such is for the prophet the starting point of something new, a new creation no less.

Israel's exile did not last forever. Deliverance came about in 539 BCE, when the Persian king Cyrus II conquered Babylonia and released all foreign captives, inaugurating a *Pax Persiaca* that lasted over two centuries. Heralding Cyrus as YHWH's anointed (45:1), the poet behind "Second Isaiah" beheld the storm clouds of salvation looming on the horizon and boldly announced a new exodus, one in which his people would not have to flee in haste but travel in triumph, homeward bound (43:16–21).

Character of Creation and Community

The prophet's view of creation reflects something of the powerful historical forces that gripped him and an exiled people. Case in point: the ancient poet views celestial creation as unfurled fabric:

> Thus says YHWH, who redeems you,
>   and who forms you in the womb:
> I am YHWH, who makes all things,
>   who alone stretches out (*nṭh*) the heavens,
>   who by myself hammers out (*rqʻ*) the earth. (44:24)

> I made the earth,
>   and created humankind upon it.
> It was my hands that stretched out (*nṭh*) the heavens,
>   and I commanded all their host. (45:12)

> Indeed, my hand laid the foundation (*ysd*) of the earth,
>   and my right hand spread out (*ṭpḥ*) the heavens;
> When I call them,
>   they stand together at attention. (48:13)

> You have forgotten YHWH, your Maker,
>   who stretches out (*nṭh*) the heavens,
>   and lays the foundations (*ysd*) of the earth. (51:13a [cf. 16])

For this poet of the exile, creation is a matter of both past and present. Not confined to the primordial past, God's creative activity spills into the present. The heavens and the earth are God's works in progress, but in different ways. Whereas the earth is "hammered out" and its foundations laid, the heavens are "spread" or "stretched out." The contrast between the celestial and the terrestrial is unmistakable, so also the divergence between Genesis and Isaiah. In Genesis 1, the heavens are associated with "hammering" in the form of a solid "firmament" (*rāqîaʻ*), firmly demarcating the realm of the transcendent (see Gen 1:6–7). In Isaiah, however, the heavens are likened to stretchable, unfurled fabric (see also Ps 104:2b), with the earth itself cast as a firmament. The poet's perspective, in fact, corresponds well to his vision of a reestablished Israel.

> It is [YHWH] who resides above the circle of the earth,
>   whose residents are like grasshoppers;

who stretches out (*nṭh*) the heavens like a curtain,
    and spreads (*mṭḥ*) them like a tent to reside in. (40:22)

Sing, O barren one who did not bear!
    Burst into song and shout,
        you who have not been in labor!
For the children of the desolate one will become more numerous
    than the children of the married one, says YHWH.
Enlarge the site of your tent,
    and let the curtains of your habitations be stretched out (*nṭh*).
Do not hold back;
    lengthen your cords and strengthen your stakes.
For you will spread out (*prṣ*) to the right and to the left,
    and your progeny will possess the nations,
        and the desolate towns you will settle. (54:1–3)

These two passages clarify the origin and purpose of the celestial metaphor: the heavens are like the canvas of a tent. Such imagery suggests that all creation was fashioned for habitation, including Israel's. In 54:1–3, the "desolate" daughter Zion is commanded to sing as a mother of many children and to stretch out "the curtains of [her] habitations," her "tent" (v. 2). Zion's actions are to replicate God's heavenly act of creation! As the heavens are a celestial tent stretched out, Zion reestablished is an extension of heaven on earth. Zion's restoration is itself an act of re-creation.

As the heavens and the earth were made for habitation, so Zion beckons the exiles to return and re-inhabit the land. The reference to God laying the earth's foundation also resonates with God's promise that the temple's "foundation" be laid, to be fulfilled by Cyrus (44:28). The temple's rebuilding mirrors the earth's founding. With temple restored and tent spread out—as the earth is founded and the heavens are stretched—daughter Zion becomes mother Zion, the bearer of a people's rebirth.

Another creation passage reflects God's carefully determined and care-filled redemption of Israel from exile.

To whom then will you compare me,
    or who is my equal? says the Holy One.
Lift up your eyes on high and see:
    who created these?
He who brings out their host and counts them,
    calling them all by name;

> Because he is great in strength,
>> mighty in power,
>>> not one is missing. (40:25–26)[9]

As the heavenly hosts are called by name, so also is exiled Israel:

> But now thus says, YHWH,
>> the one who has created you, O Jacob,
>>> the one who has formed you, O Israel:
> Do not fear, for I have redeemed you;
>> I have called you by name, you are mine. (43:1)

Each one is "called by my name, whom I created for my glory, whom I formed and made" (v. 7). Like the heavenly hosts, no one will go "missing" (40:26). Redemption, too, mirrors divine creation.

As the passage above makes clear, creation has as much to do with Israel as with the cosmos. It is the same God who created the heavens who also "created" Israel.

> Now listen up, O Jacob my servant,
>> Israel whom I have chosen!
> Thus says YHWH, who has made you,
>> who has formed you in the womb and will help you:
> Do not fear, my servant Jacob,
>> Jeshurun, whom I have chosen. (44:1–2 [see also vv. 21, 24; 54:5])

Israel's creation is bound up with Israel's election. To choose a people is, in effect, to create a community. Israel was formed from the womb of God's decision to constitute and sustain a people. In creation, cosmos and community are wedded together.

Also merging cosmos and community is the new exodus the prophet proclaims to the exiles, the one that leads Israel back to its land.

> Thus says YHWH,
>> who makes a way in the sea,
>>> a path through the mighty waters,
>> who brings out chariot and horse,
>>> army and warrior together;
>> they lie down, not able to rise,
>>> they are extinguished, quenched like a wick.

"Do not remember the former things,
>   or consider the things of old.
Look, I am doing a new thing!
>   Now it springs forth; do you not perceive it?
I will make a way in the wilderness,
>   rivers in the desert.
The wild animals will honor me,
>   the jackals and the ostriches;
for I give water in the wilderness, rivers in the desert,
>   to give drink to my chosen people." (43:16–20)

Although announced as something entirely new, this exodus is actually a mirror image of the old. The elements of sea and land remain present, but they are now reversed: whereas a dry path once split the mighty waters that extinguished Pharaoh's army in the old exodus, now rivers cut through the parched wasteland, prompting praise even from jackals and ostriches. The old is a precursor of the new; the new is patterned off of the old.

Finally, in one densely packed passage, creation reveals its essential purpose for God's people.

For thus says YHWH,
>   who creates the heavens (he is God!),
>   who forms the earth and fashions it;
He established it;
>   he did not create it to be a waste (tōhû),
>   but formed it to be inhabited!
"I am YHWH,
>   and there is no other.
I did not speak in secret,
>   in a land of darkness.
I did not say to Jacob's offspring,
>   'Seek me in waste (tōhû).'
I YHWH speak the truth,
>   I declare what is right." (45:18–19)

This compact poem divides itself evenly into two parts. The creation part (v. 18) introduces the divine declaration that follows (v. 19). God's intention for creation is habitation, not "waste" or "darkness," and seeking God can only take place in a reconstituted land, not one emptied and left dark (cf. Jer 4:23–26). For the poet of the exile, a land "emptied" or "wasted" by exile and darkened by the lengthening shadows of its ruins constitutes the very antithesis of creation. "Seek me" in life-sustaining creation is God's invitation to a displaced and despairing people. Seeking God in a wasteland, on the other hand, is simply a waste.

Character of God

God in "Second Isaiah" is preeminent creator, indomitable warrior, and impassioned savior, all wrapped into one. God's announcement in 45:19, quoted above, amounts to a declaration of divine supremacy, so much so that "there is no other" god to offer resistance. The creator's self-identification as YHWH in 45:19 is shorthand for an unequivocally monotheistic claim spelled out more fully in 45:5: "I am YHWH, and there is no other; besides me there is no god." Such is Isaiah's monotheistic refrain.[10] Its explication is given in an earlier chapter:

> Before me no god was formed,
>   nor shall there be any after me.
> I, I am YHWH,
>   and there is no savior besides me. . . .
> I am God,[11] and henceforth I remain so;[12]
>   there is no one who can deliver from my hand.
> I work, and who can hinder it? (43:10b–11, 13)

Pressed to its (theo)logical conclusion, YHWH's unparalleled claim to power excludes all other claims to divinity. As consummate savior and creator, YHWH remains alone in the divine realm. Hence, all would-be divinities are reduced to nothing, and that includes Babylon, the "Gate of the Gods" (*Bâb-ili*):

> Now therefore hear this, O luxuriant one,
>   who dwells securely,
> who says in her heart,
>   "I am; there is no one else except me;
> I shall not sit as a widow or know childlessness." (47:8)

Only once in Isaiah does the self-referential formula for divinity get pronounced apart from YHWH, and it is here. Personified as a woman, Babylon makes her own mono-imperialistic claim, succumbing to the conceit of regal supremacy and, thereby, warranting the scourge of divine judgment. Babylon, with its formidable pantheon of gods, with its seemingly indomitable military might, is nothing more than an "idol" threat to YHWH's sovereign will. Idols, the prophet defines, are gods that "cannot save" (45:20). They are "nothing" (41:24), whereas Israel's God is both "first" and "last," and everything in between (41:4; 44:6; 48:12).

The God of "Second Isaiah" negates all other powers, both earthly and divine. Princes are "naught" (40:23a), rulers and nations are "nothing" (vv. 17, 23b), and the gods are nonexistent. In Isaiah, YHWH's monotheistic status is squarely rooted in divine incomparability. Three times God confronts the reader with the question: "'To whom then will you compare me, or who is my equal?' says the Holy One" (40:25; cf. 40:18; 43:5). Most elaborate is the passage below.

> Listen to me, O Jacob's house,
>> all the remnant of Israel's house,
>>> borne (by me) since birth,
>>> carried from the womb.
>> Even in your old age, I am he;
>>> even when you turn gray, I will carry you.
>> I have made, and I will bear;
>>> I will carry, and I will rescue.
>> To whom will you liken me and make me equal,
>>> and compare me, as though we were alike? (46:3–5)

Israel's God is unmatchable in forbearance and incomparable in commitment, ever demonstrated from Israel's "birth" to "old age," from beginning to end. God's lifelong sustaining care is at the root of God's incomparability and, in turn, God's exclusive membership in the divine realm. YHWH's power is not only unmatched; it is categorically unique. Nothing else can create like YHWH; indeed, nothing else can create at all.

Because God's sovereign, unrivaled work in creation is so totalizing, it can encompass fundamentally opposing phenomena.

> I am YHWH; there is no other,
>> who forms light and creates (br') darkness;
>> who makes weal[13] and creates (br') woe;
>> I, YHWH, make all these things. (45:6b–7)

The technical term for creation used in Genesis 1 (br') is employed specifically in Isaiah to designate the creation of "darkness" and "woe." The God of Isaiah, thus, is more expansive in power than the God of Genesis 1. In Genesis, God's first act of creation is the domain of light, which separates the pre-existent darkness. In Isaiah, God creates *both* darkness and light.[14] By such a claim, the prophet brings us closer to an absolute beginning and, thus, to a more totalizing and, therefore, ambiguous rendering of divinely wrought creation. In the beginning was God alone, the creator of all.

But the prophet does not linger there. From the sheer brevity of his statement, it is clear that the poet is interested not so much in the primordium of creation as in the immediacy of Israel's exile. Isaiah's God does not cease or rest, unlike the God of the Sabbath (cf. Gen 2:1–3). The Creator of all does "not faint or grow weary" (Isa 40:28). Whereas the God of Genesis models the salutary rhythm of work and rest, the God of Isaiah works unceasingly to empower "the faint" and strengthen "the powerless" (40:29).[15] For the poet of the exile, creation is constant; it bears witness to YHWH's unrivaled claim of "only-ness." This God is the *only* God, and it just so happens that the unstoppable creator of all is YHWH, Israel's God. YHWH calls exiled Israel back home to re-inhabit and rebuild, back to a land re-created.

Flowering of Creation

How is the land re-created? The prophet's central focus on the land points to an additional aspect of creation that is often overlooked by interpreters. Such neglect is understandable. It is tempting to focus only on creation from on high, from the commanding perspective of God single-handedly creating light and life, darkness and woe. Creation is certainly all that for the poet, but creation also emerges from below, from the ground up. One need not look far in the poetry to find numerous references to botanical growth. Rich with metaphors and images drawn from the realm of horticulture, the prophet's discourse covers a remarkable range of botanical diversity, from the lowly brier (55:13) to the most majestic of trees, the cedar of Lebanon (41:19; 44:14). Indeed, "Second Isaiah" contains a veritable catalogue of flora. Creation on the ground bears a distinctly botanical twist.

Case in point: "Second Isaiah" begins with God's words of "comfort" directed to Jerusalem (40:1). What such "comfort" entails is spelled out later in 51:3:

> For YHWH will comfort Zion;
>   he will comfort all her desolate places,
> making her wilderness like Eden,
>   her desert like the garden of YHWH.

> Joy and gladness will be found in her,
>   thanksgiving and the voice of song.

More than expressing words of kindness, "comfort" dirties itself with transforming the desolate land into a veritable garden paradise. For desolate Zion, comfort is all about God's ongoing, renewing work in creation, the "new thing" that "sprouts forth" (42:9 and 43:19). And like the new exodus, this "new thing" paradoxically recalls a rather old thing, namely, primordial Eden. The prophet could have easily employed other images to describe God's "new thing," but he chose distinctly botanical imagery. "Sprouting" signals the growth of new creation and the emergence of a new community. The restored Israel shall emerge not just from the ashes of destruction but also from a soil restored and a land re-fructified. Even salvation is cast naturally:

> Shower, O heavens, from above,
>   and let the clouds rain down righteousness.
> Let the earth open so that salvation may spring up,[16]
>   and let it cause righteousness to sprout up also.
> I YHWH have created it. (45:8)

Rain showers and sprouting plants give testimony to God's imminent deliverance of exiled Israel. Such language "naturalizes" the miraculous nature of divine activity. Nature sets the context—primed as it is—for supernatural intervention. So it is with God's "word":

> For just as the rain and the snow descend from the heavens,
>   and do not return there unless they have saturated the land,
>   causing it to produce and sprout forth,
>   giving seed to the sower and food to the eater,
> So is my word that issues from my mouth;
>   it shall not return to me empty,
>   unless it has done that which I intended,
>   accomplishing that for which I sent it. (55:10–11)

The language of creation, both agricultural and meteorological, highlights the creative power of divine discourse. YHWH's word is likened to precipitation. From on high the word descends, and from below the word returns. What exactly does the prophet have in mind with his reference to God's "precipitous" word returning to its divine source? Precipitation seems to be a one-way event for the prophet. By analogy, the "return" of precipitation corresponds to the fructified land. Precipitation enables life to flourish in return for such provision. Like the rain that restores the land, the divine word restores and renews a people. The word of God is as much an enlivening force for nature as it is a sustaining one for Jacob and Zion (cf. 50:4; see also Ps 147:15–20).

With his focus set specifically on the exiled community, the poet develops the botanical analogy more fully in the following passage.

> Thus says YHWH who has made you,
>> who has formed you in the womb and will help you:
> Do not fear, my servant Jacob,
>> Jeshurun, whom I have chosen.
> For I will pour water on the thirsty land,
>> and streams on the dry ground.
> I will pour my spirit upon your descendants,[17]
>> and my blessing on your offspring.[18]
> They shall spring up like a green tamarisk,[19]
>> like willows[20] by flowing streams. (44:1–4)

In stark contrast to YHWH's withering "breath" in 40:7 and 24, YHWH's "spirit" is here likened to the refreshing rain that restores the arid land. The land thirsts for water as much as Jacob's "seed" stands in need of regeneration. Thus, the fructification of the land points to the proliferation of Israel's descendents, for they themselves constitute the garden grove!

> When the poor and needy seek water,
>> and there is none,
>>> and their tongue is parched with thirst,
> I YHWH will answer them;
>> I the God of Israel will not forsake them.
> I will open rivers on the bare heights,
>> and foundations in the midst of the valleys;
> I will make the wilderness a pool of water,
>> and the dry land springs of water.
> I will put in the wilderness the cedar,
>> the acacia, the myrtle, and the pine.[21]
> I will set in the desert the fir tree,
>> the plane and the cypress together,[22]
> so that all may see and know,
>> all may consider and understand,
> that the hand of YHWH has done this,
>> the Holy One of Israel has created it. (41:17–20)

Israel's restoration is likened to the cultivation of a garden. But it is no ordinary garden. The various kinds of trees listed, seven total, reflect a garden grove that is extraordinarily diverse, including the lowly desert acacia and the majestic mountain cedar. Their transplantation from disparate habitats signals the gathering of dispersed exiles, who together constitute Zion's new garden, Eden restored (51:3). Resilient, diverse trees cultivated together present a striking counterimage to the uniform metaphor of human beings as grass, which merely withers away (40:7; 51:12). Representative of a reconstituted people, Zion is a victory garden, and its produce is not its fruit or its wood but its people, a vivid testimony to God's horticultural prowess (41:20).

Granted the promise of restoration, Israel in exile is called upon to *grow* into the new community. To do so, according to the prophet, is first to bear witness to YHWH's "new thing" and then to live into it. To live into the newness entails being uprooted out of the old, out of captivity and chaos, and being planted into freedom and new community. The new exodus is a transplantation, but one that does not happen without the exercise of choice. In the face of God's "new thing," a crisis of decision is placed squarely upon human shoulders. One can choose either to enter the new or to remain stuck in the old. The "Israel" of exile is called upon to bear witness to the emergence of the new *and* to act accordingly. As Moses on the plains of Moab once commanded the Israelites, "I have set before you life and death, blessings and curses. Choose life so that you and your descendants may live" (Deut 30:19), so the prophet of the exile proclaims that God has set before the exiles the choice between the new and the old, between life and death. Call it comfort with a kick.

To sum up: the botanical realm for the exilic poet is a window into the social realm as much as it is a glimpse into the divine. The regeneration of the land constitutes nothing less than Zion's revival, established and sustained by God. The poet is not merely urging the exiled in Babylon to embark on a hazardous wilderness journey with the promise of an oasis to sustain them along the way. Eden's re-cultivation in 51:3 is no mere way station in the desert for those making their way to Zion. The wilderness is not just a *terra intermedia*,[23] but a *terra transformanda*, a land primed for transformation, a community ready for its reconstitution. All streams lead to Zion (49:10b), for the well-watered grove *is* the newly restored Zion, God's goal and Israel's destination. What sprouts forth from the soil is a reconstituted people, vindicated and transplanted. It is no wonder, then, that the poet performs a uniquely powerful wordplay when he recasts the creation of the heavens in 51:16. Whereas the heavens elsewhere are described as "stretched out" (*nṭh*), here they are "planted" (*nṭ'*), just like Zion, God's people.

> I have put my words in your mouth,
>> and in the shadow of my hand I have hidden you,
> planting (*nṭ*)[24] the heavens,
>> founding the earth,
>>> and saying to Zion, "You are my people." (51:16)

Science of Emergence

More than an event in the primordial past or a state of steady equilibrium, creation according to Isaiah is the arena of divine innovation, replete with novelty and renewal, as open-ended as God is freely sovereign. Creation is dynamically historical: things happen, and genuine change is the result. According to Isaiah, the community will never be the same once God is finished redeeming Zion. Neither will creation.

The word "new" (*ḥādāš*) encapsulates "Second Isaiah's" message and perspective as much as "vanity" (*hebel*) captures Qoheleth's. "There is nothing new under the sun" (Eccl 1:9) is set against "I am about to do a new thing" (Isa 43:19). Historically, Qoheleth's obituary of the new tries to bury the prophet's newsflash made a century or so earlier. Canonically, however, Isaiah has the latter word.[25] But regardless, both books are held together canonically, generating hermeneutical sparks left and right. Indeed, no greater tension is to be found within the biblical canon, a creative friction comparable to that which lies at the heart of scientific inquiry. But first, a few points of virtual contact between science and "Second Isaiah" are in order.

Fabric of the Cosmos

Of all the familiar images that could have been used to describe the celestial realm, the poet of the exile settles on fabric. So also some physicists. Einstein's theory of general relativity draws from the image of fabric to describe spacetime as warped and curved by gravity. Spacetime itself is "a stretchable fabric that underpins physical reality."[26] According to Einstein's principle of equivalence, gravity, akin to accelerated motion, provides the wrinkles in the fabric. To be stationary is to be accelerating upward relative to an object in weightless free fall. The apple that falls from the tree by gravity, hitting Newton's head, is equivalent to Newton's head rushing up in accelerated motion to meet the apple.[27] "[Y]ou feel gravity's influence only when you resist it."[28] Only those who are freely floating, "whether deep in outer space or on a collision course with the earth's surface," experience no acceleration.[29]

With this observation, Einstein concluded that gravity itself is nothing but "warps and curves in the fabric of spacetime,"[30] rather than a force between objects, à la Newton.[31] In the absence of matter or energy, spacetime remains flat and smooth. In their presence, however, space becomes warped and curved.[32] Like a marble rolling on a warped surface, a meteor traveling through space in proximity to a planet will have a curved trajectory. The same applies even to light, which can bend or deflect with gravity. The earth orbits around the sun because it follows curves in the spacetime fabric caused by the sun's gravity. Time, too, is warped. The warping of spacetime is the "manifestation—the geometrical embodiment—of a gravitational field."[33] The primordial era of inflation, the most drastic rearrangement of matter, energy, and spacetime in the universe as we know it, would have produced "a riot of gravity waves,"[34] "relic ripples in the structure of space and time."[35] Like God unfurling fabric.

As the universe continues to expand, the fabric of spacetime continues to stretch. Space and time are not fixed and rigid, but flexible and "rubbery."[36] As Edwin Hubble observed through the astronomical phenomenon of redshift, the farther away two galaxies are, the faster they speed away from each other in an "artful symmetry," stretching without a center.[37] In the poet's eyes as well, creation itself is "stretched out" but for a reason: to accommodate a chosen people living in constrained conditions. Valleys are "lifted up" and mountains are "made low" (Isa 40:4). And from this leveled landscape something new "springs forth."

Emergence of Novelty

The prophet announces the newness of salvation arising out of the punishing reality of exile, like new growth taking root on barren soil. If there is one word that captures the poet's vision of new creation, specifically from the ground up, it is emergence. Emergence also has taken its rightful place in the field of science as the counterpart and corrective to causal reductionism,[38] the view that ultimately everything can be explained by the movement of particles.[39] But science, as it has recently discovered, cannot live by reductionism alone. Pressed far enough, reductionism leads to George Wald's winsome definition of a physicist as "the atom's way of knowing about atoms."[40] Through emergence, however, the physicist has good reason to retain her consciousness.

Among the sciences, emergence acknowledges that certain properties of higher-level systems of complexity cannot be reduced to the "combined effects of lower-level causal processes."[41] Emergence is more a process than an outcome, a process by which something unanticipated appears, either gradually or suddenly, from existing conditions. Its hallmarks are complexity and novelty.[42]

Examples include the emergence of atoms out of the primordial plasma, metabolism from organic molecular structures, and consciousness from cognition. George Mason University biologist Harold Morowitz identifies no less than twenty-eight emergences that have driven the history of the cosmos to its current state, from the "Primordium" of the Big Bang to the apprehension of the spiritual.[43] Of them all, according to mineralogist Robert Hazen, "life is the quintessential emergent phenomenon."[44]

Emergence is not mechanical. It acknowledges the complex interactions of constituent parts that give rise to something greater than themselves. According to Hazen, "each emergent step increases the degree of order and complexity, and each step follows logically, sequentially from its predecessor." Nevertheless, "emergent phenomena remain elusive."[45] "The inherent novelty and layered complexity of emergent phenomena all but preclude prediction."[46] The logical sequence of emergence can only be discerned by "hindsight," even as it entails a "conceptual leap."[47] There always remains an "ontological gap between one sort of property and its emergent successors."[48]

In the process of emergence, no physical laws are broken, but new laws may be waiting in the wings. An "accidental" emergence is an oxymoron, for emergence takes place within "an orderly unfolding of the world, but an unfolding rich in novelty."[49] "Directionality" may even be apparent in the emergent history of the cosmos, according to Morowitz: "There need not be a knowable end point, but there may be an arrow."[50] However, because of the sheer complexity involved, the outcome at each stage of emergence remains unpredictable; the result is something truly new. According to Stuart Kauffman, we live in a "universe of ceaseless creativity in which life, agency, meaning, value, consciousness, and the full richness of human action have emerged."[51] The higher order forms of life that have emerged in the universe cannot be reduced to their constituent parts anymore than evolutionary biology can be reduced to particle physics.[52] "We cannot from physics deduce upwards to the evolution of the biosphere."[53] By whatever name, emergence is a master of surprise, in part because it is always a group effort. Temperature, for example, is an emergent property that only makes sense in view of a *collection* of particles. No single gas molecule has a temperature.[54] Atoms themselves are not conscious. Emergent networks range from that of "interacting proteins within cells of an organism, to the colonial organization of social insects, to the World Wide Web."[55] Emergence is all about "relationships and connections."[56]

What specifically drives emergence? Each case is special. The eukaryotic cell—the most advanced cell branch responsible for the final three kingdoms of life (plants, fungi, and animals)—came about from the "symbiotic alliance" of bacteria, according to Lynn Margulis and Dorion Sagan,[57] or, as Morowitz describes it, through "one cell engulfing another" and the two merrily co-existing, one inside the other. As a result, "redundant functions were lost and the evolving organisms changed their character."[58] Technically, this process of cellular merging is called "endosymbiosis," in which one single-cell organism (or organelle) lives within a larger cell. And so the first nucleated cell emerged, "a totally new life form."[59] Cast generally, emergence takes place when "new forms emerge from a coalescence of existing types."[60] Cast simply, emergence is the result of merging.

Land of Plants and Niches

The prophet of the exile saw that the conditions were ripe for something un-precedented to happen in behalf of his people, and botanical imagery was well suited for describing the emergence of this "new thing." From an evolutionary perspective, the conditions were ripe around 465 million years ago for plants to colonize the land.[61] Their leaves, however, came nearly 100 million years later. As nature's "flat solar panel," the leaf developed to "exploit photosynthetic pro-ficiency" in response to the depletion of CO_2 in the atmosphere. Plants spread-ing on the land dramatically changed the earth's surface, atmosphere, and their own evolution. Indeed, they changed the very course of Earth's history. For the prophet of the exile, the example of plants flourishing in the desert held the promise of a dramatic change in the course of Israel's history.

If it grows, they will come. The presence of plants on the land was a prereq-uisite for animal life emerging out of the sea and coming ashore, beginning with insects. Plants later became the necessary food source for vertebrates. The first terrestrial plants were mosslike, from which three major groups evolved. The earliest were ferns. They were followed by the gymnosperms, including conifers, around 363 million years ago, roughly the same time when the earli-est land vertebrates began to appear. Over 200 million years later, angiosperms or flowering plants took root, resulting in the greatest diversity of plants on Earth today, from towering redwoods to tiny pond plants. The forests, in par-ticular, provided a great variety of potential niches to shape and sustain the land's emerging biosphere, contributing to and sustaining the enormous diver-sity of animal life on land.[62]

It is no coincidence, then, that the ancient prophet likens the gathering of exiles to a transplanted grove (e.g., Isa 41:17–20). Taken from its native habitat, each transplanted tree is to flourish within a new environment, all happily co-existing in God's miraculous bioconservatory, the Edenic Zion. For the prophet, the diversity of the grove reflects the diversity of the gathered community, whose members had been dispersed to the land's "farthest corners" (41:9). But the metaphor runs deeper. Trees are the preeminent niche providers. They provide shelter and shade as well as food. Thus, a botanically diverse grove makes possible a thriving ecosystem. Diversity is key. For the prophet, the metaphor of the grove represents not only a flourishing community in all its diversity but also the niches and structures that sustain it. Israel's arboreal grove is the nursery of a nation.

Cosmic New Birth

The emergence of newness covers not only the biological but also the cosmological, from the formation of atoms to that of stars and planets. But there may be more. Armed with powerful mathematical models, science has entertained the possibility that the vacuum-bubble formation that threatens the very existence of the cosmos could also work in reverse in the formation of a new universe.[63] Studied by Japanese physicists in 1981, a working mathematical model demonstrates that a small bubble of "false vacuum" surrounded by "true vacuum" would inflate within a trillionth of a second as in the Big Bang. Viewed from the vantage point of the true vacuum (our universe), the ballooning baby universe would resemble a black hole, even though a hypothetical observer inside the false-vacuum balloon would see the universe swell to cosmic proportions. The analogy Paul Davies gives is of a "rubber sheet that blisters up in one place and balloons out."[64] All that connects the baby universe to the mother universe is a wormhole, like an umbilical cord, which quickly evaporates or gets "pinched off." And, *voila*, a baby universe is born. The new universe would develop its *own* spacetime, not displace ours.

No one knows whether a baby universe would inherit any of the properties of the mother universe. Lee Smolin suggests a kind of Darwinian evolution of universes that encourages the emergence of life and consciousness, but without the competition for resources.[65] This remarkable picture suggests the possibility that something of the sort has already happened and that our universe may simply be one among many universes, all part of a developing, evolving "system" of multiverses. The possibility of "new things" happening even on the most cosmic of scales would warm the heart of any scientist . . . and prophet.

Isaiah Emergent

If anything, science sets the exuberant prophet of the exile in conversation with the steely empiricist of Ecclesiastes. Together, Ecclesiastes and Isaiah parallel (virtually, of course) the two sides of scientific inquiry: reductionism and the kind of synthesis that acknowledges emergence. Reduction seeks to explain by breaking down physical phenomena into their most minimal parts, seeing only the "trees" (e.g., quarks, atoms, biochemical reactions) within the ecosystem. Synthesis, by contrast, discerns emergent properties and dynamics that are unique to higher levels of complexity, recognizing their irreducibility to simpler or lower levels. To borrow from Qoheleth, scientific inquiry swings between "a time to break down and a time to build up" (Eccl 3:3b). In science, reduction and synthesis coexist methodologically. In the Bible, Ecclesiastes and Isaiah coexist canonically: the sage of the static and the prophet of change, the entropic and the novel, the old and the new, the reductive and the emergent.

Fullness of Emptiness

New life, the prophet announces, emerges out of "emptiness" or waste (tōhû), like vegetation sprouting from the barren ground (see 45:18–19). In the physical world, however, emptiness is never truly empty. Any stretch of outer space, for example, is suffused with energy carried by subatomic particles/waves and cosmic rays, all interacting and exchanging energy. Even empty space teams with quantum "jitters."[66] Like potential energy, there is a potential fullness to the apparent emptiness. Tōhû, in other words, is never total.

In light of the scientific understanding of "emergence," we return to the prophet's message of newness within the social and theological reality of exile. Emergence reflects the coalescence of disparate factors that result in something novel and more complex. For the prophet, what constitutes the new seems to emerge miraculously from the old: from the barren soil of exile, from a people's despair over the impotence of their God, from Babylon's absolute power over the world. But, again, emptiness is never truly empty. The new salvation proclaimed by the prophet took root within an existing level of social and religious complexity primed for something truly extraordinary.

The seemingly "empty" reality of the exile is, in fact, filled with background. The historical and social situation was more complicated than the prophet lets on. Life in a foreign land under Babylonian hegemony afforded many exiles opportunities of livelihood. As the prophet Jeremiah commanded the exiles:

> Build houses and dwell in them; plant gardens and eat their produce. Take wives and have sons and daughters. . . . But seek the welfare (šālôm) of the city where I have sent you into exile, and pray to YHWH on its behalf, for in its welfare you will find your welfare.
> (Jer 29:5–7)

Foreign soil was not only the land of captivity but also the land of opportunity. Some were able to establish niches in the land of exile, which would explain why, after fifty-plus years, many ignored the prophet's call to return "home." They remained and prospered in Babylonia; their roots had sunk too deep.[67]

The complex social milieu of captivity and opportunity helped to shape the prophet's novel message. His pronouncement of new life was already evident, "germinating" as it were, among those exiles who were actually prospering. But the prophet did not want them to sink their roots too deeply in foreign soil. His repeated call to "wait" for YHWH (40:31; 49:23; 51:5) served to warn fellow exiles against fully pursuing their livelihood in Babylonia and to encourage them to remain ready for a new exodus and a new community in their land of origin. Just such an opportunity arrived in the form of Cyrus II, whose conquest of Babylon afforded the exiles the opportunity to return home and start anew. Cyrus, the prophet boldly announced, was none other than YHWH's anointed or "messiah" (45:1)! A pagan king, a Zoroastrian no less, elected by YHWH to "feed his flock like a shepherd" (40:11), was no doubt a scandalous thing in the eyes of many exiles. It was disturbingly "new." But for the prophet it made perfect sense in light of the creator's boundless, cosmic reach. This revolutionary act of God reflects, moreover, the prophet's emergent theological vision: monotheism.

Emergence of Monotheism

The trauma of exile brought on by the loss of land, temple, and king was the social catalyst for a new theological vision, one that emerged from the fertile soil of religious polytheism. The concept of emergence is helpful in understanding the prophet's theological innovation, his claim that the God of Israel, YHWH, is the one and only God, the creator of all. Absent in "Second Isaiah" is the Decalogue's prohibition: "You shall have no other gods before me" (Exod 20:3; Deut 5:7), because for this exilic prophet there *were* no other gods besides God, only idols, and to worship idols was to worship human-constructed artifacts, a mark of theological insanity (cf. Exod 20:4; Isa 44:9–20). How the prophet of newness came to this revolutionary claim is best described as emergent.

Before the exile, ancient Israel tacitly acknowledged the existence of other deities even as it worshiped its own. Up until the exile, YHWH was considered supreme among the gods, a theological notion scholars call "monolatry" in distinction from "monotheism."[68] Some of the more ancient traditions in the Bible feature vestiges of polytheism, or what Mark Smith aptly calls "the older worldview."[69]

When *Elyon*[70] apportioned the nations,
>   when he divided up humankind,
> he established the boundaries of the peoples
>   according to the number of deities.[71]
> Indeed, YHWH's portion was his people,
>   Jacob his allotted share. (Deut 32:8–9)[72]

As all the nations were given their inheritances (i.e., their territories), they were also allotted to the various gods. Israel fell to YHWH, one deity among others, all distinguished from *Elyon*, "the Most High," otherwise known as El, the chief deity and "father" of the gods. Later biblical traditions, including "Second Isaiah," effectively merged these two divine figures into a single deity.[73]

By tracing the development of biblical traditions, one finds the God of Israel taking on the characteristics of other deities, including the Canaanite deity Ba'al, the warrior god of storm and fertility. The qualities of various Canaanite deities, in effect, became incorporated into the repertoire of attributes ascribed to YHWH.[74] In biblical poetry, the warrior imagery traditionally attributed to Ba'al came to be conjoined with the eternal age and majesty conventionally ascribed to El and ultimately subsumed under ancient Israel's Godhead, YHWH.[75]

Mark Smith refers to this process of assimilation as a *"convergence* of titles and imagery of deities to the personage of Yahweh."[76] It involved the incorporation of various attributes, characteristics, images, and titles once associated with other deities but now ascribed to Israel's God. While some of these deities were condemned in ancient Israel's faith, such as Ba'al and Asherah, others were easily assimilated, such as El.[77] For monotheism to succeed, Israel's God had to absorb the qualities of the other deities, stripping the gods of their efficacy and eventually displacing them entirely.[78]

To borrow from biology, one could call the emergence of monotheism a matter of theological "endosymbiosis." By theologically "engulfing" the other deities, God became "a plurality of conceptions."[79] But Israel's God did not simply become the summation of all gods. YHWH's differentiation also emerged in the process.[80] Call it theological singularization. New divine qualities emerged not shared by any other deity, reflecting a process of theological evolution *and* revolution, of gradual development punctuated by creative innovation, culminating in the exilic poetry of "Second Isaiah."[81]

The crowning theological achievement of "Second Isaiah" was to have YHWH stand alone, but alone in manifold fullness. Stephen Geller identifies three originally separate aspects of divinity that came to be subsumed or integrated under Israel's Godhead: "God as king, as warrior, and as protector."[82] In

"Second Isaiah," however, the list grows longer and more differentiated. YHWH is depicted as a warrior (40:10; 42:13; 51:9–11), shepherd (40:12), king (5:7), comforter (40:1–2; 49:13; 51:3, 12), lover (43:4), husband (54:5), potter (45:9), father (45:10a, 11), mother (45:10b, 11; 49:15), Holy One (41:14, 16, 20; 45:11), redeemer (41:14; 43:14; 44:6, 24; 54:5), and covenant-maker (42:6; 49:8; 54:10; 55:3). Traditionally conceived as male, YHWH also incorporates female imagery:

> Can a woman forget her nursing child,
>> or show no compassion for the child of her womb?
> Even these may forget,
>> yet I will not forget you. (49:15 [cf. 42:14; 46:3])

Finally, it is the same God who creates the world who "creates" Israel (e.g., 43:1; 44:2, 21, 24; 54:5).

God's composite personality in "Second Isaiah" cannot be reduced to any one attribute. Neither is YHWH simply a compilation of all them. God's divinity is not measured simply by addition. In the fullness of divinity, the prophet's God stands utterly alone and fully transcendent, above all categories.

> It is [YHWH] who resides above the circle of the earth,
>> whose residents are like grasshoppers;
>>> who stretches out the heavens like a curtain,
> and spreads them like a tent to reside in. (40:22 [see also vv. 12–17])

> For my thoughts are not your thoughts,
>> nor are your ways my ways, says YHWH.
> For as the heavens are higher than the earth,
>> so are my ways higher than your ways,
>>> and my thoughts than your thoughts. (55:8–9)

YHWH's transcendent status rises above the myriad attributes and roles that are ascribed to the deity. "Second Isaiah's" conception of deity is more than the sum of its roles. Except for one. God's most central role is also, not coincidentally, the one that fits God's transcendent status most fully: creator. The creator of all is "above" all. God creates both darkness and light, the old and the new. YHWH is a divine singularity, incomparably and exclusively divine, whose creative reach knows no bounds. Such a view of divinity is nothing less than an emergence from the fertile soil of religious thought upended by social trauma.

Creation to be Continued

Modern science has dispelled once and for all the notion that the world is static and unchanging. The scientific notion of emergence broadens, if not deepens, "Second Isaiah's" discernment of God's *continuing* creative activity. For the prophet, original creation and continuing creation are not different categories or alternatives. The primordial past and the surprising present are part of a seamless whole of divine activity that cuts across physical and cultural domains. Israel's deliverance from exile is folded into God's ongoing work of and in creation.[83] Israel's redemption is as novel and surprising as it is natural, exemplified by falling rain and sprouting plants. More than a primordial event or a continued state of preservation, creation for the prophet is God's ongoing work of innovation.

An open, interconnected world is increasingly recognized by science. The continual emergence of new complexities and dynamics testifies to an open, complexified world, from the cosmic to the more provincial domains, including the biological and, as the prophet was most concerned about, the political. The particularity of the prophet's announcement of newness, namely, of the exiles' deliverance from Babylon, was the result of a tectonic shift of imperial power in the ancient Near East. Yet such a shift was occasioned by the ambitions of a single, new player in the international realm, Cyrus II, deemed YHWH's "anointed" in Isaiah, a title that set him in line with Israel's previous kings but at odds with much biblical tradition (cf. Deut 17:15). "Behold, Israel's new king" is the prophet's bold announcement. Here, the new not only builds on but breaks through the old.

So also in the physical realm. The future is never completely determined by the past. The world is not only regular and repetitive but also surprising and anomalous. Even as entropy increases to the drumbeat of thermodynamics, certain "systems, surprisingly, evolve from states of lesser to greater order."[84] For all its dismantling effects, entropy can also play "a catalytic role during processes of unusual complexification."[85] The processes that give rise to decay and destruction also give rise to new and greater forms of "complexity, beauty, and diversity."[86] The forces that result in earthquakes, volcanoes, and tsunamis are the same forces that once established and now sustain the biosphere's richness.[87] Creation intermixes tragedy and triumph, death and life, decay and regeneration. It is full of "woe and weal" (Isa 45:7), and necessarily so for the kind of world that we have. "The clockwork universe is dead,"[88] and in its place is "the complex, fuzzy, dynamic, and opaque real world."[89] It is a world that is continually emergent and openly innovative, in which the smallest factor, if exercised at the right moment, can bring about large-scale changes: the flap of a butterfly's wing, a quantum fluctuation, or a flash of inspiration.

Nevertheless, the new is not created ex nihilo or even de novo. Although the exilic prophet rhetorically stressed the discontinuity between the old and the new, he also acknowledged, albeit indirectly, certain trajectories of continuity by which the old prefigures the new. As science consistently observes, the new invariably comes by way of the old: higher level complexities arising from their lower-level constituents, consciousness emerging from synaptic and social complexity, nucleated cells emerging from the symbiotic interaction of more primitive cells, the new exodus patterned off the old. As new life sprouts forth from the barren soil, so redemption emerges from punishment, and hope from despair and memory (see Isa 40:2). The new is a reconfiguration of the old, such as water and land reversing their conventional roles in God's new exodus. The processes and events of the past (and present) serve as the precursors and precedents of tomorrow. Does the new emerge as part of the "autopoietic," self-sustaining processes of nature,[90] or through the theopoietic activity of God? The prophet would have found the distinction artificial. For him, *creatio continua* involves both the natural and the supernatural in a correlated, seamless effort. And that is why both Qoheleth and Isaiah belong together, regardless of the shouting match.

10

The God Allusion

Creation as Consciousness-Raiser

What is the difference between a cathedral and a physics lab? Are they not both saying: Hello?

—Annie Dillard[1]

The principle that is opposite to reduction . . . is God's love for all things, for each thing for its own sake and not for its category.

—Wendell Berry[2]

Ethics is everything.

—Edward O. Wilson[3]

Our cosmic tour has ended for now. Through the eyes of the biblical authors we have crossed heaven's expanse and plumbed the sea's teeming depths, felt the animating breath of life and inspected Leviathan's impenetrable scales. Through the eyes of scientists we have pondered the weirdness of quantum reality, wondered about dark matter, and marveled at the dramatic beginnings of the universe. Fish that crawl, primates that console, wisdom at play, and stars that dance are just a few of the astonishing phenomena discovered on our tour. We have proceeded from the cosmic temple into the howling wilderness, from the summit of dominion to life at the margins. Behemoth and bacteria have given us pause. We have read the biblical witness of God's multifaceted character, from artisan and law-giver to improviser and provider, and we have considered the scientific testimony of evolution at work, innovating, selecting, modifying, and complexifying. Our trek, nevertheless, has covered only a fraction of the strange, manifold world disclosed in the Bible and discovered by science.

Yet even as the journey remains ever incomplete, now is the time to step back and reflect on where we have been on our hermeneutical adventure. It is also time to step forward, like Job, into our local contexts with a new outlook. In this final chapter we cast one last glance upon our journey, not to repeat where we have gone but to discern a direction forward. Here we will also reflect on the interpretive dynamics that have marked our way and explore how the biblical traditions, read in conversation with science, address the greatest crisis we humans have yet to face, one of global, even biblical, proportions.

Textual Orientation

We began our hermeneutical journey by doing something quite counterintuitive: we entered every biblical text through the "backdoor" as we worked our way to the "front." That is to say, we began with the text's ancient context before exploring its contemporary significance; we proceeded from what the text possibly said to what it could mean today. Most popular engagements with the biblical text remain at the front door, standing as a guest peering in, unwilling to cross the threshold and roam about inside. But by entering through the backdoor, we have taken full advantage of the text's invitation to wander and wonder. Through the backdoor we explored the texts' ancient contexts as much as they could be ascertained and speculated about how these texts might have engaged their earliest audiences. The ancient traditions, we discovered, are not scientific but theologically and existentially oriented. They were composed to provide guidance to the community, that is, to offer wisdom. We examined the literary contours of these texts and listened carefully to their theological cadences, noting how each tradition exhibited its own coherence of meaning. But as we did so, we also found that each text presented a probing foray into the mysteries of creation and of God.

By working our way from back to front, we noted possible points of contact between the biblical text and our context, particularly as informed by science. By opening the front door, we have let the world of science co-mingle with the world of the text. And so it must be. To lock the front door is to turn the Bible into a musty museum, a temptation of many biblical scholars. But to wall in the back entrance is to turn the Bible into something of a Rorschach test.

The analogy is far too simple. Shaped as we are by our various contexts, we carry with us our filters, values, perceptions, and prejudices, our cultural and familial contexts, as well as our religious and political convictions. They all play a formative role in our interpretation of the text. As we wander through a given text, we also construct its meaning for us. Or to put it another way, as we enter through the backdoor of the biblical text, the front door remains wide open, whether we like it or not, and the draft that wafts through the house is felt in every room.

That draft is no ghostly visitor. It is the immediate world in which we live, the world that invariably intrudes upon and fills the ancient world of the text. The text's world cannot be hermetically sealed off from the interpreter's world any-more than we can sever ourselves from our world. And for this study I have explicitly welcomed science as a part of my interpretive world, in part because I do not know enough about science and will never do so. But my own deficit has given me opportunity to invite others to join me on the journey. Another reason is that I am fully convinced that science is a gift; it exercises the intellect and propels human inquiry into domains unimagined by previous generations. Given its rigorously empirical practices, science has within the last century exceeded all other fields of inquiry in reshaping our understanding. Thanks to contemporary science,[4] we see the world, including ourselves, much differently. We see it as overwhelmingly vast and complex, as intelligible, dynamic, interdependent, cau-sational, processual, living, resilient, and fragile. Through science we see our-selves as an integral part of creation, not just *in* the world but also *of* the world.

Science also presses the challenge of interpreting the world of Scripture to its limit. Precisely because the creation texts of the Bible are not scientific in any modern sense, science can push the hermeneutical envelope in new and, I believe, edifying directions. Our journey, after all, has been guided by the fol-lowing question: How do we interpret the Bible today in the light of science? In some cases, science has deepened and broadened the perspective of the ancient text. Our awareness of Earth's biodiversity has expanded significantly and con-tinues to do so. The manifold nature of creation featured in Psalm 104 has swelled astronomically. In Job 38–41 God's abiding astonishment at the crea-tures of the wild is now matched by our amazement over creatures thriving near oceanic thermal vents and under polar ice caps, in deep rainforests and across inhospitable deserts, as we continue to explore our largely unexplored planet. The developing project called the Encyclopedia of Life (www.eol.org), the most ambitious attempt ever to catalogue Earth's species, raises the reli-gious revelry of biblical poetry to a new level. But it also sounds an alarm. The current diminishment of biodiversity through human-induced extinctions is cutting short God's fanfare for the common creature. According to the psalm-ist, creation's diversity is essential for God's joy. According to science, biodiver-sity—the variety of life on Earth—is essential for the flourishing of life, including human life.[5]

Another example: The laws of physics extend Qoheleth's static sense of time. For both the sage and the physicist, time does not flow; it simply is. And yet the inexorable rise of entropy gives time both its "arrow" and its "talons," driving the cosmos toward inevitable dissolution. Of course, the ancient sage knew neither the Second Law of Thermodynamics nor the Special Theory of Relativity. Even to say that Qoheleth anticipated them, semi-empiricist that he was, would be a gross overstatement and a categorical mistake: entropy belongs to physics; *hebel* ("vanity") belongs to the cynic. Nevertheless, the parallel exists, albeit virtually: for one physicist, at least, the scientific prediction of cosmic dissolution leads to despair. The time-immune world of physics and the static world of Qoheleth's observations, the progression of entropy and the inexorable deterioration of life, resonate well, lamentably.

Other examples of positive contact between the Bible and science share a degree of ambiguity, if not mystery, such as the creation of light in Genesis 1:3 and cosmic inflation as theorized by science. The Big Bang of astrophysics and the Big Flash of Genesis both describe a spectacular cosmic beginning. Ancient tradition and modern science, moreover, credit light as a primordial and essential feature of the early universe. Nevertheless, scientists cannot as of yet reconstruct the state of the universe prior to the initial lapse of Planck time. Perhaps not coincidentally, Genesis 1:2 leaves ambiguous the state of things prior to the creation of light and time. God's "breath" fluctuating over the dark waters and the quantum state of near infinite density at $t < 10^{-43}$ seconds equally confound the mind.

An understanding of the natural sciences also supplements and transforms the interpretation of biblical texts. Cosmologists, for example, theorize a rapid and brief phase of inflation by which the early, tiny universe expanded in unprecedented fashion. This cosmic "jumpstart" is nowhere referenced in any creation tradition in Scripture. But by taking the Big Bang into account, a peculiarly dynamic dimension is added to God's first act of creation in Genesis 1. In the Priestly account, light itself is created as a domain, one that precedes all other domains—space itself. From what we know from science, the "creation" of light also entails the inflation of space. Far from diminished by science, the wonder that gripped the Priestly cosmogonist expands to the nth degree under the lens of modern cosmology.

In its encounter with science, the biblical text can also come to signify something quite different from what it says within its own context. One example is Qoheleth's depreciatory view of nature's cycles. Their incessant movements are for the sage a sign of cosmic *hebel* (Eccl 1:2–11). We know now, however, that such cycles—and many more, from citric and carbon to geological and stellar—make possible life as we know it. But little did Qoheleth know. The natural cycles the sage so disparaged as wearisome are in fact life giving.

Reading the ancient text in conversation with science also lifts up certain features of the text that might otherwise be overlooked. Case in point: the power of observation, particularly at the quantum level, highlights the creative impact of divine sight upon creation in Genesis 1. Because perception is itself an act, more than simple observation is signaled by the words "God saw. . . ." Something happens to creation when it is beheld by God throughout the creative process: creation is advanced, sharpened, brought into focus. By sight, creation attains its "goodness." Creation, according to Genesis and by analogy with quantum mechanics, requires not only a Maker but also an Observer.

Biological evolution, as well, lifts up certain features of the biblical text that at first sight seem insignificant. The orderly process depicted in Genesis 1 can be regarded as developmental, proceeding from light to human life, "from so simple a beginning," to quote Darwin, to "endless forms most beautiful and most wonderful."[6] Interpreted in conjunction with biblical faith, evolution also highlights the creative openness of the God who improvises, even experiments, in the process, as one finds in Genesis 2–3. "Earth is a laboratory wherein Nature (God, if you prefer . . .) has laid before us the results of countless experiments," observes E. O. Wilson.[7] As evolution's byproduct, biodiversity supplements the panoramic sweep of creation in Psalm 104 and Job 38–41, and it does so by also demonstrating the kinship of all life, of Job created *with* Behemoth, of the "groundling" sharing common substance with the nonhuman animals (Gen 2:7, 19).

Finally, the collisions between ancient text and scientific understanding must be acknowledged; they too must figure in the ongoing process of interpretation. The six days of creation, for example, cannot be allegorized to fit the sweeping saga of the cosmos as reconstructed by cosmologists today. Indeed, any attempt to do so leads only to more problems. The Priestly author(s) clearly had in mind a regular cycle of day and night for describing the orderly process of creation. In this case, the disparity between text and science at the temporal level leads the interpreter to explore another avenue of significance for the seven-day schema of creation in Genesis, namely, the importance of Sabbath-rhythm for human flourishing, indeed for all of life.

So also the garden story. Much of what constitutes suffering in nature, from predation and parasitism to earthquakes and disease, cannot be attributed to an original failure of responsibility on the part of humans, that is, to humanity's "Fall." Such endemic suffering existed long before humanity's appearance. Evolutionary biology exposes the limits of interpreting the garden story in line with the Christian doctrine of original sin. But science also lifts up the narrative's ecological import. The garden story is all about humanity growing up from its falling down, about its "falling forward" to moral consciousness through painful transitions. The Yahwist narrative is also about the human grasp for power, specifically divine power, and its tragic consequences. The garden story illustrates, symbolically, the adverse effects that humanity has wrought on nonhuman nature. With the help of science, the garden remains a story worth retelling while its message continues to repeat itself with ever greater vengeance.

In short, through its encounter with science, the text's meaning undergoes change: it is extended, deepened, supplemented, narrowed, and transformed. But not willy-nilly. Textual meaning *evolves* in the act of interpretation, and as it evolves, the new emerges out of the old. Meaning is created ex vetere, not ex nihilo; it remains textually based, and as such the text retains its constraints on the interpreter. Science, too, places constraints upon the interpreter. But acknowledging the constraints is only a part of the process. Engaging biblical tradition and scientific understanding also opens doors. As new discoveries about the world are reached, new discoveries about the text are made, and a greater complexity of meaning emerges in the process. The hermeneutical quest, thus, plunges us more deeply into the world of the text *and* into the world around us.

Bible and Science

Genesis 1 provides a strong precedent for inviting science to join the hermeneutical adventure.

> (1:20) Then God said, "Let the waters produce swarms of living
> beings, and let the winged creatures fly about on the earth, across the
> surface of the heavenly firmament."

> (1:21) So God created the great sea monsters and every living being
> that moves, of which the waters produced swarms, according to their
> kinds, and every winged creature, each according to its kind. And
> God saw that it was good.

In v. 20 the waters are commanded to create aquatic life. In v. 21 God takes credit for creating aquatic life, including the "great sea monsters." One finds in these two parallel verses a relatively naturalistic and a fully supernaturalistic account of the genesis of life, both coexisting quite happily. There is no "however" or "but," no adversative conjunction, separating these two verses; one flows from the other. Indeed, the waters are acknowledged in both verses for having had a hand in creation. Juxtaposed as tightly as they are, these two verses bring a little closer the theist and the evolutionist, the creationist and the scientist. They could even suggest a model of divine activity that operates naturally.

For the sake of review, the hermeneutical points of contact or "virtual parallels" between the biblical creation traditions and the scientific understandings of the natural world are listed in table 10.1 in the appendix. From this highly selective overview, one can readily see the distinctive theological accents of each tradition as they are "virtually" matched by the various perspectives of science. Together, the creation traditions touch upon a wide range of scales. Genesis 1, the most comprehensive and cosmically oriented of the traditions, divides creation into interrelated domains and fills them with agency, from the astronomical to the biological. Resonant with science, the creative process unfolds from the fruitful collaboration of chaos and order, culminating in a complexity of the highest order on Earth, *Homo sapiens*. Even a parallel with special relativity can be drawn out of the first day with the creation of light as both a domain and a marker of time.

Paired with this account is the garden story, whose opening scene is more geologically oriented and whose focus is more narrowly anthropological. Celebrating the commonality and uniqueness among all species, Psalm 104 and Job 38–41 revel in the zoological, placing *Homo sapiens* squarely within the great panorama of biodiversity. Covering both the microcosmic and macrocosmic extremes, Wisdom's lively "play" in Proverbs 8 can be found within both the quirkiness of quantum reality and the complex choreography of the cosmos, all caught up in the bewildering interplay of forces known and still unknown. But most important is Wisdom's most engaging partner (next to God), the human being, as attested in the wonder of human development and evolution, from infant to adult, from consciousness to wisdom, from nonhuman primate to *Homo sapiens*. Last but by no means least, Ecclesiastes and Isaiah present sharply contrasting yet fully interrelated aspects of reality, forces that have held sway throughout the life of the cosmos and the history of life: stasis and change, open and closed systems, the transient and the evolving, the entropic and the complex, the familiar and the novel.

Taken together, these traditions bear witness to creation's manifold nature. From the nudge of science, biblical creation attests to nature's intelligibility and unpredictability, its emergent properties and sustaining regularities, its differentiated diversity and its ordered interdependence. Like science, the Bible presents a remarkably thick depiction of the world.

Canonical Consonances

Varied as they are, the Bible's perspectives on creation are literarily and canonically bound. As Einstein observed, every measurement of an object's speed taken from a different frame of reference is equally valid. The same could be said about every creation tradition in Scripture. Each one has its own validity and integrity; each adds its distinctive contribution to the manifold nature of creation. Even so, one tradition does not stand apart from the other. They are canonically connected. Respecting the differences among the traditions, while discerning their interconnections, has been called the task of "polycentric" hermeneutics.[8] One could also call it the hermeneutics of relativity.

But is it all relative? Does the sheer diversity of the traditions undercut their authority and relevance? Are there coherent claims to be found amid the myriad differences that distinguish the traditions? Let us look again at the distinctive features of each creation tradition (see table 10.2 in the appendix). Each text or corpus of texts presents a remarkably different view of creation's integrity, of its process of becoming, of God's role, and of humanity's place. Take the first two accounts: one begins with water (Gen 1:2), the other with dry land (2:5). They part ways even farther as Genesis 1 likens creation to an unfolding temple in time and Genesis 2 to a garden flourishing in the desert. Beyond Genesis, the garden of "plenty" ('ēden) finds its antithesis in Job's wilderness of want. Whereas the animals find their place in relation to the "groundling" in the garden, God in Job sets them free to roam a land beyond human control. In the wild, Job learns, rather than gives out, their names.

Despite their differences, the creation traditions are tied together. That simple canonical fact alone gives license for relating them, varied as they are, to scientific understandings, divergent as they may be from the biblical traditions. Scripture provides the warrant and impetus to seek constructive connections, while also taking into account the irreducible distinctions. To explore the connections among the biblical traditions themselves is to do something comparable to what I have done in relating scientific understanding and biblical tradition together throughout this study: explore hermeneutical points of contact. In other words, the fruitful interactions I have tried to facilitate between Scripture and science are paralleled within the biblical corpus itself.[9] Because the intertextual connections among the various creation traditions cut across diverse genres and perspectives, they offer a basis for exploring the interdisciplinary connections between biblical theology and natural science, two very different genres in their own right.

As for the two accounts in Genesis, the divergent opening scenes of water and land, each in its own way, depict a harrowing scene of deprivation. The deep, dark waters of Genesis 1 lack form and substance; the dry land of Genesis 2 lacks fecundity. As the narratives develop, one complements other: the waters must be divided by the land to advance and sustain creation's emerging structure; the barren land requires water to provide the grounding for life. Water and land, form and fecundity—both are acknowledged as essential for creation. The Priestly account and the Yahwist story "need" each other to present a fuller picture of creation. Together they provide the central elements of land, water, and air—the elemental conditions for life.

Another natural pairing is, again, Psalm 104 and Job 38–41. Both texts stretch the boundaries of creation beyond habitat for humanity. The wilderness takes center stage along with its wild inhabitants. Lions, onagers, mountain goats, and Leviathan are shared between these two accounts, and in both God takes on the role of provider. But whereas Job 38–41 emphasizes the animals' fierce independence, Psalm 104 lifts up their habitational dependence. In the psalm the creatures of the wild inhabit their places; in Job they freely roam. In Job these denizens of the margins are terrifyingly formidable; in the psalm they are dependent and domestic in God's eyes. Such creatures are quintessentially wild and quintessentially at home in the wild. The mountain goats in the psalm claim their refuge in the "high mountains" (Ps 104:18), but in Job they, along with the deer, give birth to young who "go forth and do not return" (Job 39:4). Psalm 104 shows how the vastness of the wilderness is vastly inhabited. Hence, trees receive greater emphasis in the psalm than in Job; they are a distinct part of creation's habitational landscape. As Job and the psalm together affirm, dignity and dependence are the hallmarks of creation; the creatures that inhabit the wild share the need for habitation. If the psalm is the poet's fanfare for the common creature, Job 38–41 is God's ode to the uncommon creature. The creature in question is one and the same.

Like Psalm 104, Proverbs 8 emphasizes the habitational dimension of creation. Human beings, the poem acknowledges, inhabit the earth (Prov 8:31), but they do so only in service to the preeminent inhabitant of all creation, Wisdom. And like any good habitation, creation is made safe and secure for Wisdom's play, as well as rich and engaging for her delight. As the cosmic child, Wisdom is creation's paramount subject. And like the wild creatures in Job, she is free to roam and play. God and creature are her playmates. Though her human playmates are not prominently featured in this poem, Wisdom plays a crucial role in their behalf. She facilitates their growth in wisdom. More than any other creation tradition, Proverbs 8 emphasizes the dynamic of human growth and development, from conception to maturation. Wisdom puts the *sapientia* into *Homo sapiens*. Human ontogeny recapitulates Wisdom's ontogeny; consequently, the human self is made in the *imago sapientiae*.

A seemingly less than natural pairing is Ecclesiastes and Second Isaiah. At first glance, they appear to be diametrically opposed. Yet despite their divergent perspectives, there lies a host of interrelationships, made all the more evident by science. Whether cosmologically or biologically demonstrated, the emergence of novelty depends on the work of entropy; evolution is driven in part by dissolution. "[T]ranscience, dissolution, and death are the entropic price" for new systems and properties to emerge, for they arise only at the expense of old ones.[10] As entropy leads to disorder and decay at one level, it occasions growth and the emergence of new order at other levels.[11] Second Isaiah's proclamation of the "new," in fact, requires Qoheleth's lament of *hebel* ("vanity"). The science of entropy and evolution, of reduction and emergence, builds a vital hermeneutical bridge between Ecclesiastes and Second Isaiah.

Humanity's Place

A critically relevant way of comparing the creation traditions of the Bible is to linger over their anthropological profiles. The various traditions offer remarkably divergent portrayals of humanity. Differing geneses reflect differing identities and roles. In the opening narratives of Genesis, for example, the elevated "image of God" (Gen 1:26–27) is matched by the earthy image of humanity created out of the "dust of the ground" (Gen 2:7), the ground from which all other animals are also created (Gen 2:7, 19). In the first two chapters alone, two anthropological extremes meet and merge: the *imago Dei* and the *imago terrae*. Made from the earth yet cast in God's image, humans are bound to both God and ground. Corresponding to these two perspectives of human genesis are humanity's divergent roles. On the one hand, humanity wields a Sabbath-tempered "dominion" and, on the other, performs a divinely ordained service. Such a composite depiction finds its unity in an irreducibly complex, if not paradoxical, profile of human identity. Human beings are at once set apart from and held in common with the rest of life. Animals we are, but divergently so.

Taken together, the first two creation accounts of the Bible conjoin kingship and kinship—kingship over creation and kinship with creation—as two fundamental aspects of humanity's relation to the rest of creation. And woe to any interpreter who rends them asunder! Kingship without kinship allows for exploitation and subjugation in our partnership with creation. Kinship without kingship could devalue initiative and ingenuity in our stewardship of creation.

The anthropological relationship between Genesis 1 and Genesis 2–3, moreover, drips with irony. The "image of God" finds some resonance with the conclusion of the garden story in Genesis 3:22: "See, the groundling has become *like* one of us, knowing good and bad," so God observes, confirming the serpent's wily prediction (3:5). As "image" corresponds to "likeness" in 1:26, so "image" links itself canonically to the theme of likeness in 3:22, but with added nuance. The second creation story tells how humanity has attained "God-

likeness" in a way that is not welcomed by God. The couple is expelled to sub-
sist on the cursed ground rather than elevated to rule the earth. The Yahwist
brings God's lofty "image" painfully down to earth.

Another ironic consonance between these two traditions is found in the
language of dominion, which is, admittedly, harsh in Genesis 1: "fill the earth
and subdue it and rule over . . ." (Gen 1:28).[12] Such language is absent in the
garden, where mutuality rather than hierarchy and violence is the norm. The
woman and the man flourish in mutual intimacy, and humanity and humus
maintain a symbiotic relationship of productivity. But this picture of harmony
becomes noticeably marred. The language of rulership raises its head only
when the curse is pronounced against the woman: "Your desire shall endure
for your husband, even though he shall rule over you" (3:16b). More an etiology
than an ethical charge, the curse in Genesis 3 takes up the royal language of
Genesis 1 and perverts it.

The curse-filled picture of the man ruling over the woman is not at all what
is envisioned in Genesis 1. The curse in Genesis 3 does not live up to the high
calling of kingship universalized across gender in Genesis 1. It breaks apart
ethical ideals and shatters salutary relationships, leaving the pieces strewn
across the moral landscape to be picked up and put back together, but only
partially and with concerted, painful effort. "Power with" is twisted into "power
over," power shared into power lorded. The ground, moreover, pushes back by
resisting the farmer's labors (3:18–19). The groundling has become subject to
the ground, and greater effort is required to produce something edible. Who
will subdue whom, ground or groundling in this creational conflict? And so we
come full circle back to the beginning, back to "subduing" the earth in Genesis
1, which acknowledges "the particular harshness of subsistence agriculture."[13]
But sandwiched in between cosmology (Genesis 1) and etiology (Genesis 3) is
the ethos of mutuality and harmony embodied in the garden, an ideal that
remains to be attained. As if to say: only "by the sweat of your nostrils," only by
concerted, painful effort and resolve, shall you reclaim some semblance of the
life of mutual kinship with each other and with all creation. The moral life of
mutuality is made all the more necessary, even as it is made all the more difficult, in life outside the garden.[14]

There is, however, much more to human identity and responsibility than
what is portrayed in Genesis. Other creation traditions count humanity as one
among myriad biotic species that thrive on Earth, as in Psalm 104 and Job
38–41. In the psalm, human flourishing fits within the larger natural rhythm of
life in which all creatures flourish, each within its own niche (Psalm 104:19–23).
As diurnal creatures, human beings take the "day shift," while the lions take
the night. Humans have their homes just as lions have their lairs, coneys their

crags, and birds their trees. Humanity's place in the world is one among count-less places that accommodate the panoply of life. Nowhere does the psalm acknowledge the pride and place of human dominion; humans simply have their assigned place and rhythm in the world, and it is by no means central. Each species rules its own place and creates its own niche(s), all within given bounds. Human culture has its legitimate place on Earth, but it is one bounded by other "cultures" of the nonhuman variety.

Job, too, has his place, but it is defined entirely in relation to the wilder-ness. As the margins of creation take center stage within Job's limited purview, so Job finds himself and, by extension, all humanity on the periphery. The wil-derness, with its host of fierce and alien creatures, is no proximate place for human beings. By definition, the wilderness remains out of human reach and control, and so it should be. While humanity has no place, let alone dominion, *in* the wild, it remains inseparably linked *to* the wild. Behemoth is a creature made "with" Job, so God declares (Job 40:15; cf. Gen 3:12). In this monstrous creature, Job finds his linkage with all creation. The claim is not without war-rant elsewhere in biblical tradition. All animals, including the *'ādām*, are made from the same originating substance in Genesis 2; they all share common "ground." And yet how gloriously strange are God's creatures, including human beings.

As for Wisdom's world in Proverbs 8:22–31, humanity remains decentered in so far as it exists not for its own sake but for that of cosmic, playful Wisdom. But it is precisely Wisdom's position of primacy in creation that highlights something central about humanity, namely, its capacity and purpose to grow *in* wisdom. Humans exist for the sake of Wisdom as creatures worthy of her play in the world, for their growth depends on her even as her play and delight depend on them. As the psalm commends our praise to God for creation's marvelous diversity, so the sage commends Wisdom's wide-eyed delight over creation in all its fullness. While the psalmist profiles the human species as *Homo laudans* (the "praising human"), the sage makes a slight adjustment in favor of *Homo ludens*, the "playing human."[15]

In Qoheleth's world, however, Wisdom's game can only be "won" by los-ing. The human striving for gain and ultimate meaning invariably turns up empty. In a world destined for dissipation, humans must relearn their role. Against Genesis 1, Qoheleth deconstructs the charge to exercise dominion. Human dominion exacts too great a toll on human and creational well-being, the sage would observe, for all toil is toil done in vain. How can we control the world if we cannot even control our own lives? Qoheleth asks. Our lot, thus, is not to dominate the earth, wresting from it something of lasting value or utility. Rather, it is to receive from the earth, however temporarily, something gracious

for our fragile, transient lives, namely, the simple gifts of God found in creation.[16] Psalm 104 specifies those gifts as the grape, the olive, and the grain (v.
15), the threefold source of sustenance and joy. The gifts of God are the gifts of
the ground. In Qoheleth's world, humans are cast in the image of the recipient,
though they invariably choose to cast themselves in the image of the toiler, a
hopelessly self-enslaving image. For Qoheleth, the good life can only be a "nonprofit" enterprise.

In "Second Isaiah," humans are primarily recipients as well, but of God's
new and renewing work in creation, to which they are to bear witness (Isa 42:9;
43:19; 48:6). In the eyes of the exilic prophet, God's new work is tied historically to liberation from captivity, a liberation that enables a dispersed people to
be gathered together in unity. On a more universal front, "Second Isaiah" cracks
open Qoheleth's closed, cyclical world of entropic demise, allowing for newness to take root. But the new reality the prophet announces cannot be realized
fully except by living into that reality. Only then can the desert blossom, tama-
risks spring up, and Eden be recultivated (35:1–2; 44:4; 51:3).

The canonical composite of these various traditions yields a complex profile of human identity. Nearly divine and eminently earthly, humans are both
the theophanic creatures Genesis 1 claims they are and the fragile, dirt-bound
creatures Genesis 2 says they are. The first two chapters of the Bible present
two divergent anthropological extremes; the subsequent traditions fill in the
wide spectrum. As Wisdom's playful partners, human beings are deemed distinctly wise. As day laborers in the pursuit of life, human beings are bona fide
citizens of the kingdom *animalia*; their work and livelihood fit within the alter-
nating rhythm of life shared by all creatures. As peripheral creatures in God's
wild kingdom, humans are one among many inhabitants of Earth, all alien to
each other yet all of inalienable value to God, their creator. As central, powerful
agents, human beings are uniquely responsible for Earth's fate. As servants of
the soil, they help sustain Earth's flourishing. As grateful recipients, they are
sustained by the simple gifts of the ground, the signs of God's grace and providence. As witnesses to the truth, human beings tell of God's new work in creation and live it.

Each in its own way, the creation traditions lift up the critical value of community. So does science. Life arose and developed communally, from a network
of molecules robustly evolving into a single cell, from single cells merging and
differentiating to form multicellular life, from coordinating neurological nodes
making possible cognitive development. Symbiotic cooperation has made it all
possible, just as, according to the ancients, the symbiosis shared between
ground and groundling, between man and woman, between humans and all
animal life, indeed, between creation and creator, sustains life in all its rich

abundance. Community is the secret to any sustainable ecosystem. And as nature's cycles sustain the natural community, so covenant maintains the moral community. According to biblical lore, Noah's preservation of creation set the occasion for God's covenant with all creation, God's self-restraining order to never again unleash destructive floodwaters upon the earth (Gen 9:8–17). With the rainbow as its sign, God's covenant, like Sabbath, sets an example: it offers a model for human conduct, for only by covenant, by the resolute work of the human community working in consort, can life be sustained amid a new onslaught of destruction, this time wrought by human hands, against the community of creation.

Word for a Warming World

It is only fitting, then, to conclude our journey with a discussion of humanity's responsibility to creation, as informed by the ancient traditions and by the mounting evidence of environmental degradation. The Bible's multitextured view of creation offers a multifaceted ethical framework for addressing this crisis. I apologize to the reader if this final section sounds too sermonic or heavy-handed, but this is what I would call the "come-to-the-creator" moment of our journey. In marshaling the biblical resources to address our threat to creation's integrity, I do find it necessary to move from the descriptive and the reflective to the urgently prescriptive. Confronting our growing legacy of environmental destruction with eyes wide open, with faith seeking understanding, challenges us to move toward a new way of living in God's good and groaning world.[17]

Much to Earth's harm, we have culturally evolved within the last two and half centuries into *Homo industrialis*,[18] remaking ourselves in the *imago deletoris*, in the "image of the destroyer." Nevertheless, we remain *Homo sapiens*, made in the *imago Dei*, a distinctly wise species uniquely endowed with God-given responsibility. Science provides us the data of danger. It excels in revealing nature's interdependent mechanisms and feedback loops. As science has given fair warning about global warming and other forms of environmental degradation, it is also developing technologies that can shift our dependence from fossil fuels to renewable energy sources. But even with much of the technology in place, or soon to be, the political and economic will is not.

If there is one thing that this grave crisis teaches us about the relationship between science and faith, it is that "man" cannot live by science alone. Science can explain the crisis, identifying its root causes and projecting trends into the future; it can even suggest ways to mitigate it. But science cannot bring about the repentance, indeed conversion, necessary to chart a new way of life. Science alone cannot provide the impetus for changing human conduct. It does not provide a compelling warrant for acknowledging the intrinsic value of life or its sanctity. Even the "anthropic principle" cannot provide the compelling symbols of ultimate meaning and orientation.[19] We may recognize how improbable Earth-based life is in the universe, with or without God, but that will not prevent us from pushing much of life on Earth to the brink of extinction. If, however, we take our cue from Genesis, damaging creation is tantamount to defacing God's sanctuary, an act of utter sacrilege.

In his recent book, Richard Dawkins identifies the theory of evolution, specifically Darwinian natural selection, as a "consciousness-raiser." What he means is that evolution powerfully "explains the whole of life" and demonstrates the "power of science to explain how organized complexity can emerge from simple beginnings without any deliberate guidance."[20] The stunning simplicity and elegance of evolutionary theory cannot be gainsaid. Rather than one option among many, evolutionary theory is foundational for all the life sciences. "Nothing in biology makes sense except in the light of evolution," so Theodosius Dobzhansky famously claimed, a biologist who was also a Russian Orthodox Christian.[21] Biological evolution demonstrates most fully the power of scientific explanation.[22] And, as we have seen, evolution adds a developmental depth and open-endedness to what the Bible's creation traditions say about nature, including ourselves.

Evolutionary theory also, as Dawkins rightfully points out, exposes the "flaws" of creation and the "cruelty and wastefulness of natural selection," from predation to extinction.[23] Darwin himself noted the "clumsy, wasteful, blundering, low and horridly cruel works of nature."[24] But again, no biblical tradition ever claimed creation as perfect, not even the magisterial account of Genesis 1. While proclaimed "extremely good" in its God-given capacities to sustain and develop the diversity of life, creation is also granted the freedom to become. The earth and the waters are not preprogrammed machines; they are invited by God to further the process of creation, each according to its own natural and, yes, messy way. Both the sage and the psalmist acknowledge the intractable presence of suffering. Scientifically speaking, a *perfectly* harmonious creation could never give rise to the emergent complexity that characterizes the world as we know it.[25] The beauty and order that we know rests on the same processes that produce suffering and death. The "weal" depends on the "woe" (cf. Isa 45:7).

But, I ask, can the awareness of evolution, in all its theoretical elegance and empirical power, provide sufficient "consciousness-raising" to inspire new practices, to establish a new orientation toward the environment, that is, toward creation? Yes, global warming could dramatically disrupt the "accumulative power" of natural selection, as Dawkins puts it. But is that enough to motivate significant change in our habits of consumption? A keen awareness of the sanctity of life does not emerge unambiguously from evolution. Rather, reverence for life arises directly from discerning the world as *creation*, as the open-ended product of God's resolve and delight. In the faith spawned by the ancients, the climate chaos spawned by our imperious practices is nothing less than a breach of covenant, one that threatens a new inundation of destruction. To claim the world as created is to claim God's care for it and our responsibility to care for it. In faith, sacred responsibility meets holy passion.

If science excels in revealing the wonders of creation, then faith excels in responding to such wonders in praise, humility, and gratitude, out of which emerges the holy passion and sacred duty to "serve and preserve" creation and to address anything that would threaten its integrity. Scientifically informed faith raises both consciousness and conviction. According to Psalm 104, damaging habitat and diminishing diversity are tantamount to divesting God of joy and passion for creation. Through deforestation, greenhouse gas emissions, and overexploitation, we are systematically destroying this basic feature of creation, namely, its habitational integrity for diversity.[26] The insidious nature of global warming is that it is global; it cannot be confined to the culprits, to the gas-emitting nations. Cutting across all boundaries—economic, national, and ethnic, land and sea—climate chaos knows no limits on Earth. Unlike the waters in Job and Psalm 104, there is no natural or divinely prescribed boundary to stop the spread of greenhouse gases from inundating the earth (cf. Ps 104:9; Job 38:10–11). To be sure, some suffer from it more than others: the poor are bearing the brunt, ever more so. They are our first environmental victims and refugees, their habitats destroyed because of the energy hungry, carbon-emitting habits of developed nations.

For God so loved the world that God gave Leviathan and Behemoth, "the first of the great acts of God" (Job 40:19), so that all who would see them will gain everlasting wonder. As the God of the psalmist bursts with joy over the vastness of creation's diversity, so the God of the Joban poet swells with amazement over every wild thing. As Mary Midgley puts it, "We need the vast world, and it must be a world that does not need us; a world constantly capable of surprising us . . . since only such a world is the proper object of wonder."[27] Yes, Leviathan does not need Job; nevertheless, Job needs Leviathan, and so do we all. As the Leviathans, large and small, of land and sea are rapidly being pushed to extinction while we deplete and fight over Earth's nonrenewable resources—all carbon-based and now blood-drenched—we must lift our eyes toward that which truly sustains. As God once called the exiles to live into the new reality of freedom and set a new course of conduct, so today we are called to break the bonds of our consumptive, self-enslaving ways and live anew. As newness can emerge in nature, so it can also emerge in human culture and behavior. A new day can dawn.

And for each new day, so Qoheleth would remind us, the sun rises, the wind blows, and the streams flow into the sea (Eccl 1:5–7). Solar, wind, and water: these renewing resources have been present before the dawn of human history. As Qoheleth grew weary of them, we have ignored them outright, much to our detriment. Even as the cosmos slowly unwinds and dissipates, the wind, the sun, and the flowing streams remain the closest thing to eternal, inexhaustible sources of energy that we will ever know. Yet the first two remain largely untapped, and the third has been exploited in such a way as to destroy habitats and entire ecosystems.[28] The solutions remain ever before us. Qoheleth warns against the striving for luxury and legacy and commends a life of grateful simplicity, of acceptance and sufficiency, a life of settling for less while living for joy. Qoheleth exposes the self-destructive striving for gain, whose environmental consequences we are just beginning to face. Seeking gain, the sage points out, bears its own heavy yoke. Seeking simplicity, however, replaces greed with gratitude, enslavement with freedom, despair with joy. Qoheleth is no hedonist, for he does not seek enjoyment as one strives for gain. Qoheleth, rather, *receives* pleasure, however momentary, in gratitude to God. In his own small (and cynical) way, this sage of sages acknowledges the giftedness of creation. Creation's delicate gift of sustenance is best received and shared, not grasped and exploited. And it is in the receiving, so claims Qoheleth, that God is encountered.

And it is in the playing, according to Proverbs, that Wisdom encounters us. God so loved the world that God gave daughter Wisdom, so that everyone who plays with her may gain enlightened life. Proverbs boldly claims that human beings exist not for themselves but for Wisdom, specifically for her play and enrichment. Yet, reciprocally, Wisdom's play nurtures and enriches all conscious life. Her play is mutually edifying, and there are no losers, except those who refuse her invitation or simply quit, much to their impoverishment. Wisdom's play, moreover, is no otherworldly, mystical exercise. Both Proverbs and Psalms declare God creating the world *in* and *by* wisdom (Ps 104:24; Prov 3:19). However, more than creation's intelligibility, more than its orderliness is meant, as science so powerfully demonstrates. Creation in wisdom reflects its joie de vivre, a vitality reflected in its interactive, self-regulatory, life-sustaining processes.

Creation according to Proverbs is made for Wisdom's play, and to play is to discover and cherish creation made in wisdom. It is what scientists do best in their quest to understand the wonders of creation. It is what people of faith do best in their quest to cherish and care for creation. Wisdom takes hold of both science and faith to engage in the play of discovery and the exercise of responsibility. Wisdom is Stephen Jay Gould's "beautiful and coherent quilt" that unites the separate "magisteria" of science and religion.[29] But wisdom also calls for action, the wisdom to relinquish destructive habits and to do so joyfully. The world is Wisdom's playfield, not a battlefield, and we are her partners, not her opponents.

The morality play of Genesis 2–3 weds partnership and service in our work for creation. The rationale is simple: we are inextricably tied to creation, bound to it, in fact. Created from the "dust," we are groundlings to the ground and partners to each other. We are kin to creation. But as this ancient tale unfolds, humanity fails to acknowledge such kinship, the kinship to which evolution also points. We failed in the garden, and we continue to fail to "serve and preserve" creation (Gen 2:15). The "original sin" was and remains the failure to take responsibility. Its "originality" lies in its pervasively persistent force, manifested throughout human history. Had the primal couple openly confessed before God and expressed their willingness to take responsibility for their actions, perhaps God would not have expelled them, and they would have continued to serve the garden. But the story takes a dramatically different turn: it tells of the couple succumbing to fear, blame, and the will to power, which from Cain and Lamech to today continues to engulf the world.

The man and the woman were no longer deemed fit to serve God's garden. Today, outside the garden, only one thing is different. We have not only reached for divinelike power; we have grasped it and declared it rightfully ours in the name of economic development and freedom. The greenhouse gases we emit, along with the economic and military destruction we wreak, assert our dominant, godlike powers, and the earth becomes increasingly cursed as a consequence. There is no need for divine judgment to turn creation into a curse. We are doing it quite well ourselves.

Finally, we come back to Genesis 1, with the commanding charge: "Be fruitful and multiply; fill the earth and subdue it; and have dominion" (v. 28). The legacy of our destructive dominion, coupled with the tale of failure in Genesis 3, sheds a cautionary light on Genesis 1.[30] After two and half centuries of rampant industrialization, the "blessing" of dominion has become a baneful burden. And yet Genesis 1 reminds us that as the human species has become the problem, it is also bears the solution. We remain in charge, for only we can undo the havoc that is undoing creation. A new model of dominion must emerge, a dominion of self-restraint and natural development.[31] Technology shall remain, to be sure, a crucial part of life-sustaining "dominion," particularly as we develop ways to harness renewable energy sources efficiently. Nevertheless, in most areas of human "dominion," restraint rather than expansion and development is in order.

The kind of dominion needed today is a dominion of self-mastery, of self-restraint and respect that acknowledges creation's kinship, a dominion that points to a peaceable kingdom, or better, "kindom,"[32] and acknowledges that human beings, cast in God's image, are an integral part of creation. Biblically, dominion can play out in one of two ways, which collide in the character of Cain. For whatever reason, Cain's sacrifice was rejected by God, while his brother's was accepted. In response, Cain violently subdues Abel (Gen 4:8). But such an exercise of dominion is actually a capitulation. Before Cain kills, God issues him a challenge:

> YHWH said to Cain, "Why are you angry, and why has your face
> fallen? If you do good, will you not be accepted? And if you do not do
> good, sin is crouching at the door; its desire is for you, but you must
> master it." (Gen 4:6–7)

God calls upon Cain to master the sin that is coiled up inside him and poised to strike. He is to rule over himself and, in so doing, choose life instead of death, kinship over fratricide. But Cain succumbs, committing violence. It is his sin; it is also the Bible's first "sin."[33] Cain cannot hide it, for the ground itself, the 'ădāmāh, bears witness as it cries out for justice (4:10–11). Cain becomes the Bible's first environmental refugee. His story exemplifies the tragic failure to exercise self-mastery and charts the devastating consequences of his crime. God's challenge to Cain is also directed to us. Can we exercise sufficient self-mastery? Can we, for the sake of God's good creation, stem the impulse to dominate and destroy? Made in the *imago Dei*, we take our cue from the God of Genesis, who in the very act of creation made room for creation, limiting God's own self so that creation could flourish in freedom. Self-limitation for the sake of the world, indeed for our own sake, is paramount. Such is the secret of Sabbath, the joy of fallow time.

What does it mean for us, in the face of mounting catastrophe, to "do good," as God had challenged Cain? It means to muster all the scientific, moral, and, yes, religious resources to change our devastating habits of consumption, which have privileged luxury over livelihood, greed over good. It means to compensate the victims whose blood has already drenched the ground, like Abel's. It means to mobilize ourselves toward peaceful practices, to see ourselves as creatures of the earth, entirely dependent on God and ground. It means to work and play in life-sustaining ways with each other and with all creation. It means to speak the truth, to speak the Word on behalf of a warming world, the Word on behalf of God and of the victims of our imperious practices, to speak on behalf of those that cannot speak, including polar bears and cedar trees. It is the Word informed by faith and science, the Word of Wonder and Love.

To claim the world as creation is not to denounce evolution and debunk science. To the contrary, it is to join in covenant with science in acknowledging creation's integrity, as well as its giftedness and worth. To see the world as creation is to recommit ourselves to its care, not as the fittest, most powerful creatures on the animal planet but as a species held uniquely responsible for creation's flourishing. It is to celebrate the inalienable beauty and dignity of all living kind and bear witness to God's manifold creation. It is also to bear witness to creation's groaning as the ground suffers from deforestation, mountaintop removal, toxic dumping, and rising temperatures. To see the world as God's intricate, intelligible, surprising, sustainable creation is to return to wonder and to go forth in wisdom, such that "the mountains and the hills . . . shall burst into song, and all the trees of the field shall clap their hands" (Isa 55:12).

Appendix

Biblical Text	Scientific Understanding
Genesis 1:1–2:3	
Creation by differentiation: from formless "chaos" to order	Entropic rise of complexity: from uniformity to structure
The Big Flash: "creation" of light Day 1 and Day 4 creation of space and time	The Big Bang: "creation" of the cosmos primordial light and star formation emergence of "spacetime"
Life "filling" various domains	Biological drive to "fit" various environments
Imago Dei: humanity uniquely made in the image of God	*Homo sapiens*: cognitive and cultural uniqueness
Genesis 2:4b–3:24	
Creation by improvisation	Improvisational nature of evolution
Dry Earth watered from above and below	Sterile Earth seeded with water from outer space and volcanic "outgassing"
Life created from the ground	Life emergent from stardust and organic material
"Soul" and the breath of life	Oxygenation of the atmosphere, enabling and sustaining the development of complex life
Naming the animals by the *'ādām*	Humanity's power to domesticate
Kinship of humanity with creation	Genetic and behavioral connections among the animals
Humanity's "Fall" eyes "opened" alienation and conflict	Painful evolutionary development consciousness and conscience negative impact on the environment

(*continued*)

TABLE 10.1 *(continued)*

Biblical Text	Scientific Understanding
Proverbs 8:22–31	
Creation made in wisdom	Creation's intelligibility
Liveliness of creation	Cosmic choreography of gravity and energy
Wisdom's play in creation	"playfulness" of quantum reality
Wisdom's play with humanity	Pedagogy of play human evolution and
human growth in wisdom	development
Wisdom as mother and teacher	Maternal locus of learning
Psalm 104	
Habitat for diversity	Biodiversity for diverse environments
Common creatureliness	*Homo sapiens* among the *animalia*
Dependence of life on God	Interdependence of life on Earth
Job 38–41	
Creation filled with alien life	Extremophilic life on Earth
Polycentric creation	Earth as multiverse
Job created "with" Behemoth	Genetic linkage of *Homo sapiens* to all *animalia*
Ecclesiastes 1:2–11 (3:1–8; 12:1–7)	
Wearying cycles of creation	Nature's life-sustaining cycles
Static creation	Closed universe
Transience of creation	Entropic end of the cosmos
Static nature of time	Illusion of past, present, and future
"Nothing new under the sun"	The "new" reducible to the old
Hebel ("vanity")	"Pointlessness" of the cosmos
Isaiah 40–55	
Heavens as unfurled fabric	Spacetime fabric of the cosmos
Creation of the new	Emergence of novelty
creatio continua	dynamic, open-ended universe

TABLE 10.2. The Seven Pillars of Creation: A Field Guide

Text	God as Creator	Character of Creation	Character of Humanity
Genesis 1:1–2:3	**Creation by word** God co-opting creation Creation from formlessness to form and complexity God as king, priest, and artisan Creation completed: 7th Day cessation	***Imago templi:*** creation as cosmic temple constructed in time Water-based creation with cosmic scope Creation as form-full and filled Hierarchy of creation Creation deemed "good": self-sustaining creation	***Imago Dei:*** humanity created in God's image Royal stewardship Dominion without domination Humanity differentiated in genders Humanity on top Blessing of procreation
Genesis 2:4b–3:24	**Creation by deed** God filling creation's lacks God as potter, gardener, and improviser God as the "ground" of being	***Imago horti:*** creation as garden, arable land Land-based creation with focus on the human family Kinship of creation: groundling and ground, male and female	***Imago terrae:*** human as created out of the ground Servant stewardship to "serve and preserve" the garden, to cultivate the land Kinship partnership human partnership with the land Blessing of companionship
Job 38–41	**Creation by provision,** sustained in freedom Creation as the object of God's care and awe God the biophile	***Imago vastitatis:*** creation as a vast wilderness Creatures as aliens invested with inalienable dignity Creation as polycentric	***Imago feri:*** humanity as kin to the wild ***Homo alienus:*** human as stranger in a strange land Humanity de-centered in creation
Psalm 104	**Creation by provision,** sustained by God's joy God as passionate provider God the biophile	***Imago habitationis:*** creation as habitation Habitat for divinity and diversity All life dependent on God	***Imago animalis:*** humanity as one species among many species ***Homo laudans:*** the praising human Humans as recipients of God's bounty

(continued)

TABLE 10.2 (*continued*)

Text	God as Creator	Character of Creation	Character of Humanity
Proverbs 8:22–31 (+ 3:1–8; 12:1–8)	**Creation by construction and procreation** God as builder and architect God as progenitor and play partner (*Deus ludens*)	*Imago domus*: creation as Wisdom's playhouse Cosmos as secure and engrossing *Imago nati*: Wisdom as growing child and play partner	*Imago sapientiae*: humanity in the image of Wisdom *Homo ludens*: the playful human Human growth in wisdom
Ecclesiastes 1:3–11 (+ 3:1–8; 12:1–8)	**Creation recycling itself** God's apparent indifference to the world God as inscrutable determiner of events God imparts joy through creation	*Imago futilitatis*: creation in futility (*hebel*) "Nothing new under the sun" Creation as closed and static, dying and decaying Sustaining gifts of creation: food and drink	*Imago operarii*: human as toiler, enslaved by the quest for gain *Imago acceptoris*: human as grateful recipient of God's gifts Human life as ephemeral and unpredictable Life as a "non-profit" enterprise
Isaiah 40–55 (excerpts)	**Creation re-created** God creating ever anew God as re-creator and redeemer God as the one and only deity	*Imago mutationis*: creation ever changing, made anew Everything "new" Creation as open and dynamic, living and growing	*Imago liberti*: human as freed in God's new creation *Imago testis*: human as witness to God's new creation Life of new orientation toward God's open future.

Notes

PREFACE

1. Louis C. LaMotte, *Colored Light: The Story of the Influence of Columbia Theological Seminary, 1828–1936* (Richmond, VA: Presbyterian Committee, 1937), 189.

2. Ibid., 212.

3. Ibid., 113.

4. Quoted in Laura Diamond, "Teachers Say Covering Evolution Can Be a Trial," *Atlanta Journal-Constitution* (27 October 2008): B6; or at www.ajc.com/search/content/metro/stories/2008/10/27/evolution.html.

CHAPTER 1: INTRODUCTION

1. Quoted in Huston Smith, *Beyond the Postmodern Mind: The Place of Meaning in a Global Civilization*, 3rd ed. (Wheaton, IL: Quest Books, 2003), 98, but without attribution. Sockman (1889–1970) was the Methodist pastor of Christ Church in New York City.

2. Holmes Rolston III, *Science and Religion: A Critical Survey with a New Introduction* (Philadelphia: Templeton Foundation, 2006), ix.

3. Said at the Scopes trial in response to Clarence Darrow's question about whether Bryan "made interpretations of various things." Edward J. Larson, *Summer for the Gods: The Scopes Trial and America's Continuing Debate Over Science and Religion*, 2nd ed. (New York: BasicBooks, 2006 1997), 3.

4. See the discussion book by Philip Morrison and Phylis Morrison and The Office of Charles and Ray Eames, *Powers of Ten: About the Relative Size of Things in the Universe* (New York: Scientific American Library, 1982).

5. Brian R. Greene, *The Fabric of the Cosmos: Space, Time, and the Texture of Reality* (New York: Vintage Books, 2004), 11.

6. Ursula Goodenough, *The Sacred Depths of Nature* (Oxford: Oxford University Press, 1998).

7. See, e.g., Stuart A. Kauffman, *Reinventing the Sacred: A New View of Science, Reason, and Religion* (New York: Basic Books, 2008).

8. Melvin Konner, *The Tangled Wing: Biological Constraints on the Human Spirit*, 2nd ed. (New York: Henry Holt, 2002), 488.

9. Ibid., 486.

10. Chris Impey, *The Living Cosmos: Our Search for Life in the Universe* (New York: Random House, 2007), 10.

11. I credit my former student Zach T. Roberts for this felicitous phrase.

12. For a survey of the conflict between Christian fundamentalists and the modern, including scientific, champions of atheism, see Tina Beattie, *The New Atheists: The Twilight of Reason and the War on Religion* (London: Darton, Longman and Todd, 2007). John F. Haught argues that critics of religion such as Richard Dawkins, Sam Harris, and Christopher Hitchens succumb to a fundamentalism comparable to that which afflicts the creationists. Haught, *God and the New Atheism: A Critical Response to Dawkins, Harris, and Hitchens* (Louisville: Westminster John Knox, 2008).

13. So the definition of faith offered by Barbara Brown Taylor, in *The Luminous Web: Essays on Science and Religion* (Cambridge, MA: Cowley, 2000), 63.

14. Albert Einstein, "The World as I See It" (originally published in 1931) in Einstein, *Ideas and Opinions, Based on* Mein Weltbild, trans. Sonja Bargmann, ed. Carl Seelig (New York: Crown Publishers, 1954), 11.

15. See Eugene Wigner, "The Unreasonable Effectiveness of Mathematics in the Natural Sciences," *Communications on Pure and Applied Mathematics* 13, no. 1 (February 1960): 1–14.

16. Many thanks to George W. Fisher for this discussion of "mystery."

17. Impey, *Living Cosmos*, 212.

18. Despite the best efforts of the Templeton Foundation and the Center for Theology and the Natural Sciences in Berkeley, as well as interdisciplinary journals such as *Zygon* and *Theology and Science*.

19. And most recently Austin, Texas, where the school board voted to support textbook language questioning evolution (27 March 2009).

20. Indeed, the casualties have been many, particularly among scientists who grew up in fundamentalist households. See the painful testimony of paleontologist Stephen Godfrey and others in Jennifer Couzin, "Crossing the Divide," *Science* 319 (22 February 2008): 1034–36.

21. For a brief overview of some of the various perspectives of creation in the Hebrew Bible in their ancient Near Eastern context, see Richard J. Clifford, SJ, "Creation in the Hebrew Bible," in *Physics, Philosophy, and Theology: A Common Quest for Understanding*, ed. Robert J. Russell, William R. Stoeger, SJ, G. V. Coyne (Vatican: Vatican Observatory, 1988), 151–70; Gene M. Tucker, "Rain on a Land Where No One Lives: The Hebrew Bible on the Environment," *JBL* 116 (1997): 3–17. More extensive treatments include Richard J. Clifford, SJ, *Creation Accounts in the Ancient Near East*

and in the Bible (CBQMS 26; Washington, D.C.: Catholic Biblical Association of America, 1994); William P. Brown, *The Ethos of the Cosmos* (Grand Rapids: Eerdmans, 1999); Terence E. Fretheim, *God and World in the Old Testament: A Relational Theology of Creation* (Nashville: Abingdon Press, 2005).

22. Seeking coherence between the Bible and nature, sometimes referred to as God's "two books," has deep roots in Christian tradition, beginning most clearly with John Chrysostom (ca. 347–407) and extending to Galileo and John Calvin. For a concise historical survey, see Peter J. Hess, "'God's Two Books': Revelation, Theology, and Natural Science in the Christian West," in *Interdisciplinary Perspectives on Cosmology and Biological Evolution*, Australian Theological Forum Science and Theology Series 2 (Hindmarsh, Australia: Australian Theological Forum, 2002), 19–51.

23. Taylor, *Luminous Web*, 15.

24. Edward O. Wilson, *Consilience: The Unity of Knowledge* (New York: Vintage Books, 1998), 294.

25. See chap. 7.

26. *Babylonian Talmud Sanhedrin* 38a. For a brief summary of other possibilities explored by the early rabbis, see Michael V. Fox, *Proverbs 1–9* (AB 18A; New York: Doubleday, 2000), 297–98.

27. As evidenced in the books of Proverbs, Job, and Ecclesiastes in the Hebrew canon, as well as Sirach and Wisdom of Solomon in the apocryphal corpus. See the concluding section of this chapter (pp. 17–18).

28. The phrase is more apt than the tepid labels "global warming" and, worse, "climate change." My thanks to the American historian Jon Houghton, who introduced me to the term.

29. From *fides quaerens intellectum*, the original title to Anselm's *Proslogion* as referenced in his preface. See *Anselm of Canterbury: The Major Works*, ed. Brian Davies and G. R. Evans, Oxford World's Classics (Oxford: Oxford University Press, 1998), 83, 87.

30. A notable exception is Paul Davies, whose numerous works include *The Mind of God: The Scientific Basis for a Rational World* (New York: Simon & Schuster Paperbacks, 2005 [1992]) and *God and the New Physics* (New York: Simon & Schuster, 1983). See also the concluding chapter in Stephen Hawking with Leonard Mlodinow, *A Briefer History of Time* (New York: Bantam Dell, 2005), 138–42. Such scientific ruminations about God go back to Johannes Kepler (1571–1630), to whom is attributed the oft-quoted phrase regarding the purpose of science: "to think God's thoughts after him."

31. Ted Peters, "Introduction: What Is to Come," in *Resurrection: Theological and Scientific Assessments*, ed. Ted Peters, Robert John Russell, and Michael Welker (Grand Rapids: Eerdmans, 2002), xiii.

32. Douglas F. Ottati, "Theology among the Arts and Sciences," *The Bulletin of the Institute for Reformed Theology* 7, no. 1 (Spring/Summer 2007): 6. Indeed, James M. Gustafson makes it a "moral imperative" to do so. Gustafson, *Intersections: Science, Theology, and Ethics* (Cleveland, OH: Pilgrim Press, 1996), xvi.

33. Michael Welker, *Creation and Reality* (Minneapolis, MN: Fortress, 1999), 4.

34. Martin Rees, "Pondering Astronomy in 2009," *Science* 323 (16 January 2009): 309.

35. By distinguishing between what the biblical text "says" and what it "means," I am drawing from the exegetical insights of the great Jewish exegete Rashi (Rabbi Shelomo Ben Isaac, 1040–1105 CE). For Rashi, the *peshuto shel mikra* ("literal meaning of Scripture" or *sensus litteralis*) is distinguishable from aggadic or traditional interpretations of the text *for* the Jewish community. For convenience, I refer to the "literal meaning" of a text in its ancient context as what the text "says" (or "said"). What the text "means" *explicitly* engages the world of the interpreter, which includes the world of science.

36. Ian G. Barbour has helpfully classified four models by which science and religion have been set in relation: conflict, independence, dialogue, and integration (Barbour, *When Science Meets Religion: Enemies, Strangers, or Partners?* [New York: HarperSanFrancisco, 2000], 2–4). As a biblical theologian, I confess that my approach is not beholden to any one model. The complexities of biblical interpretation are far too messy to be confined to one or another model within a set typology. Nevertheless, I take as my point of departure "dialogue," regardless of where it may lead.

37. See Karl Barth, "The Strange New World within the Bible," in Barth, *The Word of God and the Word of Man*, trans. Douglas Horton (New York: Harper, 1957), 28–50.

38. Mary Midgley, "Science in the World," *Science Studies* 9, no. 2 (1996): 57. The term "consonance" is most fully discussed in Ernan McMullin, "How Should Cosmology Relate to Theology?" in *The Sciences and Theology in the Twentieth Century*, ed. Arthur Peacocke (Notre Dame: University of Notre Dame Press, 1981), 17–57, esp. 47–52.

39. Ted Peters, "From Conflict to Consonance: Ending the Warfare between Science and Faith." *Currents in Theology and Mission* 28, nos. 3–4 (June/August 2001), 238. See also his "Theology and the Natural Sciences," in *The Modern Theologians: An Introduction to Christian Theology in the Twentieth Century*, ed. David F. Ford (Oxford: Blackwell, 1997), 652.

40. Gustafson, *Intersections*, xi–xii. Note also the logo of the Center for Theology and the Natural Sciences in Berkeley: the Golden Gate Bridge. See Robert John Russell, "Bridging Theology and Science: The CTNS Logo," *Theology and Science* 1 (June 2003), 1–3.

41. Discerning "virtual parallels" across disciplines is an instance of what J. Wentzel van Huyssteen calls "transversal reasoning" or "performance," which "facilitates different but equally legitimate ways of viewing, or interpreting, issues, problems, traditions, or disciplines" van Huyssteen, *Alone in the World? Human Uniqueness in Science and Theology* (Grand Rapids: Eerdmans, 2006), 19.

42. Similarly, McMullin regarded the consonance between science and theology as "a tentative relation, constantly under scrutiny, in constant slight shift" ("How Should Cosmology Relate to Theology?" 52).

43. See Warren Brown, "Resonance," 113–15. Brown's discussion is cast on a more general level than mine, one that incorporates the Wesleyan Quadrilateral, with the addition of science, into his resonance model.

44. Etymologically, the name "Israel" probably means "God rules."

45. See Susan Niditch's discussion of Jacob as trickster, along with Joseph, in *Underdogs and Tricksters: A Prelude to Biblical Folklore* (San Francisco: Harper & Row, 1987) 94–125.

46. Delwin Brown, *Boundaries of Our Habitations: Tradition and Theological Construction*, SUNY Series in Religious Studies (Albany: State University of New York Press, 1994), 79.

47. The following discussion is drawn from my "Introduction," in *Engaging Biblical Authority: Perspectives on the Bible as Scripture*, ed. William P. Brown (Louisville: Westminster John Knox, 2007), xi–xiv.

48. See Jacqueline E. Lapsley, *Whispering the Word: Hearing Women's Stories in the Old Testament* (Louisville: Westminster John Knox, 2005), 13–19.

49. See also Delwin Brown's helpful distinction between authority as "binding" in Roman law and the kind of authority found in the New Testament (*exousia*), which is more generative than limiting (*Boundaries of Our Habitations*, 144).

50. As evidenced in the phenomenon of "redshift," the stretching of light's wavelength.

51. See Krister Stendahl, "Biblical Theology, Contemporary," IDB, 1.418–32; reprinted in Stendahl, *Meanings: The Bible as Document and as Guide* (Philadelphia: Fortress, 1984), 11–44. Although Stendahl's distinction between what the text "meant" and what it "means" has been criticized for being too rigid and objectified, and rightly so, I still find it helpful as a point of departure for discussing the complexity of biblical hermeneutics.

52. For the "spectacles" analogy, see John Calvin, *The Institutes of the Christian Religion: Library of Christian Classics*, trans. Ford Lewis Battles, ed. J. T. McNeill (Philadelphia: Westminster, l960), 1.6.1. (70).

53. Ian Barbour, *Myths, Models, and Paradigms: A Comparative Study in Science and Religion* (San Francisco: Harper & Row, 1974), 29–70. Take, for example, the various models used to describe the atom (from "plum pudding" to planetary system to cloud) and God (from "father" to "ground of being").

54. John Goldingay, *Models for Interpretation of Scripture* (Grand Rapids, MI: Eerdmans, 1995). 262.

55. See Nancey Murphy, "Introduction: A Hierarchical Framework for Under-standing Wisdom," in *Understanding Wisdom: Sources, Science, and Society*, ed. Warren S. Brown (Philadelphia and London: Templeton Foundation Press, 2000), 7. For detailed discussion of feedback loops and systems on the methodological level, see Nancey Murphy and George F. R. Ellis, *On the Moral Nature of the Universe: Theology, Cosmology, and Ethics* (Minneapolis: Augsburg Fortress, 1996), 74–84.

56. The interactive side of the interpreter is increasingly acknowledged even in science. The notion of objectivity, for example, has shifted significantly from nine-teenth to twenty-first century science, from one that regards the observer as merely a machine, much like a camera, to one that acknowledges (and exploits) the observer's interactive, formative role in relation to the object of study. See the perceptive historical survey and discussion of scientific atlases in Lorraine Daston and Peter Galison, *Objectivity* (New York: Zone, 2007), esp. 55–190, 309–62.

57. Contra Francis S. Collins, *The Language of God: A Scientist Presents Evidence for Belief* (New York: Free Press, 2006), 206.

58. Wilson, *Consilience*, 230.

59. Stephen Jay Gould, *Rocks of Ages: Science and Religion in the Fullness of Life* (The Library of Contemporary Thought; New York: Ballantine, 1999), 5, 178.

60. All translations are my own, unless otherwise noted.

61. Such as the Babylonian "astronomical diaries," reflecting a primitive science that is "empirical, predictive, and theoretical," according to the Assyriologist Francesca Rochberg, *The Heavenly Writing: Divination, Horoscopy, and Astronomy in Mesopotamian Culture* (Cambridge: Cambridge University Press, 2004), xv (see also 28, 96). Rochberg makes a convincing case that the roots of science can be traced prior to the rise of Greek cultural dominance in antiquity and that the sacred and the scientific in the ancient world comfortably coexisted (ibid., 12, 244). For Babylonian astronomy, see James Evans, *The History and Practice of Ancient Astronomy* (New York: Oxford University Press, 1988), esp. 5–17. For examples of how Mesopotamian science impacted biblical tradition, see Baruch Halpern, "Assyrian and Pre-Socratic Astronomies and the Location of the Book of Job," in *Kein Land für sich allein: Studien zum Kulturkontakt in Kanaan, Israel/Palästina und Ebirnâri für Manfred Weippert zum 65. Geburtstag*, ed. Ulrich Hübner and Ernst Axel Knauf (OBO 186; Fribourg: Universitätsverlag Freiburg, 2002), 255–64; and Halpern, "The Assyrian Astronomy of Genesis 1 and the Birth of Milesian Philosophy," *Eretz-Israel* 27 (2003): 74*–83*.

62. First coined by Wolfram von Soden in his study of Sumerian and Akkadian lists, which include matters of philology, deities, kings, geography, zoology, and botany, in "Leistung und Grenze sumerischer und babylonischer Wissenschaft," in Benno Landsberger and Wolfram von Soden, *Die Eigenbegrifflichkeit der babylonischen Welt. Leistung und Grenze sumerischer und babylonischer Wissenschaft* (Darmstadt: Wissenschaftliche Buchgesellschaft, 1965 [1936]), esp. 29–74. The theory was developed in biblical studies by Albrecht Alt in "Die Weisheit Salomos," *ThLZ* 76 (1951), cols. 139–44 (translated as "The Wisdom of Solomon," in *Studies in Ancient Israelite Wisdom*, ed. James L. Crenshaw [New York: Ktav, 1976], 102–12). See the critique by Michael V. Fox, "Egyptian Onomastica and Biblical Wisdom," *VT* 36 (1986): 302–10.

63. See chap. 5.

64. Cf. Gen 1:11–12, 21, 24–25; Prov 30:18–19, 24–28, 29–31.

65. See Daniel Hillel, *The Natural History of the Bible: An Environmental Exploration of the Hebrew Scriptures* (New York: Columbia University Press, 2006), 150. For a survey of agricultural practices and techniques, see Oded Borowski, *Agriculture in Iron Age Israel* (Winona Lake, IN: Eisenbrauns, 1987).

66. Greene, *Fabric of the Cosmos*, 20.

67. Impey, *Living Cosmos*, vii.

68. Edward O. Wilson, *The Creation: An Appeal to Save Life on Earth* (New York: W. W. Norton, 2006).

CHAPTER 2

1. Paraphrased in Lee Smolin, *The Life of the Cosmos* (London: Phoenix Paperback, 1997), 368.

2. For example, see the collection of extrabiblical texts, in *The Context of Scripture*, vol. 1, *Canonical Compositions from the Biblical World*, ed. William W. Hallo et al. (Leiden: Brill, 2003).

3. For an alternative dating, see Tzvi Abusch, "Marduk," in *Dictionary of Deities and Demons in the Bible*, ed. Karel van der Toorn, Bob Becking, and Pieter W. van der Horst, 2nd ed. (Leiden: Brill, 1999), 547. Abusch proposes that *Enūma Elish* was composed later during a period of Babylonian weakness, not strength: sometime during the first millennium. Little evidence, however, is given for support.

4. The "canonical" status of *Enūma elish* is demonstrated in an Assyrian version of the Babylonian epic in which the patron deity Aššur replaces Marduk as the narrative's protagonist. For an English translation, see Stephanie Dalley, "The Epic of Creation," in Dalley, *Myths from Mesopotamia: Creation, the Flood, Gilgamesh and Others* (World's Classics; Oxford: Oxford University Press, 1991), 228–77, from which portions are featured below, unless otherwise noted. See also Benjamin R. Foster, "Epic of Creation (1.111)," in Hallo, *Context of Scripture*, 391–402.

5. For the latest critical edition and translation (in French), see Philippe Talon, *Enūma Eliš: The Standard Babylonian Creation Myth* (SAACT 4; University of Helsinki: The Neo-Assyrian Text Corpus Project, 2005), esp. 79–108.

6. Tablet IV, lines 136–40; Dalley, "The Epic of Creation," 255.

7. See Stephanie Dalley's translation, "Atrahasis," in *Myths from Mesopotamia: Creation, the Flood, Gilgamesh and Others*, World's Classics (Oxford: Oxford University Press, 1991), 9–38. The story was also adapted and incorporated into the Standard Babylonian Version of the Epic of Gilgamesh, in which the Noah figure was renamed Utnapishtim ("He Who Found Life").

8. A nonviolent but comparable version of human "clay" creation is found in the Epic of Gilgamesh, in which the same mother goddess (here called "Aruru") pinches off clay and shapes it into a man, specifically a man of the wild, Enkidu (Tablet I).

9. Tablet I.v.11–12.

10. The English cleric Thomas Malthus (1766–1834) contended that God ordained the population in America to increase faster than what the land could sustain and, thereby, provided means of restraining such growth.

11. See, e.g., Num 25:3–5; Judg 2:11–13; 6:25–32; 1 Kgs 16:32; 18:18–40; Jer 7:9; Hos 2:8, 13–16; 13:1; Zeph 1:4.

12. *KTU* 1.2 IV 27. All passages are drawn from Mark S. Smith's translation of the "Ba'al Epic," in *Ugaritic Narrative Poetry*, ed. Simon B. Parker (SBLWAW 9; Atlanta: Society of Biblical Literature, 1997), 81–180.

13. *KTU* 1.4 VI 24–38. See Smith, "Ba'al Epic," 133–34.

14. See also the fuller text that ascribes the victory to Anat, Ba'al's sister, in *KTU* 1.3 III 38–42, in which Yamm is included among the list of casualties.

15. The following translations of Egyptian sources are drawn from James P. Allen in Hallo, *Context of Scripture*, 5–27.

16. So Allen's translation of Coffin Texts Spell 714 (1.2). "Evolution," of course, is used differently from the Darwinian sense of "descent with modification." In the Egyptian texts, it connotes the process of differentiation.

17. Coffin Texts Spell 714 (1.2).

18. Pyramid Texts Spell 527 (1.3).

19. Pyramid Texts Spell 600 (1.4).

20. Coffin Texts Spell 80 (1.8).

21. Coffin Texts Spell 75 (1.5).

22. Coffin Texts Spell 80 (1.8).

23. From the verbal root *ḫpr.* Papyrus Bremner-Rhind (1.9). See n. 16.

24. Coffin Texts Spell 75 (1.5).

25. "Memphite Theology" (1.15).

26. Coffin Texts Spell 647 (1.12).

27. "Memphite Theology" (1.15).

28. Papyrus Leiden I 350 (1.16).

29. Ibid.

30. Ibid.

31. The goose was sacred to Amun.

32. Papyrus Leiden I 350 (1.16).

33. See, e.g., The "Book of Nut" (1.1).

34. See Coffin Texts Spell 1130 (1.17). As Allen points out, the term "separate" derives from the language of jurisprudence. To "separate the matter" is to issue judgment.

CHAPTER 3

1. Quoted in Neil deGrasse Tyson and Donald Goldsmith, *Origins: Fourteen Billion Years of Cosmic Evolution* (New York: W. W. Norton), 247.

2. Introduction to the 1968 edition of the *Whole Earth Catalog.*

3. S. Dean McBride Jr., "Divine Protocol: Genesis 1:1–2:3 as Prologue to the Pentateuch," in *God Who Creates: Essays in Honor of W. Sibley Towner*, ed. William P. Brown and S. Dean McBride Jr. (Grand Rapids, MI: Eerdmans, 2000), 6.

4. Jon Levenson, "Translation Notes for Gen 1.1–2.3," in *The Jewish Study Bible*, ed. Adele Berlin and Marc Zvi Brettler (Oxford: Oxford University Press, 2004), 12.

5. A play on Darwin's oft-quoted definition of evolution: "descent with modification." The text's recurring cadences also suggest ancient liturgical usage. See Moshe Weinfeld, "Sabbath, Temple and the Enthronement of the Lord—The Problem of Sitz im Leben of Genesis 1:1–2:3," in *Mélanges bibliques et orientaux en l'honneur de M. Henri Cazelles*, ed. A Caquot and M. Delcor, AOAT 212 (Kevelaer: Butzon & Bercker, 1981), 501–12.

6. Literally, "In the beginning of God creating the heavens and the earth...." The Hebrew syntax conflicts with that of the KJV ("In the beginning, God created the heavens and the earth."), whose precedent is found in the Septuagint (LXX or Old Greek) translation. But as pointed out by the Jewish commentator Rashi (d. 1105), the Hebrew of v. 1 is a dependent clause, not a complete sentence, because the first word *bĕrēšît* is likely in construct with the verbal clause that follows. Hence, an absolute beginning is not assumed in the first verse. For detailed discussion, see William P. Brown, *Structure, Role, and Ideology in the Hebrew and Greek Texts of*

Genesis 1:1–2:3, SBLDS 132 (Atlanta: Scholars Press, 1993), 62–72. But see also the alternative proposal in Robert D. Holmstedt, "The Restrictive Syntax of Genesis i 1," *VT* 58 (2008): 56–67, which amounts to a comparable sense of a non-absolute beginning.

7. I am indebted to Mark S. Smith for this rendering of *tōhû wābōhû*, which matches something of the alliterative quality of the Hebrew. As with the French *Le tohu-bohu*, meaning "hubbub," the conjunction of these two similar-sounding words in Hebrew suggests a state that lacks both form and substance, something of a hodgepodge. *Tōhû* refers most generally to what is uninhabitable or empty of life, the opposite of creation (see Isa 45:18–19; Deut 32:10; Job 6:18; 12:24; Ps 107:40). Always paired with its partner in biblical Hebrew, *bōhû* is semantically wedded to *tōhû* (see Isa 34:11; Jer 4:23). See the careful linguistic study by David Tsumura, *Creation and Destruction: A Reappraisal of the Chaoskampf Theory in the Old Testament* (Winona Lake, IN: Eisenbrauns, 2005), 9–35.

8. Frequently translated "spirit" (*rûaḥ*), even though both the Septuagint and Targum Onkelos understand *rûaḥ* materially as "wind." See the parallel in Gen 8:1. In view of the divine commands that follow (1:3, 6, 9, 11, 14, 20, 24, 26), *rûaḥ* is perhaps best translated as "breath," specifically breath associated with the utterance of divine speech. See also Tsumura, *Creation and Destruction*, 74–76.

9. A close parallel is found in Deut 32:11, which describes a raptor "hovering" over its nested brood. With "breath" or "wind" as its subject, the verb in Genesis 1 generates a mixed metaphor, not uncommon in Hebrew poetry: wind does not literally "hover." Nevertheless, 1:2 seems to describe God's breath suspended over the waters, perhaps like a vulture gently riding the updraft of a warm current of air. Far from a gusty wind roiling the waters, the picture is that of God's breath poised to be released in speech, beginning in v. 3.

10. *NJPS* extends the sentence to include v. 3, thereby setting v. 2 as a parenthetical statement. A more natural interpretation is to take vv. 1–2 as a self-contained preface for what follows, compelling the reader to linger (and hold one's breath in suspense!) over the vivid profile described in v. 2 before reading on.

11. See n. 36 below.

12. Literally, "And God saw the light, that it was good."

13. I adopt the older English translation; other possibilities include "vault," "dome," and "expanse." Its verbal root (*rq'*) means to "stamp" or "hammer out" (see Ezek 6:11; 2 Sam 22:43; Exod 39:3).

14. The niphal form can be taken as either passive ("be gathered") or reflexive, the latter being more likely in context.

15. Literally, "Let the earth sprout sproutlings" (*tadšē'* ... *deše'*).

16. Literally, "whose seed is in it."

17. Literally, "signs, that is, for seasons, days and years." The *waw* prefixed to "seasons" (*ûlĕmô'ādîm*) is explicative.

18. With the possible addition "over the wild animals," as indicated in the Syriac translation (Peshitta). But note that "wild animals" do not figure in the fulfillment report given in v. 28. Though likely understood, the original Hebrew text may not have explicitly mandated human dominion over the wild animals (cf. v. 30).

19. Notable is the shift from the indefinite noun in v. 26 (*'ādām*; "humanity") to the definite noun in v. 27 (*hā'ādām*; "the human being"). For discussion of this grammatical shift, see below.

20. Read with the Samaritan Pentateuch, which adds the definite article.

21. The only day, along with the seventh day, to bear a definite article, perhaps to highlight its importance.

22. From Hebrew *šbt*, commonly translated as "rested," but whose basic meaning is "cease." See the following verse.

23. Literally, "created by doing." For the function of the *lamed* prefix on the infinitive, see J-M §1240 n.2 (438).

24. Compare 2 Macc 7:28, which represents a much later perspective that comes close to the formal doctrine of *creatio ex nihilo*. See also Isa 45:6–7.

25. See chap. 1. For a trenchant critique of the conventional claim that biblical *tĕhôm* is linked, whether figuratively or etymologically, to Tiamat of Mesopotamian lore, see Tsumura, *Creation and Destruction*, 36–41. See also the broader study of "chaos" in the Hebrew Bible in Rebecca S. Watson, *Chaos Uncreated: A Reassessment of the Theme of "Chaos" in the Hebrew Bible*, BZAW 341 (Berlin: Walter de Gruyter, 2005), esp. 269–72.

26. Of course, "sea monsters" (*tannînim*) are found later in creation (Gen 1:21), but as "great" as they are, they do not pose a threat to God or to the created order.

27. Such darkness serves as the ninth plague in the exodus story (Exod 10:21–23).

28. For connection between the divine commands in Genesis 1 and the Decalogue, see McBride, "Divine Protocol," 10, n.16.

29. The heptadic numerology of Gen 1:1–2:3 is most vigorously championed by Umberto Cassuto in his *Commentary on Genesis. Part I, From Adam to Noah: Genesis I–VI 8* (Jerusalem: Magnes Press, 1961), 14–15. See also Levenson, *Creation and the Persistence of Evil*, 67–68.

30. See Claus Schedl, *History of the Old Testament: The Ancient Orient and Ancient Biblical History*, vol. 1 (New York: Alba house, 1973), 217–18.

31. Julius Wellhausen, *Prolegomena to the History of Ancient Israel* (Gloucester: Peter Smith, 1983 [1885]), 361.

32. I.e., Sun, Moon, Venus, Mercury, Mars, Jupiter, and Saturn. It may be no coincidence that the fourth day of creation marks the central day of the account.

33. The priestly calendar in Lev 23:1–44, for example, seems based on the counting of seven (e.g., vv. 3, 5, 6, 8, 15, 16, 24, 34, 36). See, also, the calendar of offerings in Num 28:1–29:40.

34. Contrary to LXX.

35. The following discussion builds on the insightful observations given in McBride, "Divine Protocol," 12–15. For further background, see J. Richard Middleton, *The Liberating Image: The* Imago Dei *in Genesis 1* (Grand Rapids, MI: Brazos, 2005), 74–76.

36. The "creation" of light on Day 1 involves its establishment as a domain, consonant with the other domains established on Days 2–3. But whether light itself, elsewhere considered a quality of God's effulgence (e.g., Ps 104:2; Isa 60:19; Ezek 43:2; cf. Isa 45:7), is *created* in Gen 1:3 remains an open question. See Mark S. Smith,

"Light in Genesis 1:3—Created or Uncreated: A Question of Priestly Mysticism?" in *Birkat Shalom: Studies in the Bible, Ancient Near Eastern Literature, and Postbiblical Judaism Presented to Shalom M. Paul on the Occasion of His Seventieth Birthday,* ed. Chaim Cohen et al. (Winona Lake, IN: Eisenbrauns, 2008), 125–34.

37. See, also, Tsumura, *Creation and Destruction,* 34.

38. Middleton, *Liberating Image,* 278.

39. The creation of vegetation, unlike that of the animals on Days 5 and 6, lacks the divine commission to "fill" the land.

40. For example, the transition formula "and it was so" is found at the end of 1:7 rather than at the end of v. 6, which is more typical (cf. 1:9, 11). Also, the fulfillment report is entirely lacking in 1:9–10. No approbation is given in Day 2. Indeed, LXX reflects a more consistent text, reflecting either a harmonizing *Tendenz* or an older textual tradition that the Masoretes saw fit to alter. See Brown, *Structure, Role, and Ideology,* which argues for the latter. For a listing of textual anomalies, see the table in Middleton, *Liberating Image,* 281.

41. See Middleton, *Liberating Image,* 75–76.

42. The cosmos construed as a temple is not unprecedented in biblical tradition. See, e.g., Isa 66:1–2, which criticizes Israel's intent on rebuilding the temple by claiming that creation itself is God's sanctuary.

43. Note the Ugaritic parallel of Ba'al's completion of his palace in seven days (KTU 1.4 VI 24–33). See chap. 2.

44. See 1 Kgs 8:10–13; cf. Ps 132:8, 13–14.

45. For a sampling of the scholarly discussion and the ambiguity of the evidence, see Tryggve N. D. Mettinger, *No Graven Image? Israelite Aniconism in its Ancient Near Eastern Context* (Stockholm: Almqvist & Wiksell International, 1995).

46. As well as the budding rod of Aaron and a jar of manna (Num 16:33; 17:10)

47. E.g., Num 33:52; 2 Kgs 11:18; 2 Chron 23:17; Ezek 7:20; Amos 5:20. For more detailed discussion of the semantic range of this term, see Middleton, *Liberating Image,* 45–46.

48. At least compared to the second creation story in Genesis 2–3.

49. These four ways are nicely laid out by Andreas Schüle in "Made in the 'Image of God': The Concepts of Divine Images in Gen 1–3," *ZAW* 117 (2005): 5–7, from which the following discussion is drawn. Note also J. Wentzel van Huyssteen's summary of the history of interpretation of the *imago Dei* as a three-phased development: from substantial to functional to relational (*Alone in the World? Human Uniqueness in Science and Theology* [Grand Rapids: Eerdmans, 2006], 124).

50. Schüle, "Made in the 'Image of God,'" 8. An important extrabiblical parallel to this "essentialist link" comes from the Tell-Fekheriyeh statue of King Had-yiṯi of the ninth or eighth century BCE, on whose skirt is engraved an Akkadian text translated into Aramaic. The inscription refers to the king's "likeness" (*dmwt*) and "image" (*ṣlm*), namely, the statue the king had fashioned to serve as his stand-in before Hadad, his patron deity. See the study in W. Randall Garr, "'Image' and 'Likeness' in the Inscription from Tell Fakhariyeh," *IEJ* 50 (2000): 231–34. See also Ali Abou-Assaf, Pierre Bordreuil, and Alan R. Millard, *La statue de Tell Fekherye et son inscription bilingue assyro-araméene,* Etudes Assyriologiques (Paris: Editions Recherche sur les

civilisationes, 1982); A. R. Millard and P. Bordreuil, "A Statue from Syria with Assyrian and Aramaic Inscriptions," *BA* 45 (1982): 135–41.

51. For example, Middleton, *Liberating Image*, 204, which he equates with "universalizing."

52. The following discussion is prompted by W. Randall Garr's discussion in *In His Own Image and Likeness: Humanity, Divinity, and Monotheism*, Culture and History of the Ancient Near East 15 (Leiden: Brill, 2003), 202–40.

53. Sometimes called the plural of self-deliberation (*pluralis deliberationis*).

54. As Middleton points out, humanity's creation by the blood of a rebel deity in *Atrahasīs* and *Enūma elish* serves to devalue rather than elevate humanity's status (*Liberating Image*, 174).

55. See Pss 33:6, 9; 147:18; 148:5–6.

56. To borrow from Michael Welker, *Creation and Reality*, trans. John F. Hoffmeyer (Minneapolis, MN: Fortress, 1999), 40, 42, who focuses only on the productive agency of the earth.

57. Drawing from Russian theologian Nicholas Berdyaev, W. Sibley Towner rightly suggests that freedom is woven into the very process of creation (Towner, *Genesis*, Westminster Bible Companion [Louisville: Westminster John Knox, 2001], 18–19).

58. As one finds in Babylonian "astronomical diaries." For a masterful treatment of Babylonian astronomy as both ancient science and religious study, see Francesca Rochberg, *The Heavenly Writing: Divination, Horoscopy, and Astronomy in Mesopotamian Culture* (Cambridge: Cambridge University Press, 2004). For the ancient "scientific" background of the priestly account of the creation of celestial bodies, see Baruch Halpern, "Assyrian and pre-Socratic Astronomies and the Location of the Book of Job," in *Kein Land für sich allein: Studien zum Kulturkontakt in Kanaan, Israel/Palästina und Ebirnâri für Manfred Weippert zum 65. Geburtstag*, ed. Ulrich Hübner and Ernst Axel Knauf, OBO 186 (Fribourg: Universitätsverlag Freiburg, 2002), esp. 255–56; Halpern, "The Assyrian Astronomy of Genesis 1 and the Birth of Milesian Philosophy," *Eretz-Israel* 27 (2003): 74*–83*.

59. For fuller discussion, see Norbert Lohfink, *Great Themes from the Old Testament*, trans. Ronald Walls (Edinburgh: T&T Clark, 1982), 214–15.

60. As compared to their specific characterizations in *Atrahasīs* and *Enūma elish*. See chap. 2.

61. Until, of course, Genesis 6–9, the story of the Flood.

62. So Middleton, *Liberating Image*, 212.

63. See, e.g., Exod 20:8–11.

64. Gen 1:22, 28; 2:3; cf. Lev 9:2–23; Num 6:22–27.

65. Gen 2:3; cf. Ezek 44:24.

66. See, e.g., Gen 14:19; Num 6:22–26.

67. See Middleton's nuanced discussion of the royal metaphor in *Liberating Image*, 70–74.

68. See Exod 29:1–7; 31:3; 35:1; 36:2–6; 39:43.

69. As felicitously put by Middleton, *Liberating Image*, 86.

70. Welker, *Creation and Reality*, 71.

71. The permission to eat meat is immediately followed by the prohibition against eating meat with its blood and against shedding human blood (Gen 9:4–6).

72. The verb "subdue" (*kbš*) is used in other priestly material to indicate simply the allotment of tribal boundaries to the Israelites for their use and livelihood (Josh 18:1). Absent is any sense of exploitation.

73. Erich Zenger, *Gottes Bogen in den Wolken: Untersuchungen zu Komposition und Theologie der priesterschriftlichen Urgeschichte*, 2nd ed, Stuttgarter Bibelstudien 112 (Stuttgart: Katholisches Bibelwerk, 1987), 90.

74. Or in its more mythically cosmogonic order, conquer and divide!

75. For a full discussion of the social and theological trauma of exile, see Daniel L. Smith-Christopher, *A Biblical Theology of Exile*, Overtures to Biblical Theology (Minneapolis: Fortress, 2002).

76. Middleton, *Liberating Image*, 231.

77. Marilynne Robinson, *The Death of Adam: Essays on Modern Thought* (Boston: Houghton Mifflin, 1998), 39.

78. See Welker, *Creation and Reality*, 14–15.

79. Aristotle, *On the Heavens* I, 3 (270^b20–25).

80. Comets are typically composed of half water. Europa, a moon of Jupiter, may actually have more water than Earth. As astrobiologist Chris Impey suggests, planets bearing water in some form may be common throughout the universe (*The Living Cosmos: Our Search for Life in the Universe* [New York: Random House, 2007], 121–23, 215–16, 241).

81. Robert M. Hazen, *Gen·e·sis: The Scientific Quest for Life's Origins* (Washington, D.C.: Joseph Henry, 2005), 122.

82. But perhaps not for long. The Large Hadron Collider (LHC) and the Gaia Satellite, slated to launch in late 2011, may enable scientists to detect what kind of particles make up dark matter and disentangle the effects of baryonic gravitation from those of dark matter. See Gerard Gilmore, "How Cold is Dark Matter?" *Science* 322 (5 December 2008): 1476–77.

83. Michio Kaku, *Parallel Worlds: A Journey through Creation, Higher Dimensions, and the Future of the Cosmos* (New York: Anchor Books, 2005), 70.

84. See Claude-André Faucher-Giguère et al., "Numerical Simulations Unravel the Cosmic Web," *Science* 319 (4 January 2008): 52. See also the Hubble Site Newscenter: http://hubblesite.org/newscenter/archive/releases/2007/01/full/.

85. Faucher-Giguère, "Numerical Simulations," 53.

86. Eric V. Linder, "Cosmology: The Universe's Skeleton Sketched," *Nature* 445 (18 January 2007): 273.

87. Quoted in Adrian Cho, "Untangling the Celestial Strings," *Science* 319 (4 January 2008): 49.

88. Richard Massy et al., "Dark Matter Maps Reveal Cosmic Scaffolding," *Nature* 445 (18 January 2007): 286–90.

89. Physicists also speak of "dark energy," a repulsive force that drives the expansion of space. Researchers estimate that dark energy constitutes 73 percent of the universe.

90. Peter Coles, *Cosmology: A Very Short Introduction* (Oxford: Oxford University Press, 2001), 81.

91. Christopher De Pree, personal communication.

92. Lee Smolin, *The Life of the Cosmos* (London: Phoenix Paperback, 1997), 152.

93. But only to an extent. The planet Jupiter, owing to its strong gravitational field, also acts as a protective shield for the earth.

94. See chap. 9.

95. Impey, *Living Cosmos*, 41.

96. Smolin, *Life of the Cosmos*, 17.

97. Ibid., 209, 212.

98. Brian R. Greene, *The Fabric of the Cosmos: Space, Time, and the Texture of Reality* (New York: Vintage Books, 2004), 220.

99. Kaku, *Parallel Worlds*, 84.

100. For further discussion of "spontaneous symmetry breaking," see Smolin, *Life of the Cosmos*, 61–68; Kaku, *Parallel Worlds*, 95–98.

101. Quoted in Ibid., 97.

102. Ibid., 98.

103. Ibid., 96.

104. According to biologist Leslie Orgel, as quoted in Impey, *Living Cosmos*, 79.

105. For a history of how a term from Greek cosmology came to be applied to Hebrew biblical literature, see Watson, *Chaos Uncreated*, 13–19.

106. Lawrence Osborn, "Theology and the New Physics," in *God, Humanity and the Cosmos*, ed. Christopher Southgate, 2nd ed. (London / New York: T&T Clark, 2005), 149.

107. Kerry Emmanuel, "Edward N. Lorenz (1917–2008)," *Science* 320 (23 May 2008): 1025.

108. Ibid.

109. Ibid.

110. See Sjored L. Bonting, *Creation and Double Chaos: Science and Theology in Discussion*, Theology and the Sciences (Minneapolis: Augsburg, 2005), 115–17.

111. J. P. Crutchfield et al., "Chaos," in *Chaos and Complexity: Scientific Perspectives on Divine Action*, ed. Robert J. Russell, Nancey Murphy, and Arthur Peacocke (Vatican City: Vatican Observatory; Berkeley, CA: Center for Theology and the Natural Sciences, 1995), 35.

112. As summarized by Ian G. Barbour, *When Science and Religion Meet: Enemies, Strangers, or Partners?* (New York: HarperSanFrancisco, 2000), 104. See Ilya Prigogine and Isabelle Stengers, *Order out of Chaos* (New York: Bantam Books, 1984).

113. Stuart A. Kauffman, *At Home in the Universe: The Search for Laws of Self-Organization and Complexity* (Oxford: Oxford University Press, 1995), 83.

114. Ibid., 71.

115. For a general overview and assessment, see Christopher Southgate, Michael Robert Negus, and Andrew Robinson, "Theology and Evolutionary Biology," in *God, Humanity and the Cosmos*, 173; Edward O. Wilson, *Consilience: The Unity of Knowledge* (New York: Vintage, 1998), 96–97.

116. Gary Zukav, *The Dancing Wu Li Masters: An Overview of the New Physics* (New York: William Morrow and Company, 1979), 213.

117. Alan Cook, "Uncertainties of Science," in *Science Meets Faith: Theology and Science in Conversation*, ed. Fraser Watts (London: SPCK, 1998), 32–33. Cf. Heinz Georg Schuster, *Deterministic Chaos: An Introduction* (Weinheim: Physik-Verlag, 1984).

118. Osborn, "Theology and the New Physics," 151.

119. Ian Stewart, *Does God Play Dice? The New Mathematics of Chaos* (New York: Penguin Books, 1990), 1. Cf. Jon Levenson, *Creation and the Persistence of Evil: The Jewish Drama of Divine Omnipotence*, 2nd ed. (Princeton, NJ: Princeton University Press, 1994 [1988]), 17.

120. See n. 36. above.

121. Greene, *Fabric of the Cosmos*, 58.

122. Kaku, *Parallel Worlds*, 35.

123. Smolin, *Life of the Cosmos*, 345.

124. Ibid., 347.

125. Coles, *Cosmology*, 115.

126. Smolin, *Life of the Cosmos*, 348.

127. Ibid., 361.

128. Fred Hoyle, *Facts and Dogmas in Cosmology and Elsewhere* (Cambridge: Cambridge University Press, 1982), 2–3.

129. Coles, *Cosmology*, 44.

130. Except for those galaxies located in clusters, such as our Virgo Cluster, which is contracting because of the attractive force of gravity overcoming the repulsive force of dark energy.

131. This is challenged by others, including Stephen Hawking and Lee Smolin, who argue that quantum dynamics prevent the formation of singularities (Hawking with Leonard Mlodinow, *A Briefer History of Time* [New York: Bantam Book, 2005], 102–3; Smolin, *Life of the Cosmos*, 115). See n. 130.

132. Greene, *Fabric of the Cosmos*, 306.

133. Coles, *Cosmology*, 34. The theory of quantum gravity proposed by James Hartle and Stephen Hawking attempts to remedy the problematic nature of cosmic singularity by regarding time itself as "finite but unbounded." Instead of a discrete, infinitesimally small point at t=0, there is curved, three-dimensional space stripped of time, hence no singular point of creation (James B. Hartle and Stephen W. Hawking, "Wave Function of the Universe," *Physics Review* D28 [1983]: 2960–75). See the discussion by Robert John Russell of the "Hartle/Hawking model" of cosmic emergence in "Finite Creation without a Beginning: The Doctrine of Creation in Relation to Big Bang and Quantum Cosmologies," in Russell, *Cosmology From Alpha to Omega: The Creative Mutual Interaction of Theology and Science*, Theology and the Sciences (Minneapolis: Fortress, 2008), 90–103.

134. The force that holds the nucleus of an atom together despite electrical repulsion of protons.

135. Evident in certain types of radioactive decay.

136. Kaku, *Parallel Worlds*, 84.

137. Otherwise known as a "nonzero Higgs field vacuum," named after the Scottish physicist Peter Higgs (Greene, *Fabric of the Cosmos*, 257).

138. Tyson and Goldsmith, *Origins*, 84.

139. Cf. Brian Greene's description of the formation of the so-called Higgs field as a "frog" perched on a higher-energy plateau ready to jump to a lower one (*Fabric of the Cosmos*, 256–63, 282–85).

140. Michael Welker, *God the Spirit*, trans John F. Hoffmeyer (Minneapolis: Fortress, 1994), 161 n. 67.

141. Smolin, *Life of the Cosmos*, 27.

142. See n. 36 above.

143. Adam Frank, "The First Billion Years," *Astronomy* 34, no. 6 (June 2006): 31.

144. Ibid., 31.

145. Tyson and Goldsmith, *Origins*, 53.

146. Ibid., 54.

147. Frank, "The First Billion Years," 32.

148. Ibid.

149. Ibid., 33.

150. Greene, *Fabric of the Cosmos*, 94.

151. Ibid., 121.

152. Ibid., 183.

153. Ibid., 11.

154. Smolin, *Life of the Cosmos*, 54.

155. See Carl R. Woese, "A New Biology for a New Century," *Microbiology and Molecular Biology Reviews* 68 (2004): 173–86.

156. Lynn Margulis and Dorion Sagan, *What is Life?* (Berkeley: University of California Press), 240.

157. Ursula Goodenough, *The Sacred Depths of Nature* (Oxford: Oxford University Press, 1998), 78.

158. Wilson, *Consilience*, 52.

159. Jones, *Darwin's Ghost*, 92.

160. Tyson and Goldsmith, *Origins*, plate 38.

161. Jones, *Darwin's Ghost*, 92.

162. See the discussion of Terence E. Fretheim, *God and World in the Old Testament: A Relational Theology of Creation* (Nashville: Abingdon, 2005), 41.

163. Not does it apply fully to biology.

164. Hazen, *Gen·e·sis*, 155.

165. Jones, *Darwin's Ghost*, 229.

166. Ibid. See the more rigid stance articulated in Stephen J. Gould, "The Evolution of Life on the Earth," *American Scientist* 21 (April 1994): 62–69, who assumes that the idea of progress involves a value judgment.

167. See Margulis and Sagan, *What is Life?*, 9.

168. For full discussion, see Harold J. Morowitz, *The Emergence of Everything: How the World Became Complex* (Oxford: Oxford University Press, 2002).

169. Impey, *Living Cosmos*, 81.

170. Attributed to Lewis Thomas, as quoted in Impey, *Living Cosmos*, 175, from Thomas's book *The Medusa and the Snail*.

171. Impey, *Living Cosmos*, 190.

172. Margulis and Sagan, *What is Life?*, 9.

173. "Uniform progress" falsely claims progress for every organism, species, etc.

174. So Francisco J. Ayala, "The Concept of Biological Progress," in *Studies in the Philosophy of Biology: Reduction and Related Problems*, ed. Francisco Jose Ayala and Theodosius Dobzhansky (Berkeley: University of California Press, 1974), 347, who draws from the work of G. G. Simpson.

175. Ibid.

176. Ibid., 351.

177. Ibid., 352.

178. Chris Impey, *Living Cosmos*, 43.

179. Jones, *Darwin's Ghost*, 312.

180. Ibid., 313.

181. Ibid., 326.

182. J. Craig Venter et al., "The Sequence of the Human Genome," *Science* 291 (16 February 2001): 1347–48.

183. Holmes Rolston III, *Science and Religion: A Critical Survey* (Philadelphia: Templeton Foundation Press, 2006), xxv.

184. Jones, *Darwin's Ghost*, 327.

185. This would include the painted caves of Lascaux in the Périgord region of France and those of Altamira in the Cantabrian region of the Basque Country in northern Spain about 20,000 years ago. Also noteworthy are the recent discoveries of figurines at the cave site of Hohle Fels in southwestern Germany (Swabia), the oldest known examples of figurative art in the world (see van Huyssteen, *Alone in the World?*, 170–71).

186. Steven Mithen, *The Prehistory of the Mind: A Search for the Origins of Art, Religion, and Science* (London: Thames and Hudson, 1996), 70–72, 153–84.

187. See the chart in Mithen, *Prehistory of the Mind*, 67.

188. Ibid., 37–39, 55–60.

189. Ibid., 151–84.

190. Holmes Rolston III, "Creation and Resurrection," *Journal for Preachers* 32, no. 3 (Easter 2009): 29.

191. Ibid., 65–72.

192. Ibid., 153.

193. Ibid., 71, 153.

194. See Ibid., 190–92.

195. Van Huyssteen, *Are We Alone?*, 106.

196. Henry Plotkin, *Evolution in Mind: An Introduction to Evolutionary Psychology* (Cambridge, MA: Harvard University Press, 1998), 222.

197. Ibid., 223.

198. Van Huyssteen, *"Are We Alone?*, 256–68.

199. Andrew Curry, "Seeking the Roots of Ritual," *Science* 319 (18 January 2008), 278.

200. Wilson, *Consilience*, 303.

201. Charles Darwin, *The Descent of Man, and Selection in Relation to Sex* (1871) in *From So Simple Beginning: The Four Great Books of Charles Darwin*, ed. Edward O. Wilson (New York: W. W. Norton, 2006), 1248.

202. Wilson, *Consilience*, 330.

203. Paul Steinhardt, "A Cyclic Universe," *Seed* 11 (August 2007): 33–34.

204. Smolin, *Life of the Cosmos*, 109; Kaku, *Parallel Worlds*, 92–93, 134–35, 222–24.

205. Augustine, *Confessions*, 11.10–14.

206. Paul Davies, *God and the New Physics* (New York: Simon & Schuster, 1983), 38.

207. This analogy was aptly suggested to me by a twelve-year-old, Alden Daniel Glass.

208. Coles, *Cosmology*, 102.

209. Ibid., 104.

210. The letters stand for the four types of nucleotide and their nitrogenous bases: adenine, thymine, cytosine, and guanine.

211. Rolston, *Science and Religion*, xix.

212. Ibid, xvii.

213. Barbour, *When Science Meets Religion*, 106.

214. Some of God's commands, for example, are rhetorically distinguished by the cognate accusative construction in Hebrew, which bears phonetic similarities between the verb and its object (1:11, 15, 20). See William P. Brown, "Divine Act and the Art of Persuasion in Genesis 1," in *History and Interpretation: Essays in Honour of John H. Hayes*, ed. M. Patrick Graham, William P. Brown, and Jeffrey K. Kuan, JSOTSS 173 (Sheffield: Sheffield Academic Press, 1993), 19–32.

215. John Polkinghorne, *Reason and Reality* (Philadelphia: Trinity International Press, 1991), 45–47; Polkinghorne, *The Faith of a Physicist* (Princeton: Princeton University Press, 1994), 77–78.

216. More consistently employed in the Septuagint than in the Masoretic Text.

217. Carl Zimmer, "In Games, An Insight into the Rules of Evolution," *New York Times*, 31 July 2007, at www.nytimes.com/2007/07/31/science/31prof.html?scp=1&sq=Carl%20Zimmer%20%22Rules%20of%20Evolution%22&st=cse.

218. Ibid.

219. For a highly readable account, see Carl Zimmer, *Microcosm:* E. coli *and the New Science of Life* (New York: Pantheon, 2008), esp 54–58.

220. Ibid.

221. "Portfolio," *Seed* 11 (August 2007): 8.

222. E.g., Pss 32:6; 107:23–27; 144:7; Isa 17:12–13.

223. See Graeme Wistow, "Lens Crystallins: Gene Recruitment and Evolutionary Dynamism," *Trends in Biochemical Sciences* 18 (1993): 301–6. Cited in Rolston, *Science and Religion*, xxi.

224. Rolston, *Science and Religion*, xxi.

225. Lynn Margulis and Dorion Sagan, *Microcosmos: Four Billion Years of Evolution from Our Microbial Ancestors* (Berkeley: University of California Press, 1997), 29. See their emphasis on microbial "teamwork" in *What is Life?*, 236–38.

226. Hazen, *Gen·e·sis*, 144–48.

227. Margulis and Sagan, *What is Life?*, 85.

228. Hazen, *Gen·e·sis*, 148.

229. Ibid., 142.

230. Margulis and Sagan, *What is Life?*, 191.

231. For a detailed listing, see ibid., 236–37.

232. So William R. Stoeger, "Entropy, Emergence, and the Physical Roots of Natural Evil," in *Physics and Cosmology: Scientific Perspectives on the Problem of Natural Evil*, vol. 1, ed. Nancey Murphy, Robert John Russell, William R. Stoeger, SJ (Vatican: Vatican Observatory Publications/Berkeley: Center for Theology and the Natural Sciences, 2007), 94.

233. Grammatically, the verb "see" takes a second object, expressed as a separate clause.

234. See William F. Albright, "The Refrain 'And God Saw *ki tob*' in Genesis," in *Mélanges Bibliques: Rédigés en l'Honneur de André Robert*, Travaux de l'Iinstitut Catholique de Paris 4 (Paris: Bloud & Gay, 1957), 22–26.

235. So Welker, *Creation and Reality*, 9.

236. Barbour, *When Science Meets Religion*, 80.

237. This Berkeleyan thesis, that the universe requires an all-seeing Observer for its existence, is given a quantum twist by physicist Raymond Y. Chiao in "Quantum Nonlocalities: Experimental Evidence," in *Quantum Mechanics: Scientific Perspectives on Divine Action*, ed. Robert John Russell et al. (Vatican: Vatican Observatory Publications / Berkeley: Center for Theology and the Natural Sciences, 2001), 17–39.

238. Paul Davies, *Cosmic Jackpot: Why Our Universe Is Just Right for Life* (Boston: Houghton Mifflin, 2007), 1.

239. Impey, *Living Cosmos*, 139.

240. See, for example, Francis S. Collins, *The Language of God: A Scientist Presents Evidence for Belief* (New York: Free Press, 2006), 152.

241. Smolin, *Life of the Cosmos*, 193.

242. Ibid., 176.

243. Wilson, *Consilience*, 124.

244. Margulis and Sagan, *What is Life?*, 17.

245. Collins, *Language of God*, 23.

246. See chap. 4.

247. Rolston, *Science and Religion*, xxiv.

248. Wilson, *Consilience*, 142.

249. Ibid.

250. Ibid., 144.

251. Morten H. Christiansen and Simon Kirby, "Language Evolution: The Hardest Problem in Science?" in *Language Evolution*, ed. M. H. Christiansen and S. Kirby (New York: Oxford University Press, 2003), 1, 14–15.

252. For an up to date and accessible discussion of the history and theories of linguistic evolution, see Christine Kenneally, *The First Word: The Search for the Origins of Language* (New York: Viking Penguin, 2007).

CHAPTER 4

1. Barbara Brown Taylor, *Leaving Church: A Memoir of Faith* (New York: Harper-SanFrancisco, 2006), 22.

2. William Bryant Logan. *Dirt: The Ecstatic Skin of the Earth* (New York: W. W. Norton, 1995), 177.

3. Francis S. Collins, *The Language of God: A Scientist Presents Evidence for Belief* (New York: Free Press, 2006), 125.

4. Usually translated "Lord," beginning with the Greek Septuagint (LXX) and found in most modern translations, this proper name for God in the Hebrew Scriptures is not pronounced by most Jewish readers. Hence, only the consonants are transliterated from the Hebrew.

5. Andreas Schüle, "Made in the 'Image of God': The Concepts of Divine Images in Gen 1–3," *ZAW* 117 (2005): 15.

6. Michael Welker, *Creation and Reality*, trans. John F. Hoffmeyer (Minneapolis, MN: Fortress, 1999), 10.

7. Usually translated "the LORD God," YHWH Elohim is the Hebrew compound name given to God by the Yahwist (J).

8. See the translation given by Theodore Hiebert, *The Yahwist's Landscape: Nature and Religion in Early Israel* (Oxford: Oxford University Press, 1996), 32.

9. Ibid.

10. The verb in Hebrew (*'bd*) can mean "work" or "serve"; cf. 2:15.

11. Hebrew *'āpār*, which refers to the upper layer of the ground or topsoil.

12. Hebrew: *nepeš ḥayyāh*. For discussion, see pp. 109–10 for discussion.

13. To be preferred over "good and evil," which suggests Greek dualistic thinking. See Jacqueline E. Lapsley, *Whispering the Word: Hearing Women's Stories in the Old Testament* (Louisville: Westminster John Knox Press, 2005), 15, 112n.41. The "knowledge of good and bad" includes the knowledge of culture and moral discernment. See Deut 1:39; 2 Sam 19:35; Isa 7:15. According to these texts, young children do not yet have such knowledge, and the aged can lose it. In the Yahwist account, such knowledge is reserved for the gods (Gen 3:22).

14. See n. 10.

15. The verb (*šmr*) can also mean "protect" or "guard."

16. *'Ādām* in the garden narrative is not a proper name; it most frequently appears with the definite article in Hebrew.

17. See Hiebert, *Yahwist's Landscape*, 33–36.

18. See chap. 2. Another parallel is found in the Epic of Gilgamesh, in which the mother goddess ("Aruru") pinches off clay and shapes into the quintessential man of the wild, Enkidu, in Tablet I.

19. Edward O. Wilson, "General Introduction," in Charles Darwin, *From So Simple a Beginning: The Four Great Books of Charles Darwin*, ed. Edward O. Wilson (New York: W. W. Norton, 2004), 12.

20. Hiebert, *Yahwist's Landscape*, 67.

21. Cf. Gen 18:12; Ps 36:8; Neh 9:25; Jer 31:12; 51:34; Ezek 36:35; Joel 2:3. Many thanks to P. Kyle McCarter Jr. for this important insight.

22. Because the garden is planted by YHWH, the *'ādām* does not share in plowing and planting. See Hiebert, *Yahwist's Landscape*, 65. Thus, his vocation *in* the garden is not the same as his work outside it (contra ibid., 59). In the garden the *'ādām* is to tend the trees, which are watered by irrigation. Outside the garden he must cultivate the ground itself, facing all the special risks and rigors involved in dry land farming (Gen 2:5; 3:17–19).

23. NRSV: "a helper as his partner." The noun *'ēzer* refers to "help" or "helper." Beyond Genesis 2, the term most often refers to God or divine aid (e.g., Exod 18:4; Deut 33:7, 26, 29; Pss 33:20; 70:6; 115:9–11; 146:4) and, hence, does not suggest subordination of any kind. The preposition *neged* in the composite phrase *kĕnegdô* literally means "in front of" or "opposite of," here in the sense of correspondence.

24. Hebrew *nepeš ḥayyāh*, the same phrase used of the *ādām* in v. 7. See discussion pp. 109–10.

25. The enigmatic term *ṣēlāʿ* is most commonly used to designate the part of a building or hill: from "side" and "ridge" (Exod 25:12; 2 Sam 16:13) to the "planks" of a wall (1 Kgs 6:15) to the "leaves" of a door (1 Kgs 6:34). Only in Genesis 2 is the term used anatomically, designating either "side" or, less likely, "rib." The Greek translation of the Hebrew, *pleura*, is equally ambiguous. Early and medieval Jewish exegesis, such as found in the commentary work of Philo, Rashi, and Kimchi, interprets *ṣēlā* as "side." (The clearest exception is found in the Pseudo-Jonathan Targum, which identifies the man's "thirteenth rib on the left" for the woman's creation.) The midrash on Genesis *Bereshit Rabah* presents a discussion of the first human being cut in two to create the woman. In the context of Genesis 2, the common translation of "rib" is overly specific. I am indebted to the careful work of Naama Zahavi-Ely in her unpublished paper "The Better Half or a Spare Rib? A Linguistic Study of Eve's Creation" (paper given at the Society of Biblical Literature meeting in Toronto on November 26, 2002).

26. The verb *dābaq* is used widely in various contexts: bone to skin (Job 19:20), hand to sword (2 Sam 23:10), tongue to the roof of the mouth in thirst (Lam 4:4).

27. Same word for "woman" (*'iššāh*).

28. Or "wife."

29. Compare God's declaration of "good" in Gen 1:4, 10, 12, 18, 21, 25, 31.

30. See, e.g., Laban's acknowledgement of Jacob in Gen 29:14, as well as similar examples of this expression in Judg 9:2 and 2 Sam 5:1.

31. See Walter Brueggemann, "Of the Same Flesh and Bone [Gen 2,23a]," *CBQ* 32 (1970): 532–42.

32. Note the wordplay between *'ārôm* ("naked") in the previous verse and *'ārûm* ("crafty, shrewd") here. The latter term can be taken positively, as in Prov 12:23 with reference to the prudent or sensible person.

33. The verb is cast in the plural, referring to *both* the man and the woman.

34. Most translations cast the serpent's statement as a question, despite the lack of an interrogative marker in the Hebrew. The serpent actually poses an assertion so absurd as to guarantee a response.

35. Dietrich Bonhoeffer, *Creation and Fall: A Theological Interpretation of Genesis 1–3: Temptation*, trans. John C. Fletcher (London: SCM Press, 1959), 69.

36. See William E. Phipps, "Eve and Pandora Contrasted," *Theology Today* 45 (1988): 34–48.

37. The gods in *Enūma elish* wear their own clothing, "mantles of radiance."

38. The Hebrew phraseology is similar to the locative expression for the forbidden tree (cf. 2:9), suggesting that the couple's hiding place is close to the "tree of knowledge."

39. Or "voice" (*qôl*).

40. The Hebrew wordplay is unmistakable, made explicit by the Peshitta, for the clause could almost be read: "And I saw that I was naked." Seeing is fearing.

41. Note the deliberate use of "give" (*ntn*) in the verse: God "gave" the woman to the man for companionship; the woman "gave" the fruit to the man, resulting in a curse. The rhetorical effect blames God for the set up.

42. Note, by contrast, the language of desire in v. 6.

43. The syntax suggests a disjunctive sentence with the main clause in the latter half.

44. Cf. 2:15.

45. The same verb (*šmr*) is found in 2:15, where the groundling is placed in the garden to "preserve" it.

46. See Danna N. Fewell and David M. Gunn, "Shifting the Blame: God in the Garden," in *Reading Bibles, Writing Bodies: Identity and the Book*, ed. Timothy K. Beal and David M. Gunn (London: Routledge, 1997), 16–33.

47. Note that the man, in his defense, holds God responsible.

48. Claus Westermann, *Genesis* (Grand Rapids, MI: Eerdmans, 1987), 25.

49. See Jon D. Levenson, *Resurrection and the Restoration of Israel: The Ultimate Victory of the God of Life* (New Haven: Yale University Press, 2006), 32; James Barr, *The Garden of Eden and the Hope of Immortality* (London: SCM, 1992), 5–6.

50. Levenson, *Resurrection and the Restoration of Israel*, 32.

51. The walls of the temple, not fortuitously, were engraved with the figures of cherubim, palm trees, flowers, and lions (1 Kgs 6:29–35; cf. Ezek 41:18–20), to re-create the primeval garden.

52. Cf. the plant of rejuvenation featured in the Epic of Gilgamesh, Tablet XI.

53. For detailed discussion, see William P. Brown, *The Ethos of the Cosmos: The Genesis of Moral Imagination in the Bible* (Grand Rapids: Eerdmans, 1999), 152–57.

54. See chap. 6.

55. For accessible modern translations of this epic, see Maureen G. Kovacs, *The Epic of Gilgamesh* (Stanford: Stanford University Press, 1989); Stephanie Dalley, "Gilgamesh," in Dalley, *Myths from Mesopotamia: Creation, the Flood, Gilgamesh, and Others* (Oxford: Oxford University Press, 1991), 39–135; Stephen Mitchell, *Gilgamesh: A New English Version* (New York: Free Press, 2004), which is technically not a translation but a "version" developed from other English translations (66–67).

56. Julie Galambush, "'*adam* from '*adama*, '*issa* from '*is*: Derivation and Subordination in Genesis 2.4b–3.24," in *History and Interpretation: Essays in Honour of John H.*

Hayes, ed. M. Patrick Graham, William P. Brown, and Jeffrey K. Kuan, JSOTSup 173 (Sheffield: JSOT Press, 1993), 43.

57. The Song of Songs (also known as Song of Solomon, or Canticles), for example, erotically recaptures the life of mutual intimacy and desire in the garden.

58. 2 Kgs 25:4; Jer 39:4; 52:7. For archaeological evidence and insights, see Lawrence E. Stager, "Jerusalem and the Garden of Eden," *Eretz-Israel* 26 (1999): 184–94.

59. Neh 2:14; 3:15; Isa 8:6–7; see Gen 2:13.

60. For discussion, see Brown, *Ethos of the Cosmos*, 248–52.

61. For further discussion, see William P. Brown, *Seeing the Psalms: A Theology of Metaphor* (Louisville: Westminster John Knox, 2002), 68–69.

62. This is particularly clear in comparison to another garden account in Ezek 28:11–19, which typecasts the king of Tyre as a "cherub" in Eden who meets his downfall. Whereas the *'ādām* in Gen 2–3 is childlike, Ezekiel's primal man is godlike. See Levenson, *Resurrection and the Restoration of Israel*, 85.

63. Such as warfare and monumental building projects. See Oded Borowski, *Agriculture in Iron Age Israel* (Winona Lake, IN: Eisenbrauns, 1987), 9.

64. Frans de Waal, "Morally Evolved: Primate Social Instincts, Human Morality, and the Rise and Fall of 'Veneer Theory,'" in *Primates and Philosophers: How Morality Evolved*, ed. Stephen Macedo and Josiah Ober (Princeton: Princeton University Press, 2006), 24.

65. Arthur Peacocke develops the notion of improvisation as divine activity by analogy with music in his *Theology for a Scientific Age: Being and Becoming—Natural, Divine, and Human* (Theology and the Sciences; Minneapolis: Fortress, 1993), 152–57, 168–77. Peacocke draws from examples of classical music (e.g., Bach and Beethoven), although jazz would have been equally, if not more, apt.

66. Steve Jones, *Darwin's Ghost*: The Origin of Species *Updated* (New York: Ballantine, 2000), xxvi.

67. Ibid., xix.

68. Niles Eldredge and Stephen J. Gould, "Punctuated Equilibria: An Alternative to Phyletic Gradualism," in *Models in Paleobiology*, ed. T. J. M. Schopf (San Francisco: Freeman, Cooper, 1972), 82–115. See Edward O. Wilson's mediating discussion of their thesis in *The Diversity of Life* (New York: W. W. Norton, 1999), 88–89.

69. Bernard Wood, *Human Evolution: A Very Short Introduction* (Oxford: Oxford University Press, 2005), 45.

70. As geneticist Steve Jones says of evolution (*Darwin's Ghost*, 74).

71. Fred J. Ciesla, "Observing Our Origins," *Science* 319 (14 March 2008): 1488.

72. William Bryant Logan. *Dirt: The Ecstatic Skin of the Earth* (New York: W. W. Norton, 1995), 10.

73. Paul G. Falkowski and Yukoi Isozaki, "The Story of O$_2$," *Science* 322 (24 October 2008): 540.

74. Ibid.

75. Logan, *Dirt*, 10.

76. Falkowski and Isozaki, "The Story of O2," 541. See also James F. Kasting, "The Rise of Atmospheric Oxygen," in *Science* 293 (3 August 2001): 819–20.

77. Organic material, which consists of carbon compounds, takes up oxygen.

78. Falkowski and Isozaki, "The Story of O_2," 541.

79. Ibid., citing the work of geophysicist R. A. Berner.

80. A. Delsemme, "The Origin of the Atmosphere and of the Oceans," in *Comets and the Origin and Evolution of Life*, ed. Paul J. Thomas et al., 2nd ed. (Berlin: Springer, 2006), 58.

81. Ibid.

82. Latest research suggests that certain chemical reactions presumed necessary for producing RNA likely took place at "temperatures and pH levels found in ponds" that occasionally dried up. Carl Zimmer, "On the Origin of Life on Earth," *Science* 323 (9 January 2009): 199.

83. Robert M. Hazen, *Gen·e·sis: The Scientific Quest for Life's Origins* (Washington, D.C.: Joseph Henry, 2005), 141–42, 155.

84. U. J. Sofia, "Interstellar Dust: Not Just Bunnies under Your Bed," *Whitman Magazine*, December 2008, 12.

85. For more detailed discussion, see Neil DeGrasse Tyson and Donald Goldsmith, *Origins: Fourteen Billion Years of Cosmic Evolution* (New York: W. W. Norton, 2004), 151–59.

86. Ibid., 27.

87. Ibid., 159.

88. Logan, *Dirt*, 183.

89. Ibid., 184.

90. As estimated by University of Virginia researchers (ibid.).

91. George Fisher, personal communication.

92. Hazen, *Gen·e·sis*, 157.

93. Logan, *Dirt*, 125. Logan cites the work of biologist Hayman Hartman, who considers clay to be alive (*Dirt*, 123). The proposal was championed by A. G. Cairns-Smith in 1966 (Hazen, *Gen·e·sis*, 160–61).).

94. Hazen, *Gen·e·sis*, 157.

95. Logan, *Dirt*, 15.

96. Logan claims that one acre of soil produces one horsepower a day (ibid., 2).

97. Edward O. Wilson, *Consilience: The Unity of Knowledge* (New York: Vintage Books, 1998), 107.

98. Lynn Margulis and Dorion Sagan, *What is Life?* (Berkeley: University of California Press, 1995), 141.

99. Collins, *Language of God*, 125.

100. Ibid., 137. Or just a 1.23 percent difference in their genes.

101. Ibid.

102. *Atlanta Journal-Constitution*, 18 May 2006, A8. For the genetic data, see The Chimpanzee Sequencing and Analysis Consortium, "Initial Sequence of the Chimpanzee Genome and Comparison with the Human Genome," *Nature* 437 (1 September 2005): 69–87.

103. Paul Ehrlich, *Human Natures: Genes, Culture, and the Human Prospect* (New York: Penguin Books, 2002), 70.

104. Dario Maestripieri, *Macachiavellian Intelligence: How Rhesus Macaques and Humans Have Conquered the World* (Chicago: University of Chicago Press, 2007), 17–35.

105. Ibid., 173–74. Rhesus macaques, of course, are not unique in this respect. See the collection of essays in *Machiavellian Intelligence: Social Expertise and the Evolution of Intellect in Monkeys, Apes, and Humans,* ed. R. Byrne and A. Whiten (Oxford: Oxford University Press, 1988).

106. For a thorough survey of de Waal's observations and analysis, see his *Good Natured: The Origins of Right and Wrong in Humans and Other Animals* (Cambridge, MA.: Harvard University Press, 1996). His most recent engagement on the philosophical and moral issues can be found in his discussion with a panel of critics (Robert Wright, Christine M. Korsgaard, Philip Kitcher, and Peter Singer) in *Primates and Philosophers: How Morality Evolved,* ed. Stephen Macedo and Josiah Ober (Princeton: Princeton University Press, 2006).

107. De Waal,"Morally Evolved," 33–36.

108. Ibid., 21–56.

109. De Waal, "The Tower of Morality," in *Primates and Philosophers,* 168.

110. In DNA comparisons, humans and bonobos "share a microsatellite related to sociality that is absent in the chimpanzee." De Waal, "Appendix B: Do Apes Have a Theory of Mind?" in *Primates and Philosophers,* 73.

111. Ibid., 72.

112. De Waal, "Morally Evolved," 56. More recent studies support this: see Kwame Anthony Appiah, *Experiments in Ethics* (Cambridge, MA: Harvard University Press, 2008); Dan Ariely, *Predictably Irrational: The Hidden Forces That Shape Our Decisions* (New York: HarperCollins, 2008).

113. See Michael S. Gazzaniga, *Human: The Science Behind What Makes Us Unique* (New York: HarperCollins, 2008), 128.

114. Ibid., 143–44, quoting Robert Wright.

115. De Waal, "Appendix A: Anthropomorphism and Anthropodenial," in *Primates and Philosophers,* 65.

116. De Waal, "The Tower of Morality," in *Primates and Philosophers,* 161.

117. Barbara King, an anthropologist at the College of William and Mary, takes the continuity argument one step further by discerning even the precursors to religion among the great apes, evidenced in the need for "belongingness" in primate behavior. See Barbara J. King, *Evolving God: A Provocative View of the Origins of Religion* (New York: Doubleday, 2007), 5–7.

118. So Eric Delson of the American Museum of Natural History of New York as quoted in John Noble Wilford, "Lost in a Million-Year Gap, Solid Clues to Human Origins" in *New York Times,* 18 September 2007, at www.nytimes.com/2007/09/18/science/18evol.html?scp=1&sq=Noble%20Wilford%20%20%22Lost%20in%20a%20Million%20Year%20Gap%22&st=cse

119. Ehrlich, *Human Natures,* 77.

120. Ibid., 86–87.

121. So named by its discovery in the Neander Valley, near Düsseldorf.

122. See Ann Gibbons, "Ice Age No Barrier to 'Peking Man,'" *Science* 323 (13 March 2009): 1419.

123. Ehrlich, *Human Natures*, 81–82.

124. See Peter Brown et al., "A New Small-bodied Hominin from the Late Pleistocene of Flores, Indonesia," *Nature* 431 (28 October 2007): 1055–61. Also see http://news.nationalgeographic.com/news/2004/10/1027_041027_homo_floresiensis.html

125. So Maxine Sheets-Johnstone, *The Roots of Thinking* (Philadelphia: Temple University Press, 1990), 92.

126. For an accessible discussion of the evolutionary formative consequences of bipedalism, see Steve Mithen, *The Singing Neanderthals: The Origins of Music, Language, Mind, and Body* (Cambridge, MA: Harvard University Press, 2006), 144–58.

127. Ibid., 158.

128. So Sheets-Johnstone, *Roots of Thinking*, 90–92.

129. As Sheets-Johnstone puts it (ibid., 100).

130. Ehrlich, *Human Natures*, 92.

131. As Ehrlich points out, archaeological evidence of sewed skins dates to only around 20,000 years ago, but "crude clothing" was a necessity once hominids ventured forth into colder environments (ibid., 95).

132. There is some discussion as to whether the loss of body fur came before or after the use of clothing among hominids. See Mithen, *Singing Neanderthals*, 199–200. Either way, both developments are closely connected.

133. Wood, *Human Evolution*, 85.

134. Ibid., 87. In Africa, *Homo erectus* may have evolved into pre-modern *Homo heidelbergensis* (ibid., 93).

135. Hisao Baba et al., "*Homo erectus* Calvarium from the Pleistocene of Java," *Science* 299 (28 February 2003): 1384–88.

136. Ehrlich, *Human Natures*, 97.

137. Wood, *Human Evolution*, 98.

138. Mithen, *Singing Neanderthals*, 228. Mithen argues that due to the lack of symbolic artifacts, coupled with "immense stability" in Neanderthal culture, syntactic language was not a possession of *Homo neanderthalensis*. What Neanderthals did possess was a form of verbal communication that was "holistic, multi-modal, manipulative, and musical," as opposed to segmented, compositional, referential, and syntactical.

139. Ibid., 245.

140. As posited by Rebecca Cann, Mark Stoneking, and Allan Wilson. See Wood, *Human Evolution*, 105–6.

141. See, e.g., Milford H. Wolpoff et al., "Modern Human Ancestry at the Peripheries: A Test of the Replacement Theory," *Science* 291 (12 January 2001): 293–97; Milford H. Wolpoff and Rachel Caspari, *Race and Human Evolution* (New York: Simon & Schuster, 1997).

142. Wood, *Human Evolution*, 108.

143. Ehrlich, *Human Natures*, 104.

144. Jared M. Diamond, *Guns, Germs, and Steel: The Fates of Human Societies* (New York: W. W. Norton, 1997), 39–41.

145. David Lewis-Williams, *The Mind in the Cave: Consciousness and the Origins of Art* (New York: Thames and Hudson, 2002), 40.

146. Such as the stunning Chauvet Cave paintings in southern France. But the actual origin of art extends farther back in time. One recent proposal argues that symbolic art began more than 100,000 years ago, as evidenced in the "crosshatched patterns" found on pieces of red ochre discovered at Blombos Cave in South Africa. See Michael Balter, "Early Start for Human Art? Ochre May Revise Timeline," *Science* 323 (30 January 2009): 569. Others suggest that certain modified stones ostensibly resembling human figures indicate a date as early as 300,000 years ago. See Michael Balter, "On the Origin of Art and Symbolism," *Science* 323 (9 February 2009): 709–11.

147. See the discussions in Ehrlich, *Human Natures*, 159–62; J. Wentzel van Huyssteen, *Alone in the World? Human Uniqueness in Science and Theology* (Grand Rapids, MI: Eerdmans, 2006), 163–215.

148. Van Huyssteen, *Alone in the World?*, 224.

149. Richard Potts, "Sociality and the Concept of Culture in Human Origins," in *The Origins and Nature of Sociality*, ed. Robert W. Sussman and Audrey R. Chapman (New York: Aldine de Gruyter, 2004), 261–63.

150. Stuart L. Pimm et al., "What Is Biodiversity?," in *Sustaining Life: How Human Health Depends on Biodiversity*, ed. Erich Chivian and Aaron Bernstein (New York: Oxford University Press, 2008), 17–18. See also Ehrlich, *Human Natures*, 171–72.

151. Cornelia Dean, "Seeing the Risks of Humanity's Hand in Species Evolution," *New York Times*, 10 February 2009, at www.nytimes.com/2009/02/10/ science/10humans.html?scp=1&sq=Cornelia%20Dean%20%20%22Seeing%20 the%20Risks%20of%20Humanity's%20Hand%22&st=cse.

152. Ehrlich, *Human Natures*, 109.

153. So Gazzaniga, *Human*, 9, citing neuroscientist T. M. Preuss.

154. Ibid., 110–11.

155. See the discussion of Nicholas Humphrey's work in Ehrlich, *Human Natures*, 111–12.

156. Steven Mithen, *The Prehistory of the Mind: A Search for the Origins of Art, Religion, and Science* (London: Thames and Hudson, 1996), 70–71, 166. For further discussion, see chap. 3.

157. Gazzaniga, *Human*, 308 (emphasis added).

158. Research in this area began with the study of brain-damaged individuals. The most publicized case was that of railway worker Phineas Gage, who on September 13, 1848, was pierced in the head by a metal rod, causing severe frontal lobe damage. Although he miraculously survived, Gage's personality changed drastically, and he was unable to make rational decisions. See the discussion in Ehrlich, *Human Natures*, 114–15.

159. Attributed to biologist S. J. Singer. Cited in Wilson, *Consilience*, 121.

160. Nancey Murphy, *Bodies and Souls, or Spirited Bodies?*, Current Issues in Theology (Cambridge: Cambridge University Press, 2006), 99.

161. Gazzaniga, *Human*, 302 (emphasis added).

162. Ehrlich, *Human Natures*, 112–13.

163. Mary Midgley, *Beast and Man: The Roots of Human Nature*, 2nd ed. (London: Routledge, 1995), 196.

164. Ibid.

165. Ehrlich, *Human Natures*, 113.

166. Thus, as counterpart to the "Mitochondrial Eve" there is also the hypothetical "Y-Chromosomal Adam," which when traced back existed only 60,000 years ago, compared to his female counterpart of 140,000 years ago!

167. Wood, *Human Evolution*, 107.

168. Ehrlich, *Human Natures*, 179–80.

169. Ibid., 180.

170. Ibid.

171. Ibid., 181.

172. An additional human distinction from all other animals is menstruation.

173. See the discussion of theories in Ehrlich, *Human Natures*, 183–87.

174. See Ehrlich, *Human Natures*, 186, who attributes the loss of estrus to the "human sociocultural environment." Not so with Sheets-Johnstone (see n. 128 above).

175. Ibid., 187.

176. Ibid., 188.

177. Ehrlich cites evidence in studies showing contrasting patterns of promiscuity between gays and lesbians, the latter more often forming "long-term stable relationships" (ibid., 190, 194).

178. Ibid., 194.

179. Shelley Taylor, *The Tending Instinct* (New York: Times Books, 2002), cited in de Waal, "Morally Evolved," 5.

180. Ehrlich, *Human Natures*, 175.

181. Ibid.

182. Ibid., 176.

183. Ibid., 177.

184. Ibid.

185. Maestripieri, *Macachiavellian Intelligence*, 21.

186. Ibid., 160.

187. Ibid.

188. Britt Peterson, "Nice Guys Didn't Finish the Neolithic," *Discover*, August 2006. 15, which reports on the work of archaeologist Rich Schulting of Queen's University Belfast and forensics investigator Michael Wysocki.

189. Ibid.

190. See also Jones, *Darwin's Ghost*, 310–11.

191. The Yahwist narrative continues in Genesis 4 with the murder of Abel by his brother Cain, the first time the word "sin" is used in the biblical narrative (Gen 4:7). Violence escalates with Lamech (4:23–24) and culminates with the flood story (see 6:5–8).

192. Charles Darwin, *On the Origin of Species by Means of Natural Selection*, in *From So Simple a Beginning: The Four Great Books of Charles Darwin*, ed. Edward O. Wilson (New York: W. W. Norton, 2006), 760.

193. The quote is from Alfred Lord Tennyson, "In Memoriam A. H. H." LVI; 13–16, in *Tennyson: A Selected Edition*, ed. Christopher Ricks (1969; repr., Harlow: Longman, 1989), 399.

194. Christopher Southgate, *The Groaning of Creation: God, Evolution, and the Problem of Evil* (Louisville: Westminster John Knox, 2008), 137 n. 22.

195. Ibid., 29.

196. Ibid., 100.

197. Michael Boulter, *Extinction: Evolution and the End of Man* (London: HarperCollins, 2003), 9.

198. Edward O. Wilson, *The Creation: An Appeal to Save Life on Earth* (New York: W. W. Norton, 2006), 117.

199. Ibid.

200. Jan Schipper et al. "The Status of the World's Land and Marine Mammals: Diversity, Threat, and Knowledge," *Science* 322 (10 October 2008): 227.

201. See Gen 4:8–11, 23–24; 6:11–13.

202. William Sloane Coffin, *Letters to a Young Doubter* (Louisville: Westminster John Knox, 2005), 64.

203. Although more than three hundred distinct dog breeds are generally recognized. Jones, *Darwin's Ghost*, 25–26, 30.

204. Ibid., 27.

205. Ibid., 26.

206. For a fuller discussion, see Ted Kerasote, *Merle's Door: Lessons from a Freethinking Dog* (Orlando: HarcourtBooks, 2007), 26–48.

207. Jones, *Darwin's Ghost*, 23.

208. See the full discussion by Murphy, *Bodies and Souls, or Spirited Bodies?*, 56–70. See also Warren S. Brown, "Cognitive Contributions to the Soul," in *Whatever Happened to the Soul? Scientific and Theological Portraits of Human Nature*, ed. Warren S. Brown, Nancey Murphy, and H. Newton Maloney, Theology and the Sciences (Minneapolis: Fortress, 1998), 99–126.

209. As suggested in the KJV: "and man became a living soul" (Gen 2:7).

210. E.g., Isa 15:4; Pss 63:6; 107:9; Prov 27:7.

211. E.g., Ps 105:18.

212. E.g., Gen 1:20.

213. E.g., Gen 9:4–5; Lev 17:11; Deut 12:23.

214. Particularly in the context of desire (e.g., Song 1:7; 3:1–4; Eccl 6:3) or volition (Gen 23:8). See also Lev 19:28; 21:11; Num 6:6; 9:10 for the sense of "dead soul."

215. E.g., Gen 9:4–5; Lev 17:11; Deut 12:23.

216. E.g., Num 31:19; Ezek 13:19.

217. Hans Walter Wolff, *Anthropology of the Old Testament* (Philadelphia: Fortress, 1974), 20. For a more detailed discussion, see Levenson, *Resurrection and the Restoration of Israel*, 110–12.

218. Tim Flannery, *The Weather Makers: How We Are Changing the Climate and What It Means for Life on Earth* (New York: Grove Press, 2005), 12–13.

219. The story of Samuel being brought up from Sheol is about the prophet's resuscitation, not about his "soul" (1 Sam 28). See Bill T. Arnold, "Soul-Searching Questions about 1 Samuel 28: Samuel's Appearance at Endor and Christian Anthropology," in *What about the Soul? Neuroscience and Christian Anthropology*, ed. Joel B. Green (Nashville: Abingdon, 2004), 75–84.

220. The following discussion is drawn from Lapsley's perceptive analysis in *Whispering the Word*, 14–17.

221. Ibid., 14.

222. Ibid.

223. It is precisely "conscience" that best captures the ethical dimension of "wisdom" (*ḥokmāh*) as profiled in the book of Proverbs. See Michael V. Fox, *Proverbs 1–9*, AB 18A (New York: Doubleday, 2000), 29–38.

224. As the serpent rightly predicted, God did not strike the couple dead as promised.

225. So Robert W. Jenson, *Systematic Theology*, vol. 2 (New York: Oxford University Press, 1999), 59–60. Cf. Gen 5:26.

226. Charles Darwin, *The Descent of Man, and Selection in Relation to Sex*, in *From So Simple a Beginning*, 1248.

227. With apologies to Paul Tillich!

228. The tree of life was also proposed independently by Alfred Russel Wallace in a paper published in 1855. Peter J. Bowler, "Darwin's Originality," *Science* 323 (9 January 2009): 224.

229. Darwin, *On the Origin of Species*, 532–33.

230. Ibid., 533.

231. Francesca D. Ciccarelli et al., "Toward Automatic Reconstruction of a Highly Resolved Tree of Life," *Science* 311 (3 March 2006): 1286.

232. Ibid., 1285. For fuller discussion, see chap. 5.

233. See the National Science Foundation's project "Assembling the Tree of Life" at http://atol.sdsc.edu.

234. See Gen 4:1–2, 17–22, 25–26; 5:1–32; 9:18–19.

CHAPTER 5

1. From her poem, "A Summer Day," in her collection, *House of Light* (Boston: Beacon, 1990), 60.

2. Andrew Harvey, *A Journey in Ladakh* (New York: Houghton Mifflin, 1983), 93.

3. Walter Isaacson, *Einstein: His Life and Universe* (New York: Simon & Schuster, 2007), 26, 114.

4. Both Uz and Eden are located vaguely in the east. There is disagreement among the biblical traditions as to whether Uz refers to a place south of Israel in Edom (Jer 2:20; Lam 4:21; Gen 36:28) or to someplace northeast (Gen 10:23; 22:21). In light of its uncertain location, see Samuel Balentine's suggestive comparison between Uz

and Eden in *Job*, Smith & Helwys Commentary (Macon, GA: Smith & Helwys, 2006), 41–44.

5. Ibid., 42.

6. Schrödinger's cat, along with Einstein's ball in the box, was devised to expose, albeit unsuccessfully, the shortcomings of quantum theory (Isaacson, *Einstein*, 453–60).

7. See, most recently, Kathryn Schifferdecker, *Out of the Whirlwind: Creation Theology in the Book of Job*, HTR 61 (Cambridge, MA: Harvard University Press, 2008), 13–21.

8. For translations, see *ANET* 601–4; *BWL* 63–91; *BM* 2.790–98; *CS* 1.492–95.

9. For translations, see *ANET* 596–600; *BWL* 21–56; *CS* 1.486–92.

10. See C. L. Seow's arguments for dating Job in the late sixth or early fifth century based on literary parallels with Second Isaiah and Zechariah 3:1–2, as well as the historical reference to the Chaldeans in 1:17. See Seow, *Job: A Commentary*, Eerdmans Critical Commentary (Grand Rapids, MI: Eerdmans, forthcoming).

11. See, e.g., Pss 44:2, 19; 79:2; Isa 13:12; 34:13; Jer 10:22; Lam 5:18.

12. Usually translated "counsel," but in context the term (*'ēṣāh*) refers to careful planning.

13. Literally, "sons of God."

14. The plural form is specifically dual.

15. Literal reading, which conveys a sense of forcefully imposed establishment of the sea's limits.

16. Read *yišbōt gĕ'ôn* for MT *yāšît big'ôn* (due to transposition of letters between the two words). For comparable, albeit more general, imagery, see Prov 8:29.

17. Literally, "from your days."

18. MT has "they take their stand like a garment," which makes little sense. Read *tiṣṣāba'* from the verb *ṣb'*.

19. Figurative expression for water's phase transition from liquid to ice.

20. Literally, "seized" or "captured." See the same verb and its form used in Job 41:9.

21. Literally, "Herd."

22. Literally, "Fool."

23. Of uncertain identity. It is possibly a reference to constellations in general (cf. 2 Kgs 23:5).

24. Likely Ursa Major. For an alternative interpretation of the astronomical references, see A. de Wilde, *Das Buch Hiob*, OTS 22 (Leiden: Brill, 1981), 142–47.

25. Hebrew singular, but likely with collective force in reference to the "heavens."

26. Although the noun *mištār* is nowhere else attested in the Hebrew Bible, it is related etymologically to an Akkadian term that refers to celestial "writing" (*mašṭāru*), suggesting horoscopic significance.

27. See *HALOT* 374; Edouard Dhorme, *A Commentary on the Book of Job*, trans. H. Knight (London: Nelson, 1967), 593; Robert Gordis, *The Book of Job: Commentary, New Translation, and Special Studies* (New York: Jewish Theological Seminary, 1978), 452–53.

28. See *HALOT* 1327; Dhorme, *Commentary*, 593; Gordis, *Book of Job*, 453.

29. Literally, "with wisdom."

30. Literally, "cause to lie down" or "tip."

31. A metaphor for sedimentation.

32. Literal translation, referring to the pain of giving birth.

33. Onagers or wild asses were used in Sumer to pull wagons and chariots in the third and second millennia BCE.

34. Or "wild ox." The Septuagint, as a rule, translates "unicorn," and the Vulgate, "rhinoceros." For a description of the now extinct auroch, see Julius Caesar's *Gallic War*, 6.28.

35. With the proposed *BHS* change. Or, according to Qere, "that it will return your seed and gather (it) to your threshing floor."

36. Literally, "rejoice" (*ʿls*).

37. Hebrew has *ḥăsîdāh* ("stork" or "heron"), but makes no sense syntactically. Possible emendation is *ḥāsĕrāh* ("it lacks"). For an alternative rendering, see Schifferdecker, *Out of the Whirlwind*, 153.

38. Read *nōsāh* for *wĕnōsāh*.

39. Meaning is uncertain (*mrʾ*). Arabic cognates include *mry*, "to strike the ground," and *maruʾa*, "to act the man" (*HALOT* 630). The latter etymological connection suggests a confrontational gesture.

40. The form is a feminine variant of a common word for "thunder." Some scholars argue for "mane"; cf. *HALOT* 1268.

41. Hebrew meaning uncertain; see *HALOT* 848.

42. A cry of joy that can disparage another; see Isa 44:16; Ezek 25:3; Ps 70:4.

43. The Hebrew form is a *hapax legomenon* with uncertain meaning. Read *yĕlaʿlēlʿû*, from *lʿʿ*.

44. This last verse in particular identifies the raptor (*nešer*) of v. 27 as a vulture, not an eagle.

45. Usually translated "Almighty."

46. Hebrew *mišpāṭ*, which generally means "justice."

47. Or "stiffens" (*ḥpṣ*). The meaning is uncertain, since the verbal root most commonly means "take delight in." Perhaps the meaning here is related to the Arabic *ḥafaḍa*, "to make lower." If so, then the "tail" of Behemoth is not necessarily phallic, as assumed by many commentators. See *HALOT* 340.

48. Literally, "can bring near his sword." For the difficulties of this verse, see Carol A. Newsom, "The Book of Job: Introduction, Commentary, and Reflections," in *The New Interpreter's Bible Commentary*, vol. 4 (Nashville: Abingdon, 1996), 619. She recommends revocalizing the MT's active participle ("maker") to a passive participle ("made"; cf. 41:25) to render: "made to dominate his companions." She fails to note, however, a consonantal emendation made necessary by her translation: the transposition of two letters in the last word (from *ḥrbw* to *ḥbrw*). The MT is clear as it stands.

49. Job 41:1 is numbered 40:25 in the Hebrew. See also textual note *d* in NRSV.

50. The Hebrew lacks the interrogative particle but is implied by the following verses.

51. I.e., the hope of someone conquering Leviathan.

52. The presence of the interrogative particle is likely due to dittography.

53. Or "Is he not cruel when one arouses him?" (so Schifferdecker, *Out of the Whirlwind*, 177–78), which requires one to read it as a question, which is not at all clear from the Hebrew. In any case, read *yĕʾîrenû* for MT *yĕʿûrenû*.

54. MT has "me" against most other Hebrew manuscripts and LXX. The emendation to "it" is not so "extensive" as claimed by Schifferdecker (*Out of the Whirlwind*, 178), and it preserves the parallelism of the verse.

55. The shift to first person makes best sense if it is attributed to Leviathan's discourse rather than to God's, particularly in light of the reference to Leviathan's "boastings" in v. 4. See Newsom, "Job," 623; contra Schifferdecker, *Out of the Whirlwind*, 178.

56. That is, the one who dares to confront Leviathan.

57. Divine discourse resumes.

58. Hebrew *bad*; cf. Job 11:3; Jer 48:30. My translation of v. 4a (despite the difficulties of the second line) is the most straightforward. Schifferdecker, by contrast, turns the verse into a preterite question ("Did I not silence . . . ?"), claiming the opposite meaning (ibid., 174, 178–79).

59. Read *gĕbûrātô* for the Hebrew plural.

60. Hebrew uncertain, since *ḥîn* is a *hapax legomenon*, perhaps related to *ḥēn* ("favorable, graceful").

61. MT is indecipherable.

62. Read *siryōnô* for MT *risnô* ("his bridle"). See LXX.

63. Read *gēwōh* for MT *gaʾăwāh* ("pride").

64. Read *ʾōgēm* for MT *ʾagmōn* ("reed"). See *HALOT* 10–11.

65. Literally, "throat."

66. See Schifferdecker, *Out of the Whirlwind*, 176.

67. Literally, "shatterings."

68. Or "ground" (*ʿāpār*).

69. Or "there is no dominion over it."

70. The phrase is used in Job 28:8 to refer to the "proud animals."

71. For a discussion of this disputed verse, see p. 279 n.106.

72. See Newsom, "Job," 595.

73. Contra Michael V. Fox, "Job 38 and God's Rhetoric," *Semeia* 19 (1981): 58–60. See J. Gerald Janzen, *Job*, Interpretation Bible Commentary (Atlanta, John Knox, 1985), 225–28.

74. Gerhard von Rad, "Job XXXVII and Ancient Egyptian Wisdom," in *The Problem of the Hexateuch and Other Essays*, trans. E. W. Truemann Dicken (London: SCM Press, 1985 [1960]), 281–91. See also John Gray, "The Book of Job in Context of N. E. Literature," *ZAW* (1970): 251–69; Yair Hoffman, *A Blemished Perfection: The Book of Job in Context*, JSOTSup 213 (Sheffield: Sheffield Academic Press, 1996), 84–114.

75. Hoffman, *Blemished Perfection*, 85.

76. Ibid., 85–86.

77. Michael V. Fox, "Egyptian Onomastica and Biblical Wisdom," *VT* 36 (1986): 304–6.

78. See the translation by A. H. Gardiner, *Ancient Egyptian Onomastica*, 2 vols. (Oxford: Oxford University Press, 1947), 1:2. Fox argues that the claims made in the

introduction "must be taken with more than a grain of salt" (Fox, "Egyptian Onomastica," 303), but the basis for his suspicion rests on too high a bar, namely, modern science.

79. Hoffman, *Blemished Perfection*, 92.

80. Cf. Samuel Balentine's use of journey language in his summary of God's answer in *Job*, 645–46.

81. Carol A. Newsom, *The Book of Job: A Context of Moral Imaginations* (Oxford: Oxford University Press, 2003), 236, 243, 247.

82. See chap. 1, pp. xx–xx.

83. Newsom, "Job," 597; Newsom, *Book of Job*, 243.

84. Dating back to 3.5 billion years ago.

85. See the discussion of Gen 1:1–2:3 in chap. 3, pp. xxx–xxx.

86. See Schifferdecker, *Out of the Whirlwind*, 77. Cf. comparable language in Prov 30:4.

87. Michael B. Dick, "The Neo-Assyrian Royal Lion Hunt and Yahweh's Answer to Job," *JBL* 2 (2006): 255.

88. Ibid., 244–45. For an exhaustive survey of leonine imagery in the Old Testament, see Brent A. Strawn, *What is Stronger than a Lion? Leonine Image and Metaphor in the Hebrew Bible and the Ancient Near East*, OBO 212 (Fribourg, Academic Press/Göttingen: Vandenhoeck & Ruprecht, 2005).

89. Cf. Newsom, "Job," 607.

90. Christopher Southgate, *The Groaning of Creation: God, Evolution, and the Problem of Evil* (Louisville: Westminster John Knox, 2008), 63.

91. The phrase is borrowed from Edward O. Wilson, *The Creation: An Appeal to Save Life on Earth* (New York: W. W. Norton, 2006), 13, 163.

92. Hippos do not, as a rule, dwell in the mountains, nor are they to be found in the Jordan River, and crocodiles do not breathe fire. For the difficulties of pinpointing their natural counterparts, see Newsom, "Job," 618.

93. For Leviathan, see chaps. 2 and 6. Behemoth *as a proper name* is attested only in Job; otherwise it is a generic term for cattle or domestic animals. The plural form designates intensity or majesty; Behemoth is a "superbeast" (Balentine, *Job*, 83). In Egyptian religion, moreover, the hippopotamus is associated with the god of chaos, Seth, the enemy of Horus.

94. In the Hebrew Bible, Leviathan looms even larger than a chapter.

95. For an exploration of this last theme, see Kathleen O'Connor, "Wild, Raging Creativity: Job in the Whirlwind," in *Earth, Wind, and Fire: Biblical and Theological Perspectives on Creation*, ed. Carol J. Dempsey and Mary Margaret Pazdan (Collegeville, MN: Liturgical Press, 2004), 48–56.

96. See Terence E. Fretheim, *God and World in the Old Testament: A Relational Theology of Creation* (Nashville: Abingdon Press, 2005), 235, 239.

97. With the possible exception of the warhorse in 39:19–25, whose strength cannot be entirely tamed.

98. Ellen F. Davis, *Getting Involved with God: Rediscovering the Old Testament* (Cambridge, MA: Cowley, 2001), 139.

99. Edward O. Wilson, *Consilience: The Unity of Knowledge* (New York: Vintage, 1998), 231. See his full treatment in *Biophilia* (Cambridge: Harvard University Press, 1984); cf. Stephen R. Kellert and Edward O. Wilson, eds., *The Biophilia Hypothesis* (Washington, D.C.: Island Press/Shearwater Books, 1993).

100. Bill McKibben, *The Comforting Whirlwind: God, Job and the Scale of Creation* (Cambridge, MA: Cowley, 2005), 43.

101. Coined by Walter Brueggemann, *Abiding Astonishment: Psalms, Modernity, and the Making of History* (Louisville: Westminster John Knox, 1991).

102. Cf. Wilson, *Creation*, 63.

103. Many thanks to Daniel G. Conklin for this apt phrase.

104. For more detail on this point, see William P. Brown, *The Ethos of the Cosmos: The Genesis of Moral Imagination in the Bible* (Grand Rapids, MI: Eerdmans, 1999), 370–75; John G. Gammie, "Behemoth and Leviathan: On the Didactic and Theological Significance of Job 40:15–41:26," in *Israelite Wisdom: Theological and Literary Essays in Honor of Samuel Terrien*, ed. John G. Gammie et al. (Missoula: Scholars Press, 1978), 217–31; and O'Connor, "Wild, Raging Creativity," 52.

105. For example, Job exposes his friends' false confidence and pride (13:2–12; cf. 42:19–21).

106. My translation, in contrast to the NRSV and most other translations (e.g., Schifferdecker, *Out of the Whirlwind*, 184, 186–87). This verse in Hebrew is the most disputed in all the book of Job. For various options and accompanying bibliography, see Newsom, "Job," 628–29; William Morrow, "Consolation, Rejection, and Repentance in Job 42:6," *JBL* 105 (1986): 211–25. The most convincing proposal is given by Thomas Krüger, "Did Job Repent?" in *Das Buch Hiob und Seine Interpretationen*, ATANT 88 (Zürich: Theologischer Verlag Zürich, 2007), 217–29, whose translation is adapted here.

107. Alan Burdick, "The Old Men and the Sea," *New York Times Magazine*, 31 December, 2006, 56 at www.nytimes.com/2006/12/31/magazine/31hedgpeth. html?scp=1&sq=Alan%20Burdick%20%22The%20Old%20Men%20and%20the%20 Sea%22&st=cse.

108. Edward O. Wilson, *Diversity of Life* (New York: W. W. Norton, 1999), 44.

109. Wilson, *Consilience*, 4.

110. Charles Darwin, *The Voyage of the Beagle*, in *From So Simple A Beginning: The Four Great Books of Charles Darwin*, ed. Edward O. Wilson (New York / London: W. W. Norton, 2006), 429.

111. Edward O. Wilson, "Introduction to *The Voyage of the Beagle*," in *From So Simple A Beginning*, 18.

112. Darwin, *Voyage of the Beagle* in *From So Simple a Beginning*, 431.

113. Wilson, *Creation*, 123.

114. Cf. Job 24:5–8; 30:3–8, 29.

115. Attributed to S. J. Singer by Wilson in *Consilience*, 121.

116. Wilson, *Diversity of Life*, 38.

117. Stearns and Hoekstra, *Evolution*, 218.

118. Ibid.

119. Ibid.

120. Ibid., 219.

121. Wilson, *Diversity of Life*, 156 (correction: in the universe galaxies, not "stars," are moving apart).

122. Ibid.

123. Ibid.

124. Ibid., 11, 14.

125. As posited by M. E. Boraas et al. in 1998 and discussed in Stearns and Hoekstra, *Evolution*, 284.

126. Beginning with the end of the Ordovician geological period 440 million years ago and concluding with the end of the Cretaceous era, 65 million years ago. The catastrophe marking the end of the Permian era 280 million years ago came close to total destruction with an 80 percent percent reduction of marine invertebrate genera brought about from an intense cycle of global cooling and warming until Pangaea split up into separate, drifting continents (Stearns and Hoekstra, *Evolution*, 271–72).

127. Wilson, *Diversity of Life*, 15.

128. Ibid., 31.

129. Quoted in David Beerling, *The Emerald Planet: How Plants Changed Earth's History* (Oxford: Oxford University Press, 2007), vi.

130. Ibid.

131. The combined study of evolutionary development and developmental biology or "evo-devo" examines how evolution occurs through the inherited changes in genes that regulate the development of an individual organism. See Catherine Baker, *The Evolution Dialogues: Science, Christianity, and the Quest for Understanding* (Washington, D.C.: AAAS, 2006), 134.

132. Stearns and Hoekstra, *Evolution*, 239.

133. As confirmed by protein sequences extracted from the soft tissue preserved in a thighbone fossil of a *T. rex*. See Mary Higby Schweitzer et al., "Analyses of Soft Tissue from *Tyrannosaurus rex* Suggest the Presence of Protein," *Science* 316 (13 April 2007): 277–80; John M. Asara, "Protein Sequences from Mastodon and *Tyrannosaurus Rex* Revealed by Mass Spectrometry," *Science* 316 (13 April 2007): 280–85.

134. I.e., prokaryotes, protists, fungi, plants, and animals.

135. Discussed in Stearns and Hoekstra, *Evolution*, 294.

136. Ibid., 295.

137. Neil DeGrasse Tyson and Donald Goldsmith, *Origins: Fourteen Billion Years of Cosmic Evolution* (New York: W. W. Norton, 2004), 244.

138. Chris Impey, *The Living Cosmos: Our Search for Life in the Universe* (New York: Random House, 2007), 95.

139. Ibid., 104. For a highly readable account of the scientific search for life's extreme origins, see Robert M. Hazen, *Gen·e·sis: The Scientific Quest for Life's Origin* (Washington, D.C.: Joseph Henry, 2005).

140. For the full story, see Hazen, *Gen·e·sis*, 96–99.

141. See, e.g., Job 24:1–12; 30:1–8, 29; Newsom, *Book of Job*, 240. For a full account of Job's disdain of the wilderness and its denizens, see W. P. Brown, *Ethos of the Cosmos*, 333–36.

142. Quoted by Jeff Whellwright, "Captive Wilderness," *Discover*, August 2006, 46.

143. Ibid., 49.

144. Enkidu was the intimate companion and former rival of the Sumerian ruler Gilgamesh, who grew up in the wild and became acculturated. See Stephanie Dalley, "Gilgamesh," in *Myths from Mesopotamia: Creation, The Flood, Gilgamesh, and Others* (Oxford: Oxford University Press, 1991), esp. 52–59.

145. See Wilson's expressed awe over the wolverine in *Creation*, 56–58.

146. Ibid., 57–58.

147. Fretheim, *God and World*, 282.

148. Quoted in Frappa Stout, "History and Imagination," *USA Weekend*, 17 July 2005, 11. See www.usaweekend.com/05_issues/050717/050717books.html. Not surprisingly, Michael Cunningham's novel *Specimen Days* (New York: Farrar, Straus and Giroux, 2005) features extraterrestrial lizard women.

149. Munir Humayun of the National High Magnetic Field Laboratory quoted in Diane Roberts, "Cosmic Dust from Distant Comet Comes to Earth," NPR, *Weekend Edition Sunday*, 10 December 2006. See www.npr.org/templates/story/story.php?storyId=6605084.

150. Quoted in ibid.

151. Perhaps a dynamic equivalent to "gird up your loins"; see Exod 12:11.

152. But lamentably not free from toxic waste!

153. This and the following descriptions of deep sea creatures are adapted from Claire Nouvian, *The Deep: The Extraordinary Creatures of the Abyss* (Chicago: University of Chicago Press, 2007), 156, 159.

154. Ibid., 207–8.

155. Ibid., 81.

156. Ibid., 39.

157. Ibid., 134–35.

158. For an account of the discovery and its implications, see Neil Shubin, *Your Inner Fish: A Journey into the 3.5-Billion-Year History of the Human Body* (New York: Pantheon Books, 2008), 17–27. *Titktaalik* remains the best preserved transitional fossil between fish and early land-living animals: it has a shoulder, elbow, and wrist all inside a fin (23). Its specialized anatomy allowed it do "push-ups" and thereby move on land (39).

159. Loren Eisley, *The Immense Journey* (New York: Random House, 1957), 49.

160. Ibid., 50–51.

161. Author's composition.

CHAPTER 6

1. From Wendell Berry's poem, "The Old Man Climbs a Tree," in *A Timbered Choir: The Sabbath Poems 1979–1997* (New York: Counterpoint, 1998), 194.

2. From St. Thomas Aquinas, "The Mandate," translated by Daniel Ladinsky in *Love Poems from God: Twelve Sacred Voices from the East and West* (New York: Penguin Compass, 2002), 127.

3. Provided that one can imagine Job ever recovering his voice of praise after his ordeal!

4. Many thanks to my colleague Stan Saunders for casting human identity in this way vis-à-vis the psalms of praise (see, e.g., Ps 148).

5. The Masoretic pointing suggests "flaming fire." The plural predicate suggests a composite subject.

6. Read *yōsēd* for MT *yāsad* ("He established"). The Masoretic pointing renders a finite verb, signaling a change in aspect, consonant with the following section. However, the original form was likely participial, with the new section beginning in v. 6.

7. Read *kĕsûtô*, along with LXX, for MT *kissîtô* ("You covered it"), due to simple graphic confusion. In MT, the antecedent of the suffix is masculine, while "earth" (*'ereṣ*) is normally feminine. Gender agreement suggests "deep" (*tĕhôm*) as the antecedent, thereby rendering the clause as a *casus pendens*. The LXX translator, while rendering the words correctly, did not recognize the syntactical form of the clause and thereby made the astounding claim that the deep serves, like light (v. 2), as *God's* covering!

8. I.e., the flowing wadis (see v. 10).

9. The text is corrected in view of a possible dittography of the *mem*, which prefixes "oil" as a preposition in MT.

10. The Masoretic pointing features a perfect verb to indicate completed action.

11. The poetic parallelism invests a nuance of purpose to the verb.

12. Slightly emended to plural, the term literally refers to possessions, as in the case of livestock (Gen 31:18; 34:23; Josh 14:4).

13. The syntax is ambiguous, given the possible antecedents for the suffixed preposition *bô*. Thus, the text could be translated: "Leviathan, which you fashioned to play in it [the sea]." But this possibility is less likely in view of the syntactical proximity of "Leviathan" in the verse.

14. "Spirit" and "breath" (v. 29) are denoted by the same word in Hebrew (*rûaḥ*). Cf. Gen 1:2.

15. Hebrew *śîaḥ*, which bears a distinctly discursive sense. Here, it refers to the hymnic poem as it is recited or sung, not to a silent meditation.

16. Cf. Pss 74:13–15; 89:9–10; Isa 27:1; 51:9.

17. E.g., Pss 37:9–15; 109:2–20; 137:7–9.

18. Cf. Grace Jantzen, *God's World, God's Body* (Philadelphia: Westminster, 1984); Sallie McFague, *The Body of God: An Ecological Theology* (Minneapolis: Fortress, 1993).

19. See only Job 40:21–22, which like the psalm details certain flora as the setting for animal life, specifically Behemoth.

20. These cedars have been protected for more than 1,500 years by the monks of the monastery in the Kadishar Valley. In 1998 the remnants were declared a UN Natural Heritage site. See Michael S. Northcott, *A Moral Climate: The Ethics of Global Warming*, (Maryknoll, NY: Orbis Books, 2007), 107.

21. See, e.g., Patrick D. Miller, "The Poetry of Creation: Psalm 104," in *God Who Creates: Essays in Honor of W. Sibley Towner*, ed. William P. Brown and S. Dean McBride Jr. (Grand Rapids, MI: Eerdmans, 2002), 87–103; James Limburg, "Down to

Earth Theology: Psalm 104 and the Environment," *Currents in Theology and Mission* 21 (1994): 340–46.

22. Gene M. Tucker, "Rain on a Land Where No One Lives: The Hebrew Bible on the Environment," *JBL* 116 (1997): 15.

23. This section is condensed from my essay, "Joy and the Art of Cosmic Maintenance: An Ecology of Play in Psalm 104," in *"And God Saw That It Was Good": Essays on Creation and God in Honor of Terence E. Fretheim*, Word & World Supplement Series 5 (Saint Paul, MN: Luther Seminary, 2006), 23–32.

24. Erhard S. Gerstenberger, *Psalms, Part 2, and Lamentations*, FOTL 15 (Grand Rapids, MI: Eerdmans, 2001), 225.

25. On the other hand, the related verbs *ḥpṣ* ("to delight") and *rṣh* ("to be pleased") are applied to God, particularly in contexts of sacrifice and human integrity (e.g., Pss 5:4; 18:19; 35:27; 37:23; 40:13; 41:11; 51:16, 19; 147:10). The verb "rejoice" signals, however, a more heightened, ecstatic level of emotion.

26. The closest parallel is found in Zeph 3:17 ("[YHWH] will rejoice over you with gladness"). From the verb *śwś*.

27. Jon D. Levenson, *Creation and the Persistence of Evil: The Jewish Drama of Divine Omnipotence*, 2nd ed. (Princeton: Princeton University Press, 1994 [1988]), 14.

28. In Hebrew, 40:25–41:26 (see n. 49 in chap. 5).

29. Hebrew *śḥq* or *ṣḥq* in piel. The qal form denotes derisive laughter, which can be attributed to God against nations and enemies of Zion (Pss 2:4; 59:9; cf. Prov 1:26).

30. See chap. 7.

31. A fact confirmed by British pediatrician and psychoanalyst D. W. Winnicott. See his *Playing and Reality* (London/New York: Tavistock, 1982).

32. Thanks to my colleague Kathy Dawson for this suggestive analogy.

33. Charles Darwin, *On the Origin of Species by Means of Natural Selection*, in *From So Simple A Beginning: The Four Great Books of Charles Darwin*, ed. Edward O. Wilson (New York: W. W. Norton, 2006), 760.

34. See the Encyclopedia of Life now being compiled at www.eol.org.

35. Paul Davies, *Cosmic Jackpot: Why Our Universe Is Just Right for Life* (Boston: Houghton Mifflin, 2007), 2.

36. Neil DeGrasse Tyson and Donald Goldsmith, *Origins: Fourteen Billion Years of Cosmic Evolution* (New York: W. W. Norton, 2004), 103.

37. Ian G. Barbour, *When Science Meets Religion: Enemies, Strangers, or Partners?* (New York: HarperSanFrancisco, 2000), 57.

38. Tyson and Goldsmith, *Origins*, 266.

39. Freeman Dyson, *Disturbing the Universe* (New York: Harper & Row, 1979), 251.

40. As detailed by John Barrow and Frank Tipler in their now classic study, *The Anthropic Cosmological Principle* (Oxford: Oxford University Press, 1986).

41. Tyson and Goldsmith, *Origins*, 104.

42. Quoted in Davies, *Cosmic Jackpot*, 2.

43. For full discussion, see Chris Impey, *The Living Cosmos: Our Search for Life in the Universe* (New York: Random House, 2007).

44. Edward O. Wilson, *The Diversity of Life* (New York: W. W. Norton, 1999), xvi.

45. Stuart L. Pimm et al., "What Is Biodiversity?" in *Sustaining Life: How Human Health Depends on Biodiversity*, ed. Erich Chivian and Aaron Bernstein (New York: Oxford University Press, 2008), 14.

46. Catherine Baker, *The Evolution Dialogues: Science, Christianity, and the Quest for Understanding*, ed. James B. Miller (Washington, D.C.: AAAS, 2006), 132. It is estimated that there are 10^{30} microbes on Earth, the vast majority of which remain unknown (James Tiedje and Timothy Donohue, "Microbes in the Energy Grid," *Science* 320 [23 May 2008]: 985).

47. Stephen C. Stearns and Rolf F. Hoekstra, *Evolution: An Introduction* (Oxford: Oxford University Press, 2000), 214.

48. At least of complex life forms. For up to date information on endangered species and extinctions, see the "Red List of Threatened Species" developed by the International Union for the Conservation of Nature and Natural Resources at www.iucnredlist.org.

49. Quoted in David Beerling, *The Emerald Planet: How Plants Changed Earth's History* (Oxford: Oxford University Press, 2007), 174.

50. Ibid.

51. Ibid., vi.

52. Ibid., 207.

53. For numerous examples of social learning among birds and mammals that have formed their evolutionary paths, see Eytan Avital and Eva Jablonka, *Animal Traditions: Behavioural Inheritance in Evolution* (Cambridge: Cambridge University Press, 2000).

54. As Avital and Jablonka point out, even "human whistles, flute melodies and mechanical sounds" can be incorporated into the blackbird's song (ibid., 82).

55. Ibid., 83.

56. Ibid., ix.

57. Ibid., 2.

58. Ibid., ix.

59. Ibid., 5. Although Avital and Jablonka acknowledge that much of human culture depends on the distinctly symbolic transmission of information (i.e., language), they define culture more widely as "a set of behaviour patterns or products of animal activities that are socially transmitted in an animal lineage, group or population" (ibid., 22).

60. Ibid., 21. By way of example, Avital and Jablonka cite the study of a group of Japanese macaques living on a small island of Koshima, who learn to forage for food outside of their natural habitat (ibid., 10).

61. Ibid., 310.

62. Ibid.

63. Quoted in Harold J. Morowitz, *The Emergence of Everything: How the World Became Complex* (Oxford: Oxford University Press, 2002), 132.

64. Ibid., 133 (italics added).

65. Ibid.

66. Avital and Jablonka, *Animal Traditions*, 16.

67. Ibid., 28, citing the older works of Conrad Waddington and Richard Lewontin.

68. F. J. Odling-Smee, "Niche-Constructing Phenotypes," in *The Role of Behavior in Evolution*, ed. J. C. Plotkin (Cambridge, MA: MIT Press, 1988), 73–132; and Odling-Smee, K. N. Laland, and M. W. Feldman, "Niche Construction," *American Naturalist* 147 (1996): 641–48.

69. Described in Avital and Jablonka, *Animal Traditions*, 13–15, 29.

70. Charles Darwin, *The Descent of Man, and Selection in Relation to Sex* in *From So Simple A Beginning: The Four Great Books of Charles Darwin*, ed. Edward O. Wilson (New York: W. W. Norton, 2006), 837.

71. Avital and Jablonka, *Animal Traditions*, 352.

72. Albeit in a rudimentary, nonsymbolic sense

73. As reported in the *Atlanta Journal-Constitution* (29 November 2007), A7.

74. Baker, *Evolution Dialogues*, 146.

75. *Plants and Climate Change: Which Future?*, cited in Dan Charles, "The Threat to the World's Plants," *Science* 320 (23 May 2008): 1000.

76. Ibid.

77. Ricahrd Dawkins, *The God Delusion* (Boston: Houghton Mifflin Company, 2006), 143.

78. Anthropic principles, for example, do not bear predictive value from a scientific standpoint.

79. For a short list of the various ways humans are destroying habits and diminishing biodiversity, see Eric Chivian and Aaron Bernstein, "How Is Biodiversity Threatened by Human Activity?," in *Sustaining Life: How Human Health Depends on Biodiversity*, ed. Erich Chivian and Aaron Bernstein (New York: Oxford University Press, 2008), 29–74.

CHAPTER 7

1. Paul Davies, *Cosmic Jackpot: Why Our Universe Is Just Right for Life* (Boston: Houghton Mifflin, 2007), 5.

2. Personal communication.

3. For the pedagogy of wisdom in Proverbs, see Charles F. Melchert, *Wise Teaching: Biblical Wisdom and Educational Ministry* (Harrisonburg, PA: Trinity Press International, 1998), 47–73.

4. See, e.g., Prov 1:20–21; 7:6; 8:1–3; 9:1–2, 14.

5. See Leo G. Perdue, *Wisdom Literature: A Theological History* (Louisville: Westminster John Knox, 2007), 49.

6. For the well-ordered arrangement of these verses, see William P. Brown, *Character in Crisis: A Fresh Approach to the Wisdom Literature of the Old Testament* (Grand Rapids, MI: Eerdmans, 1996), 24–25.

7. E.g., Prov 3:7–8; 9:10–11; 10:27; 14:27; 19:23.

8. The voice of the mother is likely found in chap. 7 of Proverbs. See Athalya Brenner, "Proverbs 1–9: An F Voice?" in *On Gendering Texts: Female and Male Voices in*

the Hebrew Bible, ed. Athalya Brenner and Fokkelien Dijk-Hemmes (Leiden: Brill, 1993), 113–30.

9. Borrowing from the Marxist theoretician Louis Althusser, Carol Newsom suggests that this figure of the silent "son" serves as the "interpellated" subject of the reader. See Carol A. Newsom, "Woman and the Discourse of Patriarchal Wisdom: A Study of Proverbs 1–9," in *Gender and Difference in Ancient Israel*, ed. Peggy L. Day (Minneapolis: Fortress, 1989), 143.

10. E.g., Prov 1:11–14; 7:14–20; 9:13–17.

11. Wisdom's discourse resembles the impassioned aretalogies of the Egyptian goddess Isis. For texts and discussion, see Michael V. Fox, *Proverbs 1–9* (AB 18A; New York: Doubleday, 2000), 336–38.

12. See Prov 1:20–33; 4:5–9; 8:1–36; 9:1–6.

13. The verb can mean "create" or "acquire." The context suggests a procreative sense, namely, Wisdom's conception (see vv. 24, 25).

14. The verb is best derived from the root *skk*, offering a suggestive parallel to Ps 139:13, which is set in the context of gestation.

15. Same verb as in v. 24a ("birthed").

16. A much disputed line that hangs on the meaning of one word (*'āmôn*). Contrary to many English translations (e.g., NRSV "master worker"), the grammatical form of the word appears to be verbal, specifically infinitival (see the same verb with similar meaning in Esth 2:20b). The larger context, moreover, tips the scale semantically. Given the lack of reference to creative activity on Wisdom's part anywhere else in the poem and the repeated reference to play at the end, the image of Wisdom as a growing child fits the context best. For further argumentation, see Michael V. Fox, "'*Āmôn* Again," *JBL* 115 (1996): 699–702; Fox, *Proverbs 1–9*, 286–87.

17. Hebrew *tēbēl*, which can mean "mainland" or "cultivable" part of the earth (*'ereṣ*). See *HALOT* 1682–83.

18. See vv. 22, 23, 24, 25. For detailed examination of the terms of gestation and birth, see William P. Brown, *The Ethos of the Cosmos: The Genesis of Moral Imagination in the Bible* (Grand Rapids, MI: Eerdmans, 1999), 272–73, along with accompanying bibliography.

19. Terence E. Fretheim, *God and World in the Old Testament: A Relational Theology of Creation* (Nashville: Abingdon, 2005), 206 (italics original).

20. See Prov 1:8, 15, 2:1; 3:1, 11; 4:1, 20; 5:1; 6:1; 7:1, 4, 24; 8:32.

21. Elsewhere in Proverbs 1–9, wisdom is cast as an object of understanding, teaching, possession, and even love (e.g., 2:2–4; 3:18; 4:5–8, 11; 5:1).

22. The term is borrowed from Stephen Toulmin, *Cosmopolis: The Hidden Agenda of Modernity* (New York: Free Press, 1990).

23. Brian R. Greene, *The Fabric of the Cosmos: Space, Time, and the Texture of Reality* (New York: Vintage Books, 2004), 222.

24. For a critical discussion of the "anthropic principle," see Lawrence Osborn, "Theology and the New Physics," in *God, Humanity and the Cosmos*, ed. Christopher Southgate (New York/London: T&T Clark, 2005), 142–48; John Polkinghorne, "Beyond the Big Bang," in *Science Meets Faith*, ed. Fraser Watts (London: SPCK, 1998), 19–22.

25. From the Greek word *sophia* = wisdom.

26. Greek: *harmozousa* (or "arranging in harmony").

27. Greene, *Fabric of the Cosmos*, 11.

28. Ibid., 80.

29. The measurement refers to the experiment conducted by Alain Aspect in the early 1980s in which two photons emitted by a calcium atom are perfectly correlated. See Ibid., 102–15.

30. Ibid., 114.

31. Cited in Dennis Overbye, "Remembrance of Things Future: The Mystery of Time," *New York Times*, 28 June 2005, at www.nytimes.com/2005/06/28/science/28time.html?scp=1&sq=Dennis%20Overbye%20%20%22The%20Mystery%20of%20Time%22&st=cse.

32. Peter Coles, *Cosmology: A Very Short Introduction* (Oxford: Oxford University Press, 2000), 115.

33. That is, their velocities and positions cannot be determined simultaneously.

34. Coles, *Cosmology*, 109–10.

35. Kitt Peak National Observatory exhibit, July 27, 2006.

36. Coles, *Cosmology*, 99–100.

37. Neil deGrasse Tyson and Donald Goldsmith, *Origins: Fourteen Billion Years of Cosmic Evolution* (New York: W. W. Norton, 2004), 210.

38. Ibid., 213.

39. Quoted from Dennis Overbye, "The Cosmos According to Darwin," *New York Times Magazine*, 13 July 1997, 26, at www.nytimes.com/1997/07/13/magazine/the-cosmos-according-to-darwin.html?scp=1&sq=Dennis%20Overbye%20%20%22The%20Cosmos%20According%20to%20Darwin%22%20&st=cse.

40. Philo, *De opificio mundi*, §§17–20, 24, 143. For translation and commentary, see Philo of Alexandria, *On the Creation of the Cosmos According to Moses*, trans. and ed. David T. Runia, PACS 1 (Brill: Leiden, 2001), 50–51, 84–85.

41. Smolin is well known for his distinctly Darwinian view of cosmic evolution. See his *The Life of the Cosmos* (London: Phoenix Paperback, 1997), esp. 93–172.

42. Warren S. Brown, "Preface," in *Understanding Wisdom: Sources, Science, and Society*, ed. Warren S. Brown (Philadelphia: Templeton Foundation Press, 2000), ix.

43. Indeed, it is the latter point to which most of the book of Proverbs is devoted.

44. Warren S. Brown, "Wisdom and Human Neurocognitive Systems: Perceiving and Practicing the Laws of Life," in *Understanding Wisdom*, 200.

45. Ibid., 198.

46. Ibid., 203.

47. Ibid., 204–10. Warren Brown bases his conclusions on the work of Antonio Damasio in *Descartes's Error: Emotion, Reason, and the Human Brain* (New York: G. P. Putnam's Sons, 1994).

48. Jeffrey P. Schloss, "Wisdom Traditions as Mechanisms for Organismal Integration: Evolutionary Perspectives on Homeostatic 'Laws of Life,'" in *Understanding Wisdom*, 164.

49. Henry Plotkin, *Darwin Machines and the Nature of Knowledge* (Cambridge, MA: Harvard University Press, 1993), 150. Plotkin defines intelligence as "an adaptation that allows animals, including ourselves, to track and accommodate to change that

occurs at a certain frequency" (ibid.). See also Plotkin, *Evolution in Mind: An Introduction to Evolutionary Psychology* (Cambridge: MA: Harvard University Press, 1998), esp. 121–260.

50. Even though the secondary heuristic is "nested" in the first (Plotkin, *Darwin Machines*, 161, 181).

51. Schloss, "Wisdom Traditions," 163–64.

52. Plotkin, *Darwin Machines*, 177.

53. Ibid., 225.

54. Ibid., 227.

55. Schloss, "Wisdom Traditions," 168.

56. Ibid., 168, 181.

57. Ibid., 168.

58. See chap. 3, pp. 64–65.

59. Lawrence M. Hinman, "Seeing Wisely—Learning to Become Wise," in *Understanding Wisdom*, 414.

60. Plotkin, *Darwin Machines*, 207.

61. Nancy Sherman, "Wise Emotions," in *Understanding Wisdom*, 324–27.

62. Ibid., 326.

63. Ibid., 328.

64. See the discussion in Eytan Avital and Eva Jablonka, *Animal Traditions: Behavioural Inheritance in Evolution* (Cambridge: Cambridge University Press, 2000), 344.

65. Ibid.

66. Ibid., 345.

67. Ibid., 346.

68. Ibid., 347.

69. Ellen Dissanayake, *Art and Intimacy: How the Arts Began* (Seattle: University of Washington Press, 2000), xi.

70. Ibid., 14.

71. Ibid., 8.

72. Ibid., 49–50.

73. Ibid., 134.

74. Ibid., 145 (italics added).

75. Ibid., 15.

76. In ancient Israel, the nursing time of infants was three years or longer. See 2 Macc 7:27; cf. Gen 21:8; 1 Sam 1:23–24.

CHAPTER 8

1. Quoted in Paul Davies, *The Last Three Minutes: Conjectures about the Ultimate Fate of the Universe* (New York: BasicBooks, 1994), 13.

2. Bertolt Brecht, *Galileo*, ed. Eric Bentley, trans. Charles Laughton (New York: Grove Press, 1966), 32.

3. R. B. Y. Scott, *Proverbs, Ecclesiastes* (AB 18; Garden City: Doubleday, 1965), 191.

4. For the sense behind Qoheleth's contradictions, see Michael V. Fox, *Qohelet and His Contradictions* (JSOTSup 71; Sheffield: Sheffield Academic Press, 1987 [repr. 1989]); Fox, *A Time to Tear Down and a Time to Build Up: A Rereading of Ecclesiastes* (Grand Rapids: Eerdmans, 1999), esp. 1–4, 14, 69, 138.

5. See Eccl 1:1; cf. v. 12; 2:9.

6. Hence the NRSV translation "Teacher."

7. See Michael V. Fox, "Wisdom in Qoheleth," in *In Search of Wisdom: Essays in Memory of John G. Gammie*, ed. Leo G. Perdue, Bernard Brandon Scott, and William Johnston Wiseman (Louisville: Westminster John Knox, 1993), 119–21.

8. E.g. Eccl 1:14; 2:13; 2:24; 3:16, 22; 4:1, 4, 7, 15; 8:9, 10, 17; 9:11.

9. Qoheleth's search for enduring meaning can be compared to Gilgamesh's unsuccessful journey for immortal life. See William P. Brown, *Ecclesiastes* (Interpretation Bible Commentary; Louisville: Westminster John Knox, 2000), 1–7.

10. See Eccl 4:2–3; 6:3 (but cf. 9:4). Quote from Mel Gussow, "For Saul Bellow, Seeing the Earth with Fresh Eyes," *New York Times* (26 May 1997) at www.nytimes.com/1997/05/26/books/for-saul-bellow-seeing-the-earth-with-fresh-eyes.html?scp=1&sq=Gussow%20%22Seeing%20the%20Earth%22&st=cse.

11. For detailed discussion of Qoheleth's economic context, see Choon-Leong Seow, "The Socioeconomic Context of 'The Preacher's' Hermeneutic," *PSB* 17 (1996): 168–95.

12. As well as his "personal" testimony as Solomon in 1:12–2:11.

13. Except, evidently, when it comes to women; see 7:26.

14. Literally, "what is left over" (*yitrôn*), which can mean gain or advantage in the sense of surplus (3:9; 5:16; cf. 5:9–10). See Choon-Leong Seow, *Ecclesiastes* (AB 18C; New York: Doubleday, 1997), 103–4. But contra Seow, the term in this context does seem to take on the nuance of commercial value.

15. Hebrew *šô'ēp* (see, e.g., Jer 2:24; 14:6; Isa 42:14; Job 7:2). For a contrasting image of the sun, see Ps 19:4b–6.

16. The subject of the verb is not divulged until the end of the line.

17. Quoted in *The Home Book of Humorous Quotations*, ed. A. K. Adams (New York: Dodd, Mead, 1969), 25.

18. See 2:22; 3:9; 6:11.

19. For an in depth study of the various nuances of "vapor" in Ecclesiastes, see Douglas B. Miller, *Symbol and Rhetoric in Ecclesiastes: The Place of Hebel in Qohelet's Work* (Academia Biblica 2; Atlanta: Society of Biblical Literature, 2002).

20. The distinction between the "light" and the celestial bodies recalls the comparable distinction made in Gen 1:3, 14–16. See Seow, *Ecclesiastes*, 353.

21. Unlikely "wheel." See Seow, *Ecclesiastes*, 367.

22. Hebrew *rûaḥ*.

23. See, e.g., Rachel Z. Dulin, "'How Sweet is the Light': Qoheleth's Age-Centered Teachings," *Interpretation* 55 (2001): 267–69.

24. See Seow's discussion of the eschatological tenor of this section in *Ecclesiastes*, 372–82.; idem, "Qohelet's Eschatological Poem," *JBL* 118 (1999): 209–34.

25. Difficult form in Hebrew. Perhaps, as Seow suggests, the *lamed* is asseverative ("surely"), as in 9:4, and the verb is finite rather than infinitival (*Ecclesiastes*, 167).

Another possibility is that the verbal root is not *brr* at all ("choose" or "test") but *br'* with an elided final aleph: "God has created them" in such as a way as "to show them...."

26. Hebrew: "to see," which makes little sense. The form is best taken as a hiphil (causative) with an elided *h* (so ibid.).

27. Hebrew *hēmmāh* ("they") is redundant, perhaps for emphasis.

28. Hebrew *'ēt wāpega'*, which can be translated as a hendiadys: "(un)timely accident."

29. It is critical to note that Qoheleth's most famous passage is entirely descriptive, as opposed to prescriptive. The infinitives that populate the poem can just as easily be translated as gerunds: "a time of bearing and a time of dying...." See Seow, *Ecclesiastes*, 158–62.

30. The verb is transitive.

31. Hebrew *hā'ōlām* more conventionally means "eternity." The disputed meaning has prompted some commentators to emend the word. However, I have chosen to make sense of the Masoretic rendering. See also Seow, *Ecclesiastes*, 163.

32. E.g., 2 Sam 13:1; 1 Kgs 1:4; Song 1:16; Ezek 33:32; see also Eccl 5:17.

33. Fox, *Time to Tear Down*, 136.

34. Quoted in David Beerling, *The Emerald Planet: How Plants Changed Earth's History* (Oxford: Oxford University Press, 2007), 112.

35. Ibid., 113.

36. Neil Shubin, *Your Inner Fish: A Journey into the 3.5-Billion-Year History of the Human Body* (New York: Pantheon Books, 2008), 201.

37. Ibid., 185–86.

38. Ibid., 184–85.

39. Ibid., 190–92.

40. In sharks and other primitive creatures, the gonads reside next to the heart (ibid., 193–96).

41. This is true even in the case of the ice ages, by which much of the earth's water supply becomes locked up in ice caps, thereby decreasing the level of water in the oceans. The opposite effect is now evident from global warming. See James Trefil, *The Nature of Science: An A–Z Guide to the Laws and Principles Governing Our Universe* (Boston: Houghton Mifflin, 2003), 416.

42. See Daniel Hillel, *The Natural History of the Bible: An Environmental Exploration of the Hebrew Scriptures* (New York: Columbia University Press, 2006), 36, 48.

43. Davies, *Last Three Minutes*, 49.

44. Neil DeGrasse Tyson, *Death by Black Hole and Other Cosmic Quandaries* (New York: W. W. Norton, 2007), 247. See also William R. Stoeger, SJ, "Scientific Accounts of Ultimate Catastrophes in our Life-Bearing Universe," in *The End of the World and the Ends of God*, ed. John Polkinghorne and Michael Welker (Harrisburg, PA: Trinity Press International, 2000), 19–28.

45. Stoeger, "Scientific Accounts of Ultimate Catastrophes," 26.

46. Brian R. Greene, *The Fabric of the Cosmos: Space, Time, and the Texture of Reality* (New York: Vintage Books, 2004), 143.

47. Ibid., 144–45 (italics original).

48. Ibid., 51, 58, 453. See also chap. 3.

49. Einstein quoted in Greene, *Fabric of the Cosmos*, 139.

50. Ibid.

51. But the distance is actually shortening: in about 7 billion years, the Milky Way and the Andromeda Galaxy will actually collide.

52. Greene, *Fabric of the Cosmos*, 246.

53. As evidenced in the homogeneity of microwave background radiation and the relative uniform distribution of galaxies.

54. Greene, *Fabric of the Cosmos*, 234.

55. Ibid., 141.

56. Ibid.

57. Ibid.

58. Ibid., 127.

59. For greater detail on the distinction, see Robert John Russell, "Finite Creation without a Beginning: The Doctrine of Creation in Relation to Big Bang and Quantum Cosmologies," in Russell, *Cosmology From Alpha to Omega: The Creative Mutual Interaction of Theology and Science*, Theology and the Sciences (Minneapolis: Fortress, 2008), 92.

60. Peter Coles, *Cosmology: A Very Short Introduction* (Oxford: Oxford University Press, 2000), 16.

61. Greene, *Fabric of the Cosmos*, 11.

62. Davies, *Last Three Minutes*, 31.

63. Ibid.

64. Ibid.

65. Ibid., 143.

66. Ibid., 156.

67. Chris Impey, *The Living Cosmos: Our Search for Life in the Universe* (New York: Random House, 2007), 78.

68. Edward O. Wilson, "Afterward," in *From So Simple a Beginning: The Four Great Books of Charles Darwin*, ed. Edward O. Wilson (New York: W. W. Norton, 2006), 1480.

69. So Robert John Russell, "Entropy and Evil: The Role of Thermodynamics in the Ambiguity of Good and Evil in Nature," in *Cosmology*, 227. Russell also refers to the "talons of time" (ibid.)

70. Davies, *Last Three Minutes*, 11.

71. Greene, *Fabric of the Cosmos*, 173, 478. For the difference between "gravitational entropy" and "nongravitational entropy," see William R. Stoeger, SJ, "Entropy, Emergence, and the Physical Roots of Natural Evil," in *Physics and Cosmology: Scientific Perspectives on the Problem of Natural Evil*, vol. 1, ed. Nancey Murphy, Robert John Russell, William R. Stoeger, SJ (Vatican City State: Vatican Observatory Publications/ Berkeley: Center for Theology and the Natural Sciences, 2007), 99.

72. See Yudhijit Bhattacharjee, "Do Black Holes Seed the Formation of Galaxies?" *Science* 323 (16 January 2009): 323.

73. To borrow from John Barrow and Frank Tipler, *The Anthropic Cosmological Principle* (Oxford: Clarendon Press, 1986), 166.

74. Davies, *Last Three Minutes*, 129–34.

75. Ibid., 133.

76. Or as Steven Weinberg tells of a "joke" circulated among scientists in Moscow about how the anthropic principle explains the misery of life: "There are many more ways for life to be miserable than happy; the anthropic principle only requires that the laws of nature should allow the existence of intelligent beings, not that these beings should enjoy themselves" (*Dreams of a Final Theory* [New York: Pantheon Books, 1992], 221–22).

77. Davies, *Last Three Minutes*, 9.

78. According to Klaus-Peter Schroeder of the University of Guanajuato in Mexico and Robert Connon Smith of the University of Sussex in the U.K. See Constance Holden, "Out of the Frying Pan," *Science* 319 (14 March 2008): 1465.

79. Ibid.

80. Davies, *Last Three Minutes*, 55.

81. Ibid., 56.

82. Ibid., 63–64.

83. Ibid., 55.

84. Ibid., 86.

85. Ibid., 88, 98.

86. Ibid., 84.

87. Greene, *Fabric of the Cosmos*, 299; Steve Nadis, "Tales from the Dark Side," *Astronomy* 34, no. 9 (September 2006): 32–33.

88. Greene, *Fabric of the Cosmos*, 296.

89. Dennis Overbye, "The Universe, Expanding Beyond All Understanding," *New York Times*, 5 June 2007, at www.nytimes.com/2007/06/05/science/space/05essa.html?scp=1&sq=Overbye%20%22The%20Universe,%20Expanding%20Beyond%22&st=cse.

90. Ibid.

91. Ibid.

92. Davies, *Last Three Minutes*, 99.

93. Coined by Lynn Margulis and Dorion Sagan, "autopoiesis" refers to life's capacity to generate itself in evolution. See *What Is Life?* (Berkeley: University of California Press, 1995), 17.

94. Steven Weinberg, *The First Three Minutes: A Modern View of the Origin of the Universe* (New York: Basic Books, 1977), 154.

95. Ibid., 154–55.

96. Eccl 2:24–25; 3:13, 22; 5:18–19; 8:15; 9:7–10; 11:9.

97. Indeed, Qoheleth refers to enjoyment as humanity's "lot" (*ḥēlek*) from God: 3:22; 5:17–18 (English vv. 18–19).

98. See Albert Camus, "The Myth of Sisyphus," in *The Myth of Sisyphus and Other Essays by Albert Camus* (New York: Vintage Books [Alfred A. Knopf and Random House], 1955), 88–90.

CHAPTER 9

1. Albert-László Barabási, *Linked: The New Science of Networks* (Cambridge, MA: Perseus, 2002), 6. The full quotation makes reference to a child taking apart a toy and crying over not being able to "put all the pieces back together."

2. *De resurrectione carnis* 12. Translated in Kevin J. Madigan and Jon D. Levenson, *Resurrection: The Power of God for Christians and Jews* (New Haven: Yale University Press, 2008), 231.

3. In the Christian canon, only the Song of Songs separates these two books. In the Hebrew canon, Isaiah and Ecclesiastes are distanced by nineteen books.

4. For an accessible discussion of the literary, sociological, and theological issues that helped to shape this corpus, see Jill Middlemas, *The Templeless Age: An Introduction to the History, Literature, and Theology of the "Exile"* (Louisville: Westminster John Knox, 2007), 93–111.

5. But not without precedent; see Gen 12:1–3, in which Abram is blessed by God to be a blessing to "all the families of the earth."

6. The Deuteronomist counts "ten thousand captives" for the first deportation (2 Kgs 24:14) with the "rest of the people ... all the rest of the population," except for "some of the poorest people of the land," exiled in the second deportation (25:11–12).

7. Rainer Albertz, *Israel in Exile: The History and Literature of the Sixth Century B.C.E.* (Atlanta: Society of Biblical Literature, 2003), 88.

8. Isa 40:17, 23; 41:29; 45:18–19; 49:4.

9. See also 40:12.

10. See, e.g., 45:22; 46:9.

11. Hebrew *'ēl*.

12. Literally, "also from this day I am he."

13. Hebrew *šālôm*.

14. For lexical comparison, see Michael Fishbane, *Biblical Interpretation in Ancient Israel* (Oxford: Clarendon Press, 1985), 325; Benjamin D. Sommer, *A Prophet Reads Scripture: Allusion in Isaiah 40–66* (Stanford: Stanford University Press, 1998), 143.

15. See Fishbane, *Biblical Interpretation*, 326; Sommer, *Prophet Reads Scripture*, 144.

16. Read, along with Qumran, *yiprah* for MT *yiprû* ("they bear fruit").

17. Literally, "seed."

18. Literally, "emergent things" (*seṣā'îm* from the verb *yṣ'*).

19. MT has "in between grass," which indicates textual corruption. Given the Aramaic and Arabic cognates, the word *bēn* probably refers to a species of tree, most probably the *tamarix jordansis*, which once lined the banks of the Jordan in plentiful groves. For further discussion, see William P. Brown, *The Ethos of the Cosmos: The Genesis of Moral Imagination in the Bible* (Grand Rapids, MI: Eerdmans, 1999), 239.

20. Likely, *salix acmophylla* or *salix alba*, a streamside tree that thrives along the Jordan.

21. Specifically, the Aleppo Pine. See Michael Zohary, *Plants of the Bible* (London: Cambridge University Press, 1982), 114.

22. For detailed discussion of the tree species listed in this passage, see W. P. Brown, *Ethos of the Cosmos*, 241–43.

23. So Claire R. Mathews, *Defending Zion* (BZAW 236; Berlin: de Gruyter, 1995), 125, 129.

24. Most emend the text to read "stretched out" (*nṭh*). But MT is clearly the more difficult text.

25. As defined in the Christian ordering of the canon.

26. Adam Frank, "A Wrinkle in Space-Time," *Astronomy* 34, no. 9 (September 2006), 39.

27. Greene, *Fabric of the Cosmos*, 67.

28. Ibid.

29. Ibid., 68.

30. Ibid., 69.

31. Frank, "A Wrinkle in Space-Time," 39.

32. Greene, *Fabric of the Cosmos*, 69.

33. Ibid., 70.

34. Frank, "A Wrinkle in Space-Time," 39.

35. Ibid., 36.

36. Greene, *Fabric of the Cosmos*, 230.

37. Ibid., 232–33.

38. See Nancey Murphy, "Reductionism: How Did We Fall Into It and Can We Emerge From It?" in *Evolution and Emergence: Systems, Organisms, Persons*, ed. Nancey Murphy and William R. Stoeger, SJ (Oxford: Oxford University Press, 2007), 19–39; and more generally Stuart A. Kauffman, *Reinventing the Sacred: A New View of Science, Reason, and Religion* (New York: Basic Books, 2008), esp. 1–43.

39. See Steven Weinberg, *Dreams of a Final Theory: The Scientist's Search for the Ultimate Laws of Nature* (New York: Random House, 1994), esp. 19–64. Cf. Kauffman's critique in *Reinventing the Sacred*, 31–43.

40. Quoted in Michio Kaku, *Parallel Worlds: A Journey through Creation, Higher Dimensions, and the Future of the Cosmos* (New York: Anchor Books, 2006), 351.

41. Murphy, "Reductionism," 27.

42. Robert M. Hazen, *Gen·e·sis: The Scientific Quest for Life's Origin* (Washington, D.C.: Joseph Henry, 2005), 14.

43. Harold J. Morowitz, *The Emergence of Everything: How the World Became Complex* (Oxford: Oxford University Press, 2002).

44. Ibid., 13.

45. Hazen, *Gen·e·sis*, 245.

46. Ibid.

47. Ibid.

48. Keith Ward, *Divine Action: Examining God's Role in an Open and Emergent Universe* (West Conshohocken, PA: Templeton Foundation Press, 2007 [1990]), 63.

49. Morowitz, *Emergence of Everything*, 20.

50. Ibid., 22.

51. Kauffman, *Reinventing the Sacred*, 2–3.

52. Ibid., 3.

53. Ibid.

54. See Ibid., 24.

55. Martinez J. Hewlett refers to these examples as "scale-free networks," including even the "network of human sexual interactions." See Hewlett, "Biological Models of Origin and Evolution," in *Emergence and Evolution*, 170.

56. Ibid., 171.

57. Lynn Margulis and Dorion Sagan, *What is Life?* (Berkeley: University of California Press, 1995), 114, 116.

58. Morowitz, *Emergence of Everything*, 87.

59. Ibid.; see also Margulis and Sagan, *What is Life?*, 114.

60. Morowitz, *Emergence of Everything*, 88.

61. David Beerling, *The Emerald Planet: How Plants Changed Earth's History* (Oxford: Oxford University Press, 2007), 16.

62. Morowitz, *Emergence of Everything*, 88.

63. Paul Davies, *The Last Three Minutes: Conjectures about the Ultimate Fate of the Universe* (New York: BasicBooks, 1994), 135.

64. Ibid., 137.

65. Lee Smolin, "Did the Universe Evolve?" *Classical and Quantum Gravity* 9 (1992): 173–91.

66. Greene, *Fabric of the Cosmos*, 330.

67. Evidence comes from a variety of extrabiblical sources. Numerous tablets from the sixth century BCE refer to the location āl-Yāhūdu ("the city of Judah" or better "Judahville"), a settlement of exiles likely somewhere in southeastern Babylonia by 572 BCE, fifteen years after the 587 deportation. The tablets indicate business transactions among Judeans and Babylonians. For a preliminary account of these yet-to-be published documents, see Laurie E. Pearce, "New Evidence for Judeans in Babylonia," in *Judah and the Judeans in the Persian Period*, ed. Oded Lipschits and Manfred Oeming (Winona Lake, IN: Eisenbrauns, 2006), 399–411. In addition, a collection of business documents from a prominent Jewish family in the fifth and fourth centuries BCE, the "Murashu Documents," reflects a thriving agricultural business in Babylonia among some Jews. (See Middlemas, *Templeless Age*, 23.)

68. See Middlemas, *Templeless Age*, 106.

69. Mark S. Smith, *God in Translation: Deities in Cross-Cultural Discourse in the Biblical World*, FAT 57 (Tübingen: Mohr Siebeck, 2008), 143. For extended discussion of this passage in relation to the various versions, see ibid., 195–212.

70. The epithet is usually translated "the Most High." For other uses of the term as a deity other than YHWH, see Gen 14:18–22 and Num 24:16.

71. Literally "sons of God": so 4QDeut[j] from Qumran (*bny 'lhym*), against MT ("sons of Israel") and LXX ("angels of God"), both of which are theological corrections.

72. See also Deut 4:19; 29:26 (Hebrew v. 25); Mic 4:5; Judg 11:24.

73. For detailed discussion, see Mark S. Smith, *The Early History of God: Yahweh and the Other Deities in Ancient Israel* (2nd ed; Grand Rapids. MI: Eerdmans: Dove Booksellers, 2002), 32–43; Smith, *The Origins of Biblical Monotheism: Israel's Polytheistic Background and the Ugaritic Texts* (Oxford: Oxford University Press, 2001), 142–45.

74. See Smith's discussion of Gen 49:24–26 in *Early History of God*, 49, which draws from Bruce Vawter, "The Canaanite Background of Genesis 49," *CBQ* 17 (1955): 12–17.

75. E.g., Deut 33:26–27; Psalm 18:13–15 (Hebrew vv. 14–16). See Smith, *Early History of God*, 55–56.

76. Ibid., 58 (italics added).

77. See Smith, *Origins of Biblical Monotheism*, 143–44.

78. A particularly dramatic take on this development toward monotheism is found in Psalm 82, in which God (*'ĕlōhîm* in the singular) sentences the other deities (*'ĕlōhîm* in the plural) to death for having failed to rule justly. Here, theodicy becomes the expressed impetus and rationale for God's takeover of the divine realm. See Smith, *God in Translation*, 130–39.

79. Stephen A. Geller, "The God of the Covenant," in *One God or Many? Concepts of Divinity in the Ancient World*, ed. Barbara Nevling Porter (Transactions of the Casco Bay Assyriological Institute 1; Casco Bay Assyriological Institute, 2000), 275.

80. Smith, *Early History of God*, 189–90, 196.

81. Isaiah, specifically "Second Isaiah," is not the only corpus in the Hebrew Scriptures that develops a monotheistic conception of divinity. Geller argues that monotheism is also presupposed in the *Shema* (Deut 6:4–5), to which one could also add Deut 4:35, 39; 2 Sam 7:22; and Ps 86:10, among others ("The God of the Covenant," 290–302). For more texts and discussion on monotheism, see Smith, *Origins of Biblical Monotheism*, 151–54.

82. Geller, "The God of the Covenant," 280.

83. E.g., Isa 43:1, 14–15; 44:24; 50:2–3.

84. Mark William Worthing, *God, Creation, and Contemporary Physics*, Theology and the Sciences (Minneapolis: Fortress, 1996), 152, who draws from the work of Ilya Prigogine,

85. Robert John Russell, "Entropy and Evil: The Role of Thermodynamics in the Ambiguity of Good and Evil in Nature," in Russell, *Cosmology From Alpha to Omega: The Creative Mutual Interaction of Theology and Science*, Theology and the Sciences (Minneapolis: Fortress, 2008), 240.

86. Christopher Southgate, *The Groaning of Creation: God, Evolution, and the Problem of Evil* (Louisville: Westminster John Knox, 2008), 29.

87. Ibid., 38.

88. John Polkinghorne, *Science and Providence* (London: SPCK, 1989), 33.

89. Ward, *Divine Action*, 100.

90. To borrow from Margulis and Sagan, *What Is Life?*, 17.

CHAPTER 10

1. Annie Dillard, *Teaching a Stone to Talk* (New York: HarperPerennial, 1992), 89.

2. Wendell Berry, *Life is a Miracle: An Essay against Modern Superstition* (New York: Counterpoint, 2000), 103.

3. Edward O. Wilson, *Consilience: The Unity of Knowledge* (New York: Vintage Books, 1998), 325.

4. To be distinguished from earlier science, which regarded nature as nothing more than a machine. See, e.g., Wendell Berry's critique of E. O. Wilson's *Consilience* in Berry, *Life is a Miracle*, 46–55.

5. See Eric Chivian and Aaron Bernstein, eds., *Sustaining Life: How Human Health Depends on Biodiversity* (New York: Oxford University Press, 2008), which explores from a variety of contexts how biodiversity and human livelihood are deeply intertwined. Paul Ehrlich and Anne Ehrlich offer the alarming analogy of Earth as a spaceship whose rivets are popping. The rivets are those species on the brink of extinction (*Extinction: The Causes and Consequences of the Disappearance of Species* [New York: Random House, 1981], xi–xiv).

6. Charles Darwin, *On the Origin of Species by Means of Natural Selection*, in *From So Simple A Beginning: The Four Great Books of Charles Darwin*, ed. Edward O. Wilson (New York: W. W. Norton, 2006), 760.

7. Edward O. Wilson, *The Creation: An Appeal to Save Life on Earth* (New York: W. W. Norton, 2006), 36.

8. So Michael Welker (personal communication).

9. For a technical discussion of the inner connections among various traditions of the Hebrew Bible, see Michael Fishbane, *Biblical Interpretation in Ancient Israel* (Oxford: Clarendon, 1985).

10. William R. Stoeger, SJ, "Entropy, Emergence, and the Physical Roots of Natural Evil," in *Physics and Cosmology: Scientific Perspectives on the Problem of Natural Evil*, vol. 1, ed. Nancey Murphy, Robert John Russell, William R. Stoeger, SJ (Vatican City State: Vatican Observatory/Berkeley: Center for Theology and Natural Sciences, 2007), 93.

11. Robert John Russell helpfully distinguishes between "negative entropy" and "dynamic entropy" in his "Entropy and Evil: The Role of Thermodynamics in the Ambiguity of Good and Evil in Nature," in Russell, *Cosmology from Alpha to Omega: The Creative Mutual Interaction of Theology and Science*, Theology and the Sciences (Minneapolis: Fortress, 2008), 239–40.

12. See chap. 3, as well as Theodore Hiebert, "The Human Vocation: Origins and Transformations in Christian Traditions," in *Christianity and Ecology: Seeking the Well-being of Earth and Humans*, ed. Dieter T. Hessel and Rosemary Radford Ruether, Harvard University Center for the Study of World Religions Publications (Cambridge, MA: Harvard University Press, 2000), 137.

13. Ibid.

14. The prime example of conjoining both kingship and kinship is found in the biblical figure of Noah, the new man "of the ground" (Gen 5:29), who in obedience to God's instructions implements the first endangered species act.

15. The expression was coined by the Dutch historian Johan Huizinga in his book *Homo Ludens; A Study of the Play-Element in Culture* (Boston: Beacon Press, 1955).

16. See, e.g., Eccl 2:24–25; 3:13; 5:18–19.

17. The science of climate change or global warming will not be rehearsed here. For accessible resources, see The National Academies, *Understanding and Responding to Climate Change: Highlights of National Academies Reports, 2008 Edition*, (The National Academy of Sciences, 2008) at http://nationalacademies.org/climatechange; Intergovernmental Panel on Climate Change, *Climate Change 2007: Synthesis Report* at http://www.ipcc.ch/pdf/assessment-report/ar4/syr/ar4_syr.pdf; Tim Flannery, *The Weather Makers: How We Are Changing the Climate and What It Means for Life on Earth* (New York: HarperCollins, 2005).

18. As coined by Michael S. Northcott, *A Moral Climate: The Ethics of Global Warming* (Maryknoll, NY: Orbis Books, 2007), 59.

19. See William R. Stoeger, SJ, "Cultural Cosmology and the Impact of the Natural Sciences on Philosophy and Culture," in *The End of the World and the Ends of God*, ed. John Polkinghorne and Michael Welker (Harrisburg, PA: Trinity Press International, 2000), 75.

20. Richard Dawkins, *The God Delusion* (Boston: Houghton Mifflin, 2006), 116.

21. Quoted in Michael Ruse, *Darwin and Design: Does Evolution Have a Purpose?* (Cambridge, MA: Harvard University Press, 2003), 284.

22. The best recent defense of evolution is Jerry A. Coyne, *Why Evolution is True* (New York: Viking Penguin, 2009).

23. Dawkins, *God Delusion*, 134. See also, in more graphic detail, Francisco J. Ayala, *Darwin's Gift to Science and Religion* (Washington, D.C.: Joseph Henry Press, 2007), 154–59.

24. In Darwin's letter to Joseph Hooker (dated 13 July, 1856), quoted in ibid., 158. See also www.darwinproject.ac.uk, where the letter is catalogued as Letter No. 1924.

25. For argumentation, see, e.g., Nancey Murphy, "Science and the Problem of Evil: Suffering as a By-product of a Finely Tuned Cosmos," in *Physics and Cosmology*, 131–54; and Stoeger, "Entropy, Emergence, and the Physical Roots of Natural Evil," 93–108. The best recent theodicy that takes account of evolution is Christopher Southgate, *The Groaning of Creation: God, Evolution, and the Problem of Evil* (Louisville: Westminster John Knox, 2008).

26. See Eric Chivian and Aaron Bernstein, "How Is Biodiversity Threatened by Human Activity?," in *Sustaining Life*, 29–74.

27. Mary Midgley, *Beast and Man: The Roots of Human Nature*, 2nd ed. (London: Routledge, 1995), 362. For a picture of the world without human beings, see Alan Wiesman, *The World Without Us* (New York: Thomas Dunne Books, 2007).

28. As an alternative to hydroelectric dams, tidal power remains a largely untapped resource.

29. Stephen Jay Gould, *Rocks of Ages: Science and Religion in the Fullness of Life*, The Library of Contemporary Thought (New York: Ballantine, 1999), 178.

30. As Lynn T. White Jr. has pointed out more than thirty years ago in "The Historical Roots of Our Ecological Crisis," *Science* 144 (10 March 1967): 1203–07. See, more recently, Norman Habel, "Playing God or Playing Earth? An Ecological Reading of Genesis 1:26–28," in *"And God Saw That It Was Good": Essays on Creation and God in Honor of Terence E. Fretheim*, ed. Frederick J. Gaiser and Mark A. Throntveit, Word & World Supplement Series 5 (Saint Paul, MN: Luther Seminary, 2006), 33–41.

31. Good examples of "natural" development come from the field of medicine. See David J. Newman et al., "Medicines from Nature," in *Sustaining Life*, 117–62; Eric Chivian, Aaron Bernstein, and Joshua P. Rosenthal, "Biodiversity and Biomedical Research," in *Sustaining Life*, 163–202; and Eric Chivian and Aaron Bernstein, "Threatened Groups of Organisms Valuable to Medicine," in *Sustaining Life*, 203–86.

32. I thank Alan Jenkins, director of Earth Covenant Ministry, for this neologism in a sermon he gave at North Decatur Presbyterian Church in Decatur, Georgia (19 October 2008).

33. The term is not used prior to this point in the Genesis narrative.

Bibliography

Abusch, Tzvi. "Marduk." In *Dictionary of Deities and Demons in the Bible*, edited by Karel van der Toorn, Bob Becking, and Pieter W. van der Horst, 543–49. 2nd ed. Leiden: Brill/Grand Rapids, MI: Eerdmans, 1999.

Albertz, Rainer. *Israel in Exile: The History and Literature of the Sixth Century B.C.E.* Atlanta: Society of Biblical Literature, 2003.

Albright, William F. "The Refrain 'And God Saw *ki tob*' in Genesis." In *Mélanges Bibliques: Rédigés en l'Honneur de André Robert*, 22–26. Travaux de l'Institut Catholique de Paris 4. Paris: Bloud & Gay, 1957.

Allen, James P., trans. "Canonical Compositions (Egyptian) 1.1–1.17." In *The Context of Scripture*. Vol. 1, *Canonical Compositions from the Biblical World*, edited by William W. Hallo et al., 5–27. Leiden: Brill, 2003.

Alt, Albrecht. "Die Weisheit Salomos," *ThLZ* 76 (1951): cols 139–44. Translated as "The Wisdom of Solomon." In *Studies in Ancient Israelite Wisdom*, edited by James L. Crenshaw, 102–12. New York: Ktav, 1976.

Saint Anselm. *Proslogion* in *Anselm of Canterbury: The Major Works*, edited by Brian Davies and G. R. Evans, 82–104. Oxford World's Classics. Oxford: Oxford University Press, 1998.

Appiah, Kwame Anthony. *Experiments in Ethics*. Cambridge, MA: Harvard University Press, 2008.

Ariely, Dan. *Predictably Irrational: The Hidden Forces That Shape Our Decisions*. New York: HarperCollins, 2008.

Arnold, Bill T. "Soul-Searching Questions about 1 Samuel 28: Samuel's Appearance at Endor and Christian Anthropology." In *What about the Soul? Neuroscience and Christian Anthropology*, edited by Joel B. Green, 75–83. Nashville: Abingdon, 2004.

Asara, John M. "Protein Sequences from Mastodon and *Tyrannosaurus Rex* Revealed by Mass Spectrometry." *Science* 316 (13 April 2007): 280–85.

Assaf, Ali Abou, Pierre Bordreuil, and Alan R. Millard. *La statue de Tell Fekherye et son inscription bilingue assyro-araméene.* Etudes Assyriologiques. Paris: Editions Recherche sur les civilisationes, 1982.

Avital, Eytan, and Eva Jablonka. *Animal Traditions: Behavioural Inheritance in Evolution.* Cambridge: Cambridge University Press, 2000.

Ayala, Francisco J. "The Concept of Biological Progress." In *Studies in the Philosophy of Biology: Reduction and Related Problems,* edited by Francisco Jose Ayala and Theodosius Dobzhansky, 339–55. Berkeley: University of California Press, 1974.

———. *Darwin's Gift to Science and Religion.* Washington, D.C.: Joseph Henry Press, 2007.

Baba, Hisao et al. "*Homo erectus* Calvarium from the Pleistocene of Java," *Science* 299 (28 February 2003): 1384–88.

Baker, Catherine. *The Evolution Dialogues: Science, Christianity, and the Quest for Understanding,* edited by James B. Miller. Washington, DC: AAAS, 2006.

Balentine, Samuel. *Job.* Smith & Helwys Commentary. Macon, GA: Smith & Helwys, 2006.

Balter, Michael. "Early Start for Human Art? Ochre May Revise Timeline." *Science* 323 (30 January 2009): 569.

———. "On the Origin of Art and Symbolism." *Science* 323 (9 February 2009): 709–11.

Barabási, Albert-László. *Linked: The New Science of Networks.* Cambridge, MA: Perseus, 2002.

Barbour, Ian G. *Myths, Models, and Paradigms: A Comparative Study in Science and Religion.* San Francisco: Harper & Row, 1974.

———. *When Science Meets Religion: Enemies, Strangers, or Partners?* New York: HarperSanFrancisco, 2000.

Barr, James. *The Garden of Eden and the Hope of Immortality.* London: SCM, 1992.

Barrow, John, and Frank Tipler. *The Anthropic Cosmological Principle.* Oxford: Clarendon Press, 1986.

Barth, Karl. "The Strange New World within the Bible." In Barth, *The Word of God and the Word of Man,* translated by Douglas Horton, 28–50. New York: Harper, 1957.

Beattie, Tina. *The New Atheists: The Twilight of Reason and the War on Religion.* London: Darton, Longman, and Todd, 2007.

Beerling, David. *The Emerald Planet: How Plants Changed Earth's History.* Oxford: Oxford University Press, 2007.

Berry, Wendell. *A Timbered Choir: The Sabbath Poems 1979–1997.* New York: Counterpoint, 1998.

———. *Life Is a Miracle: An Essay against Modern Superstition.* New York: Counterpoint, 2000.

Bhattacharjee, Yudhijit. "Do Black Holes Seed the Formation of Galaxies?" *Science* 323 (16 January 2009): 323.

Birch, Charles. "Chance, Necessity, and Purpose." In *Studies in the Philosophy of Biology: Reduction and Related Problems*, edited by Francisco Jose Ayala and Theodosius Dobzhansky, 225–39. Berkeley: University of California Press, 1974.

Blenkinsopp, Joseph. "The Structure of P." *CBQ* 38 (1976): 275–92.

Bonhoeffer, Dietrich. *Creation and Fall: A Theological Interpretation of Genesis 1–3: Temptation*, translated by John C. Fletcher. London: SCM Press, 1959.

Bonting, Sjored L. *Creation and Double Chaos: Science and Theology in Discussion*. Theology and the Sciences. Minneapolis: Augsburg, 2005.

Borowski, Oded. *Agriculture in Iron Age Israel*. Winona Lake, IN: Eisenbrauns, 1987.

Boulter, Michael. *Extinction: Evolution and the End of Man*. London: HarperCollins, 2003.

Bowler, Peter J. "Darwin's Originality." *Science* 323 (9 January 2009): 223–26.

Brecht, Bertolt. *Galileo*. Edited by Eric Bentley and translated by Charles Laughton. New York: Grove Press, 1966.

Brenner, Athalya. "Proverbs 1–9: An F Voice?" In *On Gendering Texts: Female and Male Voices in the Hebrew Bible*, edited by Athalya Brenner and Fokkelien Dijk-Hemmes, 113–30. Leiden: Brill, 1993.

Brown, Delwin. *Boundaries of Our Habitations: Tradition and Theological Construction*. SUNY Series in Religious Studies; Albany: State University of New York Press, 1994.

Brown, Peter, et al. "A New Small-bodied Hominin from the Late Pleistocene of Flores, Indonesia," *Nature* 431 (28 October 2007): 1055–61.

Brown, Warren S. "Cognitive Contributions to the Soul." In *Whatever Happened to the Soul? Scientific and Theological Portraits of Human Nature*, edited by Warren S. Brown, Nancey Murphy, and H. Newton Maloney, 99–126. Theology and the Sciences. Minneapolis: Fortress, 1998.

———. "Preface." In *Understanding Wisdom: Sources, Science, and Society*, edited by Warren S. Brown, ix-xii. Philadelphia: Templeton Foundation Press, 2000.

———. "Wisdom and Human Neurocognitive Systems: Perceiving and Practicing the Laws of Life." In *Understanding Wisdom: Sources, Science, and Society*, edited by Warren S. Brown, 193–214. Philadelphia: Templeton Foundation Press, 2000.

———. "Resonance: A Model for Relating Science, Psychology, and Faith," *JPC* 23/2 (2004): 110–20.

Brown, William P. "Divine Act and the Art of Persuasion in Genesis 1." In *History and Interpretation: Essays in Honour of John H. Hayes*, edited by M. Patrick Graham, William P. Brown, and Jeffrey K. Kuan, 19–32. JSOTSup 173. Sheffield: Sheffield Academic Press, 1993.

———. *Structure, Role, and Ideology in the Hebrew and Greek Texts of Genesis 1:1–2:3*. SBLDS 132; Atlanta: Scholars Press, 1993.

———. *Character in Crisis: A Fresh Approach to the Wisdom Literature of the Old Testament*. Grand Rapids, MI: Eerdmans, 1996.

———. *The Ethos of the Cosmos: The Genesis of Moral Imagination in the Bible*. Grand Rapids, MI: Eerdmans, 1999.

———. *Ecclesiastes*. Interpretation Bible Commentary. Louisville: Westminster John Knox, 2000.

———. *Seeing the Psalms: A Theology of Metaphor*. Louisville: Westminster John Knox, 2002.

———. "Joy and the Art of Cosmic Maintenance: An Ecology of Play in Psalm 104." In *"And God Saw That It Was Good": Essays on Creation and God in Honor of Terence E. Fretheim*, 23–32. Word & World Supplement Series 5. Saint Paul, MN: Luther Seminary, 2006.

Brueggemann, Walter. "Of the Same Flesh and Bone [Gen 2,23a]." *CBQ* 32 (1970): 532–42.

———. *Abiding Astonishment: Psalms, Modernity, and the Making of History*. Louisville: Westminster John Knox, 1991.

Burdick, Alan. "The Old Men and the Sea." *New York Times Magazine*, 31 December 2006, 56–57, at www.nytimes.com/2006/12/31/magazine/31hedgpeth. html?scp=1&sq=Alan%20Burdick%20%22The%20Old%20Men%20and%20 the%20Sea%22&st=cse.

Byrne, R., and A. Whiten, eds. *Machiavellian Intelligence: Social Expertise and the Evolution of Intellect in Monkeys, Apes, and Humans*. Oxford: Oxford University Press, 1988.

Calvin, John. *The Institutes of the Christian Religion: Library of Christian Classics*. Translated by Ford Lewis Battles. Edited by J. T. McNeill. 2 vols. Philadelphia: Westminster, 1960.

Camus, Albert. "The Myth of Sisyphus." In *The Myth of Sisyphus and Other Essays by Albert Camus*, 88–90. New York: Vintage Books, 1955.

Cassuto, Umberto. *Commentary on Genesis. Part I, From Adam to Noah: Genesis I—VI 8*. Jerusalem: Magnes Press, 1961.

Charles, Dan. "The Threat to the World's Plants." *Science* 320 (23 May 2008): 1000.

Chiao, Raymond Y. "Quantum Nonlocalities: Experimental Evidence." In *Quantum Mechanics: Scientific Perspectives on Divine Action*, edited by Robert John Russell, Philip Clayton, Kirk Wegter-McNelly, and John Polkinghorne, 17–39. Vatican: Vatican Observatory Publications/Berkeley, CA: Center for Theology and the Natural Sciences, 2001.

Childs, Brevard S. *Biblical Theology in Crisis*. Philadelphia: Westminster, 1970.

The Chimpanzee Sequencing and Analysis Consortium. "Initial Sequence of the Chimpanzee Genome and Comparison with the Human Genome." *Nature* 437 (1 September 2005): 69–87.

Chivian, Eric, and Aaron Bernstein, eds. *Sustaining Life: How Human Health Depends on Biodiversity*. New York: Oxford University Press, 2008.

———. "Threatened Groups of Organisms Valuable to Medicine." In *Sustaining Life: How Human Health Depends on Biodiversity*, edited by Eric Chivian and Aaron Bernstein, 203–86. New York: Oxford University Press, 2008.

———. "How Is Biodiversity Threatened by Human Activity?" In *Sustaining Life: How Human Health Depends on Biodiversity*, edited by Erich Chivian and Aaron Bernstein, 29–74. New York: Oxford University Press, 2008.

Chivian, Eric, Aaron Bernstein, and Joshua P. Rosenthal. "Biodiversity and Biomedical Research." In *Sustaining Life: How Human Health Depends on Biodiversity*, edited by Eric Chivian and Aaron Bernstein, 163–202. New York: Oxford University Press, 2008.

Cho, Adrian. "Untangling the Celestial Strings," *Science* 319 (4 January 2008): 47–49.

Christiansen, Morten H., and Simon Kirby. "Language Evolution: The Hardest Problem in Science?" in *Language Evolution*, edited by Morten H. Christiansen and Simon Kirby, 1–15. New York: Oxford University Press, 2003.

Ciccarelli, Francesca D., et al. "Toward Automatic Reconstruction of a Highly Resolved Tree of Life," *Science* 311 (3 March 2006): 1283–87.

Ciesla, Fred J. "Observing Our Origins," *Science* 319 (14 March 2008): 1488–89.

Clifford, Richard, SJ. "Creation in the Hebrew Bible." In *Physics, Philosophy, and Theology: A Common Quest for Understanding*, edited by Robert J. Russell, William R. Stoeger, SJ, G. V. Coyne, 151–70. Vatican: Vatican Observatory, 1988.

———. *Creation Accounts in the Ancient Near East and in the Bible*. CBQMS 26; Washington, D.C.: Catholic Biblical Association of America, 1994.

Coffin, William Sloan. *Letters to a Young Doubter*. Louisville: Westminster John Knox, 2005.

Coles, Peter. *Cosmology: A Very Short Introduction*. Oxford: Oxford University Press, 2001.

Collins, Francis S. *The Language of God: A Scientist Presents Evidence for Belief*. New York: Free Press, 2006.

Cook, Alan. "Uncertainties of Science." In *Science Meets Faith: Theology and Science in Conversation*, edited by Fraser Watts, 32–33. London: SPCK, 1998.

Couzin, Jennifer. "Crossing the Divide." *Science* 319 (22 February 2008): 1034–36.

Coyne, Jerry A. *Why Evolution is True*. New York: Viking Penguin, 2009.

Crutchfield, James P., et al. "Chaos." In *Chaos and Complexity: Scientific Perspectives on Divine Action*, edited by Robert J. Russell, Nancey Murphy, and Arthur Peacocke, 35–48. Vatican City: Vatican Observatory/Berkeley, CA: Center for Theology and the Natural Sciences, 1995.

Cunningham, Michael. *Specimen Days*. New York: Farrar, Straus and Giroux, 2005.

Curry, Andrew. "Seeking the Roots of Ritual," *Science* 319 (18 January 2008): 278–80.

Dalley, Stephanie, trans. "Gilgamesh." In Dalley, *Myths from Mesopotamia: Creation, The Flood, Gilgamesh and Others*, 35–139. World's Classics. Oxford: Oxford University Press, 1991.

———. "The Epic of Creation." In Dalley, *Myths from Mesopotamia: Creation, the Flood, Gilgamesh and Others*, 228–77. World's Classics. Oxford: Oxford University Press, 1991.

——— "Atrahasis." In Dalley, *Myths from Mesopotamia: Creation, the Flood, Gilgamesh and Others*, 9–38. World's Classics. Oxford: Oxford University Press, 1991.

Damasio, Antonio. *Descartes's Error: Emotion, Reason, and the Human Brain*. New York: G. P. Putnam's Sons, 1994.

Darwin, Charles. *On the Origin of Species by Means of Natural Selection*. In *From So Simple a Beginning: The Four Great Books of Charles Darwin*, edited by Edward O. Wilson, 441–760. New York: W. W. Norton, 2006.

————. *The Descent of Man, and Selection in Relation to Sex.* In *From So Simple a Beginning: The Four Great Books of Charles Darwin*, edited by Edward O. Wilson, 767–1248. New York: W. W. Norton, 2006.

————. *The Voyage of the Beagle.* In *From So Simple a Beginning: The Four Great Books of Charles Darwin*, edited by Edward O. Wilson, 21–432. New York: W. W. Norton, 2006.

Daston, Lorraine, and Peter Galison, *Objectivity.* New York: Zone. 2007.

Davies, Paul. *God and the New Physics.* New York: Simon & Schuster, 1983.

————. *The Last Three Minutes: Conjectures about the Ultimate Fate of the Universe.* New York: BasicBooks, 1994.

————. *The Mind of God: The Scientific Basis for a Rational World.* 1992. Reprint, New York: Simon & Schuster Paperbacks, 2005.

————. *Cosmic Jackpot: Why Our Universe Is Just Right for Life.* Boston: Houghton Mifflin, 2007.

Davis, Ellen F. *Getting Involved with God: Rediscovering the Old Testament.* Cambridge, MA: Cowley, 2001.

Dawkins, Richard. *The God Delusion.* Boston: Houghton Mifflin, 2006.

Dean, Cornelia. "Seeing the Risks of Humanity's Hand in Species Evolution." *New York Times* (10 February 2009) at www.nytimes.com/2009/02/10/science/10humans.html?scp=1&sq=Cornelia%20Dean%20%20%22Seeing%20the%20Risks%20of%20Humanity's%20Hand%22&st=cse.

Delsemme, A. "The Origin of the Atmosphere and of the Oceans." In *Comets and the Origin and Evolution of Life*, edited by Paul J. Thomas, Christopher F. Chyba, and Christopher P. McKay, 29–68. 2nd ed. Berlin/Heidelberg: Springer, 2006.

Dhorme, Edouard. *A Commentary on the Book of Job.* Translated by H. Knight. London: Nelson, 1967.

Diamond, Jared M. *Guns, Germs, and Steel: The Fates of Human Societies.* New York: W. W. Norton, 1997.

Diamond, Laura. "Teachers Say Covering Evolution Can Be a Trial," *Atlanta Journal-Constitution* (27 October 2008) at www.ajc.com/search/content/metro/stories/2008/10/27/evolution.html.

Dick, Michael B. "The Neo-Assyrian Royal Lion Hunt and Yahweh's Answer to Job." *JBL* 2 (2006): 243–70.

Dillard, Annie. *Teaching a Stone to Talk.* New York: HarperPerennial, 1992.

Dissanayake, Ellen. *Art and Intimacy: How the Arts Began.* Seattle: University of Washington Press, 2000.

Dulin, Rachel Z. "'How Sweet is the Light': Qoheleth's Age-Centered Teachings." *Interpretation* 55 (2001): 260–70.

Dyson, Freeman. *Disturbing the Universe.* New York: Harper & Row, 1979.

Ehrlich, Paul. *Human Natures: Genes, Culture, and the Human Prospect.* New York: Penguin Books, 2002.

Ehrlich, Paul, and Anne Ehrlich. *Extinction: The Causes and Consequences of the Disappearance of Species.* New York: Random House, 1981.

Einstein, Albert. "The World as I See It." In Einstein, *Ideas and Opinions, Based on "Mein Weltbild,"* translated by Sonja Bargmann, edited by Carl Seelig, 8–11. New York: Crown Publishers, 1954.

Eisley, Loren. *The Immense Journey*. New York: Random House, 1957.

Eldredge, Niles, and Stephen J. Gould. "Punctuated Equilibria: An Alternative to Phyletic Gradualism." In *Models in Paleobiology*, edited by T. J. M. Schopf, 82–115. San Francisco: Freeman, Cooper, 1972.

Elnes, Eric. "Creation and Tabernacle: The Priestly Writer's Environmentalism." *HBT* 16 (1994): 144–55.

Emmanuel, Kerry. "Edward N. Lorenz (1917–2008)." *Science* 320 (23 May 2008): 1025.

Evans, James. *The History and Practice of Ancient Astronomy*. Oxford: Oxford University Press, 1988.

Falkowski, Paul G., and Yukoi Isozaki, "The Story of O_2." *Science* 322 (24 October 2008): 540–42.

Faucher-Giguère, Claude-André, et al. "Numerical Simulations Unravel the Cosmic Web," *Science* 319 (4 January 2008): 52–55.

Fewell, Danna N., and David M. Gunn. "Shifting the Blame: God in the Garden." In *Reading Bibles, Writing Bodies: Identity and the Book*, edited by Timothy K. Beal and David M. Gunn, 16–33. London: Routledge, 1997.

Fishbane, Michael. *Biblical Interpretation in Ancient Israel*. Oxford: Clarendon, 1985.

Flannery, Tim. *The Weather Makers: How We Are Changing the Climate and What It Means for Life on Earth*. New York: Grove Press, 2005.

Foster, Benjamin R., trans. "The Poem of the Righteous Sufferer (1.153)." In *The Context of Scripture*. Vol. 1, *Canonical Compositions from the Biblical World*, edited by William W. Hallo et al., 486–92. Leiden: Brill, 2003.

———. "The Babylonian Theodicy (1.154)." In *The Context of Scripture*. Vol. 1, *Canonical Compositions from the Biblical World*, edited by William W. Hallo et al., 492–95. Leiden: Brill, 2003.

———. "Epic of Creation (1.111)." In *The Context of Scripture*. Vol. 1, *Canonical Compositions from the Biblical World*, edited by William W. Hallo et al., 391–402. Leiden: Brill, 2003.

Fox, Michael V. "Job 38 and God's Rhetoric." *Semeia* 19 (1981): 53–61.

———. "Egyptian Onomastica and Biblical Wisdom." *VT* 36 (1986): 302–10.

———. *Qohelet and His Contradictions*. JSOTSup 71. Sheffield: Sheffield Academic Press, 1987.

———. "Wisdom in Qoheleth." In *In Search of Wisdom: Essays in Memory of John G. Gammie*, edited by Leo G. Perdue, Bernard Brandon Scott, and William Johnston Wiseman, 115–32. Louisville: Westminster John Knox, 1993.

———. "'Āmôn Again." *JBL* 115 (1996): 699–702.

———. *A Time to Tear Down and a Time to Build Up: A Rereading of Ecclesiastes*. Grand Rapids, MI: Eerdmans, 1999.

———. *Proverbs 1–9*. AB 18A; New York: Doubleday, 2000.

Frank, Adam. "The First Billion Years." *Astronomy* 34/6 (June 2006): 30–35.

———. "A Wrinkle in Space-Time." *Astronomy* 34, no. 9 (September 2006): 36–41.

Fretheim, Terence E. *God and World in the Old Testament: A Relational Theology of Creation*. Nashville: Abingdon, 2005.

Galambush, Julie. "'*adam* from '*adama*, '*issa* from '*is*: Derivation and Subordination in Genesis 2.4b–3.24." In *History and Interpretation: Essays in Honour of*

John H. Hayes, edited by M. Patrick Graham, William P. Brown, and Jeffrey K.
Kuan, 33–46. JSOTSup 173. Sheffield: JSOT Press, 1993.

Gammie, John G. "Behemoth and Leviathan: On the Didactic and Theological
Significance of Job 40:15–41:26." In *Israelite Wisdom: Theological and Literary
Essays in Honor of Samuel Terrien*, edited by John G. Gammie et al., 217–31.
Missoula, MT: Scholars Press, 1978.

Gardiner, A. H. *Ancient Egyptian Onomastica*. 2 vols. Oxford: Oxford University Press,
1947.

Garr, W. Randall. "'Image' and 'Likeness' in the Inscription from Tell Fakhariyeh." *IEJ*
50 (2000): 227–34.

———. *In His Own Image and Likeness: Humanity, Divinity, and Monotheism*. Culture
and History of the Ancient Near East 15. Leiden: Brill, 2003.

Gazzaniga, Michael S. *Human: The Science behind What Makes Us Unique*. New York:
HarperCollins, 2008.

Geller, Stephen A. "The God of the Covenant." In *One God or Many? Concepts of
Divinity in the Ancient World*, edited by Barbara Nevling Porter, 273–319.
Transactions of the Casco Bay Assyriological Institute 1. Casco Bay Assyriological
Institute, 2000.

Gerstenberger, Erhard S. *Psalms, Part 2, and Lamentations*. FOTL 15. Grand Rapids, MI:
Eerdmans, 2001.

Gibbons, Ann. "Ice Age No Barrier to 'Peking Man.'" *Science* 323 (13 March 2009):
1419.

Gilmore, Gerard. "How Cold is Dark Matter?" *Science* 322 (5 December 2008):
1476–77.

Goldingay, John. *Models for Interpretation of Scripture*. Grand Rapids, MI: Eerdmans,
1995.

Goodenough, Ursula. *The Sacred Depths of Nature*. Oxford: Oxford University Press,
1998.

Gordis, Robert. *The Book of Job: Commentary, New Translation, and Special Studies*.
New York: Jewish Theological Seminary, 1978.

Gould, Stephen Jay. *Rocks of Ages: Science and Religion in the Fullness of Life*. The Library
of Contemporary Thought. New York: Ballantine, 1999.

Gray, John. "The Book of Job in Context of N. E. Literature." *ZAW* (1970): 251–69.

Greene, Brian R. *The Fabric of the Cosmos: Space, Time, and the Texture of Reality*.
New York: Vintage Books, 2004.

Gussow, M. "For Saul Bellow, Seeing the Earth with Fresh Eyes." *New York Times*,
26 May 1997, at www.nytimes.com/1997/05/26/books/for-saul-bellow-seeing-
the-earth-with-fresh-eyes.html?scp=1&sq=Gussow%20%22Seeing%20the%20
Earth%22&st=cse.

Gustafson, James M. *Intersections: Science, Theology, and Ethics*. Cleveland, OH: Pilgrim
Press, 1996.

Habel, Norman. "Playing God or Playing Earth? An Ecological Reading of Genesis
1:26–28." In *"And God Saw That It Was Good": Essays on Creation and God in Honor
of Terence E. Fretheim*, edited by Frederick J. Gaiser and Mark A. Throntveit, 33–41.
Word & World Supplement Series 5. Saint Paul, MN: Luther Seminary, 2006.

Hallo, William W., et al., eds. *The Context of Scripture*. Vol. 1, *Canonical Compositions from the Biblical World*. Leiden: Brill, 2003.

Halpern, Baruch. "Assyrian and pre-Socratic Astronomies and the Location of the Book of Job." In *Kein Land für sich allein: Studien zum Kulturkontakt in Kanaan, Israel/Palästina und Ebirnâri für Manfred Weippert zum 65. Geburtstag*, edited by Ulrich Hübner and Ernst Axel Knauf, 255–64. OBO 186. Freiburg: Universitäts-verlag Freiburg: Vandenhoeck & Ruprecht, 2002.

———. "The Assyrian Astronomy of Genesis 1 and the Birth of Milesian Philosophy." In *Eretz-Israel* 27 (2003): 74*–83*.

Hartle, James B, and Stephen W. Hawking, "Wave Function of the Universe." *Physics Review* D28 (1983): 2960–75.

Haught, John F. *God and the New Atheism: A Critical Response to Dawkins, Harris, and Hitchens*. Louisville: Westminster John Knox, 2008.

Hawking, Stephen W., with Leonard Mlodinow, *A Briefer History of Time*. New York: Bantam Book, 2005.

Hazen, Robert M. *Gen·e·sis: The Scientific Quest for Life's Origin*. Washington, D.C.: Joseph Henry, 2005.

Hess, Peter J. "'God's Two Books': Revelation, Theology, and Natural Science in the Christian West." In *Interdisciplinary Perspectives on Cosmology and Biological Evolution*, 19–51. Australian Theological Forum Science and Theology Series 2. Hindmarsh, Australia: Australian Theological Forum, 2002.

Hewlett, Martinez J. "True to Life?: Biological Models of Origin and Evolution." In *Evolution and Emergence: Systems, Organisms, Persons*, edited by Nancey Murphy and William R. Stoeger, SJ, 158–72. Oxford: Oxford University Press, 2007.

Hiebert, Theodore. *The Yahwist's Landscape: Nature and Religion in Early Israel*. Oxford: Oxford University Press, 1996.

———. "The Human Vocation: Origins and Transformations in Christian Traditions." In *Christianity and Ecology: Seeking the Well-being of Earth and Humans*, edited by Dieter T. Hessel and Rosemary Radford Ruether, 135–54. Harvard University Center for the Study of World Religions Publications. Cambridge, MA: Harvard University Press, 2000.

Hillel, Daniel. *The Natural History of the Bible: An Environmental Exploration of the Hebrew Scriptures*. New York: Columbia University Press, 2006.

Hinman, Lawrence M. "Seeing Wisely—Learning to Become Wise." In *Understanding Wisdom: Sources, Science, and Society*, edited by Warren S. Brown, 413–24. Philadelphia: Templeton Foundation Press, 2000.

Hoffman, Yair. *A Blemished Perfection: The Book of Job in Context*. JSOTSup 213. Sheffield: Sheffield Academic Press, 1996.

Holden, Constance. "Out of the Frying Pan." *Science* 319 (14 March 2008): 1465.

Holmstedt, Robert D. "The Restrictive Syntax of Genesis i 1." *VT* 58 (2008): 56–67.

Hoyle, Fred. *Facts and Dogmas in Cosmology and Elsewhere*. Cambridge: Cambridge University Press, 1982.

Huizinga, Johan. *Homo Ludens; A Study of the Play-Element in Culture*. Boston: Beacon Press, 1955.

Huyssteen, J. Wentzel van. *Alone in the World? Human Uniqueness in Science and Theology*. Grand Rapids. MI: Eerdmans, 2006.

Impey, Chris. *The Living Cosmos: Our Search for Life in the Universe*. New York: Random House, 2007.

Intergovernmental Panel on Climate Change. *Climate Change 2007: Synthesis Report*. www.ipcc.ch/pdf/assessment-report/ar4/syr/ar4_syr.pdf.

Isaacson, Walter. *Einstein: His Life and Universe*. New York: Simon & Schuster, 2007.

Isham, Chris J. "Creation of the Universe as a Quantum Process." In *Physics, Philosophy, and Theology: A Common Quest for Understanding*, edited by Robert J. Russell, William R. Stoeger, and George V. Coyne, 375–408. Rome: Vatican Observatory, 1988.

———. "Quantum Theories of the Creation of the Universe." In *Quantum Cosmology and the Laws of Nature: Scientific Perspectives on Divine Action*, edited by Robert John Russell, Nancey Murphy, and C. J. Isham, 49–89. Series on Divine Action in Scientific Perspective 1. Rome: Vatican Observatory/Berkeley, CA: Center for Theology and the Natural Sciences, 1993.

Jantzen, Grace. *God's World, God's Body*. Philadelphia: Westminster, 1984.

Jenson, Robert W. *Systematic Theology*. Vol. 2. New York: Oxford University Press, 1999.

Jones, Steven. *Darwin's Ghost: "The Origin of Species" Updated*. New York: Ballantine, 2000.

Kaku, Michio. *Parallel Worlds: A Journey through Creation, Higher Dimensions, and the Future of the Cosmos*. New York: Anchor Books, 2006.

Kasting, James F. "The Rise of Atmospheric Oxygen." *Science* 293 (3 August 2001): 819–20.

Kauffman, Stuart A. *At Home in the Universe: The Search for Laws of Self-Organization and Complexity*. Oxford: Oxford University Press, 1995.

———. *Reinventing the Sacred: A New View of Science, Reason, and Religion*. New York: Basic Books, 2008.

Kellert, Stephen R., and Edward O. Wilson, eds. *The Biophilia Hypothesis*. Washington, D.C.: Island Press/Shearwater Books, 1993.

Kenneally, Christine. *The First Word: The Search for the Origins of Language*. New York: Viking Penguin, 2007.

Kerasote, Ted. *Merle's Door: Lessons from a Freethinking Dog*. Orlando: HarcourtBooks, 2007.

King, Barbara J. *Evolving God: A Provocative View of the Origins of Religion*. New York: Doubleday, 2007.

Konner, Melvin. *The Tangled Wing: Biological Constraints on the Human Spirit*. 2nd ed. New York: Henry Holt, 2002.

Kovacs, Maureen G. *The Epic of Gilgamesh*. Stanford, CA: Stanford University Press, 1989.

Krüger, Thomas. "Did Job Repent?" In *Das Buch Hiob und Seine Interpretationen*, 217–29. ATANT 88. Zürich: Theologischer Verlag Zürich, 2007.

LaMotte, Louis C. *Colored Light: The Story of the Influence of Columbia Theological Seminary, 1828–1936*. Richmond: Presbyterian Committee, 1937.

Lapsley, Jacqueline E. *Whispering the Word: Hearing Women's Stories in the Old Testament.* Louisville: Westminster John Knox Press, 2005.

Larson, Edward J. *Summer for the Gods: The Scopes Trial and America's Continuing Debate Over Science and Religion.* 1997. 2nd ed. Reprint, New York: Basic Books, 2006.

Levenson, Jon D. *Creation and the Persistence of Evil: The Jewish Drama of Divine Omnipotence.* 1988. 2nd ed. Reprint, Princeton, NJ: Princeton University Press, 1994.

———. "Translation Notes for Gen 1.1–2.3." In *The Jewish Study Bible,* edited by Adele Berlin and Marc Zvi Brettler, 12. Oxford: Oxford University Press, 2004.

———. *Resurrection and the Restoration of Israel: The Ultimate Victory of the God of Life.* New Haven: Yale University Press, 2006.

Lewis-Williams, David. *The Mind in the Cave: Consciousness and the Origins of Art.* New York: Thames and Hudson, 2002.

Limburg, James. "Down to Earth Theology: Psalm 104 and the Environment." *Currents in Theology and Mission* 21 (1994): 340–46.

Linder, Eric V. "Cosmology: The Universe's Skeleton Sketched." *Nature* 445 (18 January 2007): 273.

Logan, William Bryant. *Dirt: The Ecstatic Skin of the Earth.* New York: W. W. Norton, 1995.

Lohfink, Norbert. *Great Themes from the Old Testament.* Translated by Ronald Walls. Edinburgh: T&T Clark, 1982.

Macedo, Stephen, and Josiah Ober, eds. *Primates and Philosophers: How Morality Evolved.* Princeton: Princeton University Press, 2006.

Madigan, Kevin J., and Jon D. Levenson. *Resurrection: The Power of God for Christians and Jews.* New Haven: Yale University Press, 2008.

Maestripieri, Dario. *Macachiavellian Intelligence: How Rhesus Macaques and Humans Have Conquered the World.* Chicago: University of Chicago Press, 2007.

Margulis, Lynn, and Dorion Sagan, *What is Life?* Berkeley: University of California Press, 1995.

———. *Microcosmos: Four Billion Years of Evolution from Our Microbial Ancestors.* Berkeley: University of California Press, 1997.

Massey, Richard, et al. "Dark Matter Maps Reveal Cosmic Scaffolding." *Nature* 445 (18 January 2007): 286–90.

Mathews, Claire R. *Defending Zion.* BZAW 236; Berlin: Walter de Gruyter, 1995.

The National Academies. *Understanding and Responding to Climate Change: Highlights of National Academies Reports, 2008 Edition.* National Academy of Sciences, 2008, http://nationalacademies.org/climatechange.

McBride, S. Dean, Jr. "Divine Protocol: Genesis 1:1–2:3 as Prologue to the Pentateuch." In *God Who Creates: Essays in Honor of W. Sibley Towner,* edited by William P. Brown and S. Dean McBride Jr., 3–41. Grand Rapids, MI: Eerdmans, 2000.

McFague, Sallie. *The Body of God: An Ecological Theology.* Minneapolis: Fortress, 1993.

McKibben, Bill. *The Comforting Whirlwind: God, Job and the Scale of Creation.* Cambridge, MA: Cowley, 2005.

McMullin, Ernan. "How Should Cosmology Relate to Theology?" In *The Sciences and Theology in the Twentieth Century,* edited by Arthur Peacocke, 17–57. Notre Dame: University of Notre Dame Press, 1981.

Melchert, Charles F. *Wise Teaching: Biblical Wisdom and Educational Ministry.*
Harrisonburg, PA: Trinity Press International, 1998.

Merchant, Carolyn. *The Death of Nature: Women, Ecology, and the Scientific Revolution.*
San Francisco: HarperCollins, 1980.

Mettinger, Tryggve N. D. *No Graven Image? Israelite Aniconism in its Ancient Near
Eastern Context.* Stockholm: Almqvist & Wiksell International, 1995.

Midgley, Mary. *Beast and Man: The Roots of Human Nature.* London: Routledge,
1995.

——— "Science in the World." *Science Studies* 9, no. 2 (1996): 49–58.

Middlemas, Jill. *The Templeless Age: An Introduction to the History, Literature, and
Theology of the "Exile."* Louisville: Westminster John Knox, 2007.

Middleton, J. Richard. *The Liberating Image: The* Imago Dei *in Genesis 1.* Grand Rapids,
MI: Brazos, 2005.

Millard, A. R., and P. Bordreuil, "A Statue from Syria with Assyrian and Aramaic
Inscriptions." *Biblical Archeologist* 45 (1982): 135–41.

Miller, Douglas B. *Symbol and Rhetoric in Ecclesiastes: The Place of Hebel in Qohelet's
Work.* Academia Biblica 2. Atlanta: Society of Biblical Literature, 2002.

Miller, Patrick D. "The Poetry of Creation: Psalm 104." In *God Who Creates: Essays in
Honor of W. Sibley Towner,* edited by William P. Brown and S. Dean McBride Jr.,
87–103. Grand Rapids, MI: Eerdmans, 2002.

Mitchell, Stephen. *Gilgamesh: A New English Version.* New York: Free Press, 2004.

Mithen, Steve. *The Prehistory of the Mind: A Search for the Origins of Art, Religion, and
Science.* London: Thames and Hudson, 1996.

———. *The Singing Neanderthals: The Origins of Music, Language, Mind, and Body.*
Cambridge, MA: Harvard University Press, 2006.

Morowitz, Harold J. *The Emergence of Everything: How the World Became Complex.*
Oxford: Oxford University Press, 2002.

Morrison, Philip, Phylis Morrison, and the Office of Charles and Ray Eames. *Powers
of Ten: About the Relative Size of Things in the Universe.* New York: Scientific
American Library, 1982.

Morrow, William. "Consolation, Rejection, and Repentance in Job 42:6." *JBL* 105
(1986): 211–25.

Murphy, Nancey. *Theology in the Age of Scientific Reasoning.* Ithaca, NY: Cornell
University Press, 1990.

———. "Introduction: A Hierarchical Framework for Understanding Wisdom." In
Understanding Wisdom: Sources, Science, and Society, edited by Warren S. Brown,
1–11. Philadelphia: Templeton Foundation Press, 2000.

———. *Bodies and Souls, or Spirited Bodies?* Current Issues in Theology. Cambridge:
Cambridge University Press, 2006.

———. "Science and the Problem of Evil: Suffering as a By-product of a Finely Tuned
Cosmos." In *Physics and Cosmology: Scientific Perspectives on the Problem of Natural
Evil,* Vol. 1, edited by Nancey Murphy, Robert John Russell, William R. Stoeger, SJ,
131–54. Vatican City State: Vatican Observatory/Berkeley, CA: Center for Theology
and Natural Sciences, 2007.

———. "Reductionism: How Did We Fall Into It and Can We Emerge From It?" In *Evolution and Emergence: Systems, Organisms, Persons*, edited by Nancey Murphy and William R. Stoeger, SJ, 19–39. Oxford: Oxford University Press, 2007.

———, and George F. R. Ellis, *On the Moral Nature of the Universe: Theology, Cosmology, and Ethics*. Minneapolis: Augsburg Fortress, 1996.

Nadis, Steve. "Tales from the Dark Side." *Astronomy* 34. no. 9 (September 2006): 30–35.

The National Academies. *Understanding and Responding to Climate Change: Highlights of National Academies Reports, 2008 Edition*. The National Academy of Sciences, 2008.

Newman, David J., et al. "Medicines from Nature." In *Sustaining Life: How Human Health Depends on Biodiversity*, edited by Eric Chivian and Aaron Bernstein, 117–62. New York: Oxford University Press, 2008.

Newsom, Carol A. "Woman and the Discourse of Patriarchal Wisdom: A Study of Proverbs 1–9." In *Gender and Difference in Ancient Israel*, edited by Peggy L. Day, 142–60. Minneapolis: Fortress, 1989.

———. "The Book of Job: Introduction, Commentary, and Reflections." In *The New Interpreter's Bible Commentary*. Vol. 4, 319–637. Nashville: Abingdon, 1996.

———. *The Book of Job: A Context of Moral Imaginations*. Oxford: Oxford University Press, 2003.

Niditch, Susan. *Underdogs and Tricksters: A Prelude to Biblical Folklore*. San Francisco: Harper & Row, 1987.

Northcott, Michael S. *A Moral Climate: The Ethics of Global Warming*. Maryknoll, NY: Orbis Books, 2007.

Nouvian, Claire. *The Deep: The Extraordinary Creatures of the Abyss*. Chicago: The University of Chicago Press, 2007.

O'Connor, Kathleen. "Wild, Raging Creativity: Job in the Whirlwind." In *Earth, Wind, and Fire: Biblical and Theological Perspectives on Creation*, edited by Carol J. Dempsey and Mary Margaret Pazdan, 48–56. Collegeville, MN: Liturgical Press, 2004.

Odling-Smee, F. J. "Niche-Constructing Phenotypes." In *The Role of Behavior in Evolution*, edited by J. C. Plotkin, 73–132. Cambridge, MA: MIT Press, 1988.

Odling-Smee, F. J., K. N. Laland, and M. W. Feldman, "Niche Construction." *American Naturalist* 147 (1996): 641–48.

Osborn, Lawrence. "Theology and the New Physics." In *God, Humanity and the Cosmos*, edited by Christopher Southgate, 119–53. New York / London: T&T Clark, 2005.

Ottati, Douglas. "Theology among the Arts and Sciences." *Bulletin of the Institute for Reformed Theology* 7/1 (Spring/Summer 2007): 1, 3–9, 14, 16.

Overbye, Dennis. "The Cosmos According to Darwin." *New York Times Magazine*, 13 July 1997, 26–27, at www.nytimes.com/1997/07/13/magazine/the-cosmos-according-to-darwin.html?scp=1&sq=Dennis%20Overbye%20%22The%20Cosmos%20According%20to%20Darwin%22%20&st=cse.

———. "Remembrance of Things Future: The Mystery of Time." *New York Times*, 28 June 2005, at www.nytimes.com/2005/06/28/science/28time.html?

scp=1&sq=Dennis%20Overbye%20%20%22The%20Mystery%20of%20
Time%22&st=cse.

———. "The Universe, Expanding Beyond All Understanding." *New York
Times*, 5 June 2007, at www.nytimes.com/2007/06/05/science/space/05essa.
html?scp=1&sq=Overbye%20%22The%20Universe,%20Expanding%20
Beyond%22&st=cse.

Peacocke, Arthur. *Theology for a Scientific Age: Being and Becoming—Natural, Divine,
and Human*. Theology and the Sciences. Minneapolis: Fortress, 1993.

Pearce, Laurie E. "New Evidence for Judeans in Babylonia." In *Judah and the Judeans in
the Persian Period*, edited by Oded Lipschits and Manfred Oeming, 399–411.
Winona Lake, IN: Eisenbrauns, 2006.

Perdue, Leo G. *Wisdom Literature: A Theological History*. Louisville: Westminster John
Knox, 2007.

Peters, Ted. "From Conflict to Consonance: Ending the Warfare between Science and
Faith." *Currents in Theology and Mission* 28, nos. 3–4 (June/August 2001), 238–47.

——— "Introduction: What Is to Come," in *Resurrection: Theological and Scientific
Assessments*, edited by Ted Peters, Robert John Russell, and Michael Welker,
viii–xvii. Grand Rapids, MI: Eerdmans, 2002.

Peterson, Britt. "Nice Guys Didn't Finish the Neolithic." *Discover* (August 2006): 15.

Philo of Alexandria. *On the Creation of the Cosmos according to Moses*. Translated and
edited by David T. Runia. PACS 1. Brill: Leiden, 2001.

Phipps, William E. "Eve and Pandora Contrasted." *Theology Today* 45 (1988): 34–48.

Pimm, Stuart L., et al., "What Is Biodiversity?" In *Sustaining Life: How Human Health
Depends on Biodiversity*, edited by Erich Chivian and Aaron Bernstein, 3–28. New
York: Oxford University Press, 2008.

Plotkin, Henry. *Darwin Machines and the Nature of Knowledge*. Cambridge, MA:
Harvard University Press, 1993.

———. *Evolution in Mind: An Introduction to Evolutionary Psychology*. Cambridge: MA:
Harvard University Press, 1998.

Polkinghorne, John. *Science and Providence*. London: SPCK, 1989.

———. *Reason and Reality*. Philadelphia: Trinity International Press, 1991.

———. *The Faith of a Physicist*. Princeton: Princeton University Press, 1994.

Potts, Richard. "Sociality and the Concept of Culture in Human Origins." In *The
Origins and Nature of Sociality*, edited by Robert W. Sussman and Audrey
R. Chapman, 249-69. New York: Aldine de Gruyter, 2004.

Prigogine, Ilya, and Isabelle Stengers. *Order out of Chaos*. New York: Bantam Books,
1984.

Rad, Gerhard von. "Job XXXVII and Ancient Egyptian Wisdom." In *The Problem of the
Hexateuch and Other Essays*, translated by E.W. Truemann Dicken, 281–91.
London: SCM Press, 1985.

Rees, Martin. "Pondering Astronomy in 2009." *Science* 323 (16 January 2009): 309.

Roberts, Diane. "Cosmic Dust from Distant Comet Comes to Earth." *National Public
Radio* (NPR), Weekend Edition Sunday, 10 December 2006. See www.npr.org/
templates/story/story.php?storyId=6605084.

Robinson, Marilynne. *The Death of Adam: Essays on Modern Thought*. Boston: Houghton Mifflin, 1998.

Rochberg, Francesca. *The Heavenly Writing: Divination, Horoscopy, and Astronomy in Mesopotamian Culture*. Cambridge: Cambridge University Press, 2004.

Rolston, Holmes, III. *Science and Religion: A Critical Survey*. 2nd ed.; Philadelphia: Templeton Foundation Press, 2006.

———. "Creation and Resurrection." *Journal for Preachers* 32, no. 3 (Easter 2009): 25–32.

Ruse, Michael. *Darwin and Design: Does Evolution Have a Purpose?* Cambridge, MA: Harvard University, 2003.

Russell, Robert John. "Entropy and Evil: The Role of Thermodynamics in the Ambiguity of Good and Evil in Nature." In Russell, *Cosmology from Alpha to Omega: The Creative Mutual Interaction of Theology and Science*, 226–48. Theology and the Sciences. Minneapolis: Fortress, 2008.

——— "Finite Creation without a Beginning: The Doctrine of Creation in Relation to Big Bang and Quantum Cosmologies." In Russell, *Cosmology From Alpha to Omega: The Creative Mutual Interaction of Theology and Science*, 77–109. Theology and the Sciences. Minneapolis: Fortress, 2008.

——— "Bridging Theology and Science: The CTNS Logo." *Theology and Science* 1 (June 2003): 1–3.

Sandars, N.K. *The Epic of Gilgamesh: An English Version with an Introduction*. 3rd ed. London: Penguin Classics, 1972.

Sandoval, Timothy J. "Revisiting the Prologue of Proverbs" *JBL* 126 (2007): 455–73.

Schama, Simon. *Landscape and Memory*. New York: Knopf, 1995.

Schedl, Claus. *History of the Old Testament: The Ancient Orient and Ancient Biblical History*. Vol. 1. New York: Alba house, 1973.

Schifferdecker, Kathryn. *Out of the Whirlwind: Creation Theology in the Book of Job*. HTS 61; Cambridge, MA: Harvard University Press, 2008.

Schipper, Jan, et al. "The Status of the World's Land and Marine Mammals: Diversity, Threat, and Knowledge." *Science* 322 (10 October 2008): 225–30.

Schloss, Jeffrey P. "Wisdom Traditions as Mechanisms for Organismal Integration: Evolutionary Perspectives on Homeostatic 'Laws of Life.'" In *Understanding Wisdom: Sources, Science, and Society*, edited by Warren S. Brown, 153–92. Philadelphia: Templeton Foundation Press, 2000.

Schüle, Andreas. "Made in the 'Image of God': The Concepts of Divine Images in Gen 1–3." *ZAW* 117 (2005): 1–20.

Schuster, Heinz George. *Deterministic Chaos: An Introduction*. Weinheim: Physik-Verlag, 1984.

Schweitzer, Mary Higby, et al. "Analyses of Soft Tissue from *Tyrannosaurus rex* Suggest the Presence of Protein." *Science* 316 (13 April 2007): 277–80.

Scott, R.B.Y. *Proverbs, Ecclesiastes*. AB 18. Garden City: Doubleday, 1965.

Seow, Choon-Leong. "The Socioeconomic Context of 'The Preacher's' Hermeneutic." *PSB* 17 (1996): 168–95.

———. *Ecclesiastes*. AB 18C. New York: Doubleday, 1997.

———. "Qohelet's Eschatological Poem." *JBL* 118 (1999): 209–34.

———. *Job: A Commentary*. Eerdmans Critical Commentary. Grand Rapids, MI: Eerdmans, forthcoming.

Sheets-Johnstone, Maxine. *The Roots of Thinking*. Philadelphia: Temple University Press, 1990.

Sherman, Nancy. "Wise Emotions." In *Understanding Wisdom: Sources, Science, and Society*, edited by Warren S. Brown, 319–38. Philadelphia: Templeton Foundation Press, 2000.

Shubin, Neil. *Your Inner Fish: A Journey into the 3.5-Billion-Year History of the Human Body*. New York: Pantheon Books, 2008.

Smith-Christopher, Daniel L. *A Biblical Theology of Exile*. Overtures to Biblical Theology. Minneapolis: Fortress, 2002.

Smith, Mark S., trans. "Ba'al Epic." In *Ugaritic Narrative Poetry*, edited by Simon B. Parker, 81–180. SBLWAW 9. Atlanta: Society of Biblical Literature, 1997.

———. *The Origins of Biblical Monotheism: Israel's Polytheistic Background and the Ugaritic Texts*. Oxford: Oxford University Press, 2001.

———. *The Early History of God: Yahweh and the Other Deities in Ancient Israel*. 2nd ed. Grand Rapids, MI: Eerdmans/Dove Booksellers, 2002.

———. *God in Translation: Deities in Cross-Cultural Discourse in the Biblical World*. FAT 57. Tübingen: Mohr Siebeck, 2008.

———. "Light in Genesis 1:3—Created or Uncreated: A Question of Priestly Mysticism?" In *Birkat Shalom: Studies in the Bible, Ancient Near Eastern Literature, and Postbiblical Judaism Presented to Shalom M. Paul on the Occasion of His Seventieth Birthday*, edited by Chaim Cohen, et al., 125-34. Winona Lake, IN: Eisenbrauns, 2008.

Smolin, Lee. "Did the Universe Evolve?" *Classical and Quantum Gravity* 9 (1992): 173–91.

———. *The Life of the Cosmos*. London: Phoenix Paperback, 1997.

———. *The Trouble with Physics: The Rise of String Theory, the Fall of a Science, and What Comes Next*. Boston: Houghton Mifflin, 2006.

Soden, Wolfram von, "Leistung und Grenze sumerischer und babylonischer Wissenschaft." In *Die Eigenbegrifflichkeit der babylonischen Welt. Leistung und Grenze sumerischer und babylonischer Wissenschaft*, edited by Benno Landsberger and Wolfram von Solden, 21–123. 1936. Reprint, Darmstadt: Wissenschaftliche Buchgesellschaft, 1965.

Sofia, U.J. "Interstellar Dust: Not Just Bunnies under Your Bed." *Whitman Magazine* (December 2008): 12–13.

Sommer, Benjamin D. *A Prophet Reads Scripture: Allusion in Isaiah 40–66*. Stanford, CA: Stanford University Press, 1998.

Southgate, Christopher. *The Groaning of Creation: God, Evolution, and the Problem of Evil*. Louisville: Westminster John Knox, 2008.

Stager, Lawrence E. "Jerusalem and the Garden of Eden." *Eretz-Israel* 26 (1999): 184–94.

Stearns, Stephen C., and Rolf F. Hoekstra. *Evolution: An Introduction*. Oxford: Oxford University Press, 2000.

Steinhardt, Paul. "A Cyclic Universe." *Seed* 11 (August 2007): 33–34.

Stendahl, Krister. "Biblical Theology, Contemporary," IDB, 1.418–32. Reprinted in Stendahl, *Meanings: The Bible as Document and as Guide*, 11–44. Philadelphia: Fortress, 1984.

Stewart, Ian. *Does God Play Dice? The New Mathematics of Chaos*. New York: Penguin Books, 1990.

Stoeger, William R., SJ. "Cultural Cosmology and the Impact of the Natural Sciences on Philosophy and Culture." In *The End of the World and the Ends of God*, edited by John Polkinghorne and Michael Welker, 65–77. Harrisburg, PA: Trinity Press International, 2000.

———. "Scientific Accounts of Ultimate Catastrophes in our Life-Bearing Universe." In *The End of the World and the Ends of God*, edited by John Polkinghorne and Michael Welker, 19–28. Harrisburg, PA: Trinity Press International, 2000.

———. "Entropy, Emergence, and the Physical Roots of Natural Evil." In *Physics and Cosmology: Scientific Perspectives on the Problem of Natural Evil*. Vol. 1, edited by Nancey Murphy, Robert John Russell, and William R. Stoeger, SJ, 93–108. Vatican City State: Vatican Observatory/Berkeley, CA: Center for Theology and Natural Sciences, 2007.

Stout, Frappa. "History and Imagination." *USA Weekend*, 15–17 July 2005, 11–12. See www.usaweekend.com/05_issues/050717/050717books.html.

Strawn, Brent A. *What is Stronger than a Lion? Leonine Image and Metaphor in the Hebrew Bible and the Ancient Near East*. OBO 212. Fribourg: Academic Press/Göttingen: Vandenhoeck & Ruprecht, 2005.

Talon Philippe. *Enūma Eliš: The Standard Babylonian Creation Myth: Introduction, Cuneiform Text, Transliteration, and sign List with a Translation and Glossary in French*. SAACT 4. University of Helsinki: The Neo-Assyrian Text Corpus Project, 2005.

Taylor, Barbara Brown. *Leaving Church: A Memoir*. New York: HarperSanFrancisco, 2006.

———. *The Luminous Web: Essays on Science and Religion*. Cambridge: Cowley, 2000.

Taylor, Shelley. *The Tending Instinct*. New York: Times Books, 2002.

Tennyson, Alfred Lord. "In Memoriam A.H.H." LVI; 13–16. In *Tennyson: A Selected Edition*, edited by Christopher Ricks, 399. Harlow: Longman, 1989.

Toulmin, Stephen. *Cosmopolis: The Hidden Agenda of Modernity*. New York: Free Press, 1990.

Towner, W. Sibley. *Genesis*. Westminster Bible Companion. Louisville: Westminster John Knox, 2001.

Trefil, James. *The Nature of Science: An A–Z Guide to the Laws and Principles Governing Our Universe*. Boston: Houghton Mifflin, 2003.

Tsumura, David. *Creation and Destruction: A Reappraisal of the Chaoskampf Theory in the Old Testament*. Winona Lake: Eisenbrauns, 2005.

Tucker, Gene M. "Rain on a Land Where No One Lives: The Hebrew Bible on the Environment." *JBL* 116 (1997): 3–17.

Tyson, Neil DeGrasse. *Death by Black Hole and Other Cosmic Quandaries*. New York: W.W. Norton, 2007.

Tyson, Neil DeGrasse, and Donald Goldsmith. *Origins: Fourteen Billion Years of Cosmic Evolution*. New York: W.W. Norton, 2004.

Vawter, Bruce. "The Canaanite Background of Genesis 49." *CBQ* 17 (1955): 1–18.

Venter, J. Craig, et al. "The Sequence of the Human Genome." *Science* 291 (16 February 2001): 1304–51.

Waal, Frans de. *Good Natured: The Origins of Right and Wrong in Humans and Other Animals*. Cambridge, MA: Harvard University Press, 1996.

———. "Appendix A: Anthropomorphism and Anthropodenial." In *Primates and Philosophers: How Morality Evolved*, edited by Stephen Macedo, Josiah Ober, and Robert Wright, 59–67. Princeton, NJ: Princeton University Press, 2006.

———. "Appendix B: Do Apes Have a Theory of Mind?" In *Primates and Philosophers: How Morality Evolved*, edited by Stephen Macedo, Josiah Ober, and Robert Wright, 69–73. Princeton, NJ: Princeton University Press, 2006.

———. "Morally Evolved: Primate Social Instincts, Human Morality, and the Rise and Fall of 'Veneer Theory.'" In *Primates and Philosophers: How Morality Evolved*, edited by Stephen Macedo, Josiah Ober, and Robert Wright, 1–58. Princeton, NJ: Princeton University Press, 2006.

———. "The Tower of Morality." In *Primates and Philosophers: How Morality Evolved*, edited by Stephen Macedo, Josiah Ober, and Robert Wright, 161–82. Princeton, NJ: Princeton University Press, 2006.

Ward, Keith. *Divine Action: Examining God's Role in an Open and Emergent Universe*. 1990. Reprint, West Conshohocken, PA: Templeton Foundation Press, 2007.

Watson, Rebecca S. *Chaos Uncreated: A Reassessment of the Theme of "Chaos" in the Hebrew Bible*. BZAW 341; Berlin: Walter de Gruyter, 2005.

Weinberg, Steven. *The First Three Minutes: A Modern View of the Origin of the Universe*. New York: Basic Books, 1977.

———. *Dreams of a Final Theory: The Scientist's Search for the Ultimate Laws of Nature*. New York: Random House, 1994.

Weinfeld, Moshe. "Sabbath, Temple and the Enthronement of the Lord—The Problem of Sitz im Leben of Genesis 1:1–2:3." In *Mélanges bibliques et orientaux en l'honneur de M. Henri Cazelles*, edited by A Caquot and M. Delcor, 501–12. AOAT 212; Kevelaer: Butzon & Bercker; Neukirchen-Vluyn: Neukirchener, 1981.

Welker, Michael. *God the Spirit*. Translated by John F. Hoffmeyer. Minneapolis, MN: Fortress, 1994.

———. *Creation and Reality*. Translated by John F. Hoffmeyer. Minneapolis, MN: Fortress, 1999.

Wellhausen, Julius. *Prolegomena to the History of Ancient Israel*. Gloucester: Peter Smith, 1983 (1885).

Westermann, Claus. *Genesis*. Grand Rapids, MI: Eerdmans, 1987.

Whellwright, Jeff. "Captive Wilderness." *Discover* (August 2006): 42–49.

White, Lynn T., Jr. "The Historical Roots of Our Ecological Crisis." *Science* 144 (10 March, 1967): 1203–07.

Wiesman, Alan. *The World Without Us*. New York: Thomas Dunne Books, 2007.

Wigner, Eugene. "The Unreasonable Effectiveness of Mathematics in the Natural Sciences." *Communications on Pure and Applied Mathematics* 13, no. 1 (February 1960): 1–14

de Wilde, A. *Das Buch Hiob*. OTS 22; Leiden: Brill, 1981.

Wilford, John Noble. "Lost in a Million-Year Gap, Solid Clues to Human Origins." *New York Times*, 18 September 2007, at www.nytimes.com/2007/09/18/science/18evol.html?scp=1&sq=Noble%20Wilford%20%20%22Lost%20in%20a%20Million%20Year%20Gap%22&st=cse.

Wilson, Edward O. *Biophilia*. Cambridge, MA: Harvard University Press, 1984.

———. *Consilience: The Unity of Knowledge*. New York: Vintage Books, 1998.

———. *The Diversity of Life*. New York: W.W. Norton, 1999.

———. "Afterword." In *From So Simple a Beginning: The Four Great Books of Charles Darwin*, edited by Edward O. Wilson, 1479–83. New York: W.W. Norton, 2006.

———. *The Creation: An Appeal to Save Life on Earth*. New York: W.W. Norton, 2006.

———. "General Introduction." In *From So Simple a Beginning: The Four Great Books of Charles Darwin*, edited by Edward O. Wilson, 11–13. New York: W.W. Norton, 2006.

———. "Introduction to *The Voyage of the Beagle*." In *From So Simple a Beginning: The Four Great Books of Charles Darwin*, edited by Edward O. Wilson, 17–19. New York: W.W. Norton, 2006.

Winnicott, D.W. *Playing and Reality*. London: Tavistock, 1982.

Wistow, Graeme. "Lens Crystallins: Gene Recruitment and Evolutionary Dynamism." *Trends in Biochemical Sciences* 18 (1993): 301–6.

Woese, Carl R. "A New Biology for a New Century." *Microbiology and Molecular Biology Reviews* 68 (2004): 173–86.

Wolff, Hans Walter. *Anthropology of the Old Testament*. Philadelphia: Fortress, 1974.

Wolpoff, Milford H., et al. "Modern Human Ancestry at the Peripheries: A Test of the *Replacement* Theory." *Science* 291 (12 January 2001): 293–97.

Wolpoff, Milford H., and Rachel Caspari. *Race and Human Evolution*. New York: Simon & Schuster, 1997.

Wood, Bernard. *Human Evolution: A Very Short Introduction*. Oxford: Oxford University Press, 2005.

Worthing, Mark William. *God, Creation, and Contemporary Physics*. Theology and the Sciences. Minneapolis: Fortress, 1996.

Wright, G. Ernest. *God Who Acts: Biblical Theology as Recital*. London: SCM Press, 1952.

Zahavi-Ely, Naama. "The Better Half or a Spare Rib? A Linguistic Study of Eve's Creation." Paper given at the Society of Biblical Literature meeting in Toronto, 26 November 2002.

Zenger, Erich. *Gottes Bogen in den Wolken: Untersuchungen zu Komposition und Theologie der priesterschriftlichen Urgeschichte*. 2nd ed. Stuttgarter Bibelstudien 112; Stuttgart: Katholisches Bibelwerk, 1987.

Zgoll, Annette. *Der Rechtsfall der En-hedu-Ana im Lied nin-me-shara*. AOAT 246; Münster: Ugarit-Verlag, 1997.

Zimmer, Carl. "In Games, An Insight into the Rules of Evolution," *New York Times*, 31 July 2007, at www.nytimes.com/2007/07/31/science/31prof.html?scp=1&sq=Carl%20Zimmer%20%22Rules%20of%20Evolution%22&st=cse.

———. *Microcosm*: E. coli *and the New Science of Life*. New York: Pantheon, 2008.

———. "On the Origin of Life on Earth," *Science* 323 (9 January 2009): 198–199.

Zohary, Michael. *Plants of the Bible*. London: Cambridge University Press, 1982.

Zukav, Gary. *The Dancing Wu Li Masters: An Overview of the New Physics*. New York: William Morrow and Company, 1979.

Index

Index of Scripture Passages